DATE			

COLLECTED STUDIES SERIES

Studies in the Culture of Science in France and Britain Since the Enlightenment

Professor Maurice Crosland

Maurice Crosland

————

Studies in the Culture of Science in France and Britain Since the Enlightenment

————

VARIORUM
1995

This edition copyright © 1995 by Maurice Crosland.

Published by VARIORUM
Ashgate Publishing Limited
Gower House, Croft Road,
Aldershot, Hampshire GU11 3HR
Great Britain

Ashgate Publishing Company
Old Post Road,
Brookfield, Vermont 05036
USA

ISBN 0–86078–498–3

British Library CIP Data
Crosland, Maurice P.
Studies in the Culture of Science in France
and Britain Since the Enlightenment.
(Variorum Collected Studies Series; CS501)
I. Title. II. Series.
509. 44

US Library of Congress CIP Data
Crosland, Maurice P.
Studies in the Culture of Science in France and Britain
Since the Enlightenment / Maurice Crosland
p. cm. -- (Collected Studies Series; CS501). Includes Index
ISBN 0–86078–498–3 (alk. Paper)
1. Science--France--History--18th Century. 2. Science--France--
History--19th Century. 3. Science--Great Britain--History--18th
Century. 4. Science--Great Britain--History--19th Century.
I. Title. II. Series: Collected Studies; CS501.
Q127. F8C76 1995   95–15900
509. 44' 09' 033--dc20   CIP

Printed by Galliard (Printers) Ltd, Great Yarmouth, Norfolk, Great Britain

COLLECTED STUDIES SERIES CS501

CONTENTS

NATIONAL AND INTERNATIONAL SCIENCE

This volume contains xxii + 291 pages

PREFACE

Historians usually consider that a knowledge of the past is worthwhile for its own sake. Non-historians are often less convinced and require some further justification, such as the argument that the past is relevant to the present. Although any historical paper must be judged *as history*, it is arguable, without taking a Whig perspective, that many parts of this collection are fully relevant today. Thus in the first section, where the image of science in the 1790s is discussed, one may agree that a potentially hostile public image of science is still topical today. Again in a modern meritocratic society we are increasingly concerned with qualifications, a subject dealt with in the second section. The slogan "publish or perish" has been applied to the modern university world. The very idea of publications as credentials for appointment or promotion in higher education can be traced back to practices within the early nineteenth-century French Academy of Sciences. Another subject to be considered is the national context of science. The nineteenth and twentieth centuries have seen the rise of nationalism which, in its extreme form, has led to wars. Those who think of science as essentially international will find some relevant evidence both for and against in section three.[1] The application of science to war is usually thought of as a twentieth-century phenomenon but here some antecedents are discussed. We can surely learn from what has gone before, even if it is only in the limited sense of avoiding some of the mistakes of the past.[2] Those who ignore history may find themselves claiming to have reinvented the wheel.

In the title of this book France has been placed before Britain because more of the author's work has been connected with France. When asked why he has not done more research on British science, he has usually replied that there is no shortage of historians of British science. On the other hand, in France history of science is not officially recognised as a subject in higher education. France has the distinction of having made a major contribution to the emergence of modern science (no less than Britain) and there are important documentary sources in that country waiting to be studied, yet the native labourers are few. This explains why a large part of the work on the history of French science since the 1960s has been by scholars from Britain and North America.[3]

In the eighteenth century France and Britain were the two great international powers. The population of France of approximately 27 million by 1800 greatly exceeded Britain's 11 million. Its language had become the *lingua franca* of European civilisation and diplomacy so that even the Prussian Academy in Berlin published its memoirs in French. What France has given to the world in the way of literature, music and art is well known. What France gave the world in science is less well known. The metric system, for example, was a product not only of the French Revolution but of the French Enlightenment. As Burke reminds us, the French have long had a reputation for science and learning which contrasts with the British empirical tradition, which has continued, even into the twentieth century, to value the practical approach and to distrust intellectuals. Britain was comparatively late in obtaining state help for science. The British laissez-faire tradition has emphasised individual initiative, as can be seen in such lives as those of Priestley and Faraday, neither of whom owed anything to their government. By way of contrast France provides an ideal case study of early state-sponsored science.

From a European perspective France has a special place in the growth of the nation state. Although there were a few minor boundary changes in the late eighteenth century, France was a single country with a national identity long before either Germany or Italy. The centralised administration of Louis XIV was further developed in the Revolutionary and Napoleonic period so that one can without apology speak of a 'French system', 'French education', and so on. One aspect of the centralised system was the enormous importance of the capital city. In science and higher education in particular, Paris tended to monopolise talent and resources for most of the nineteenth century.

It is interesting that a reviewer of the author's 1992 book on the French Academy of Sciences should have found reason to complain of the brevity of the treatment of prizes.[4] Since the author had written at length on this subject, it would have been all too easy for him to have doubled the length of the relevant chapter by simply repeating material already published. It seemed more reasonable to give only a brief summary of what was already - in principle - in the public domain. But there is obviously the world of difference between articles hidden in back issues of journals consulted by a minority of specialists and making the articles more visible in a collection. A further justification for putting together this collection is the diversity of journals and conference proceedings in which the articles originally appeared. Apart from the author, no one person, not even a specialist in the field, is likely to be aware of the existence of more than a few of the articles included. One might expect articles on the history of science to appear mainly in history of science journals.   But this is to ignore the cultural and social dimensions of the subject. Thus if one were writing about science in the context of the Enlightenment, a journal centred on the Enlightenment might seem the most appropriate medium.   The journal *Studies on Voltaire and the Eighteenth Century* not only publishes articles by specialists in eighteenth-century French literature but also material of a more interdisciplinary kind. On the other hand, a considerable part of the author's research has been concerned with institutions, notably the nineteenth-century Paris Academy of Sciences and, to a lesser extent, the Royal Society of London, two institutions of comparable importance in their respective countries, yet both exerting influence internationally. In so far as the study of institutions deals with the social rather than the cognitive aspect of science, this suggests publication in a journal concerned more with the social and political context; this explains why several of the author's articles on the Academy were originally published in the two journals: *Minerva. A Review of Science, Learning and Policy* and *Social Studies of Science*. Indeed the fact that science has a social context is now widely regarded as a truism, although at the time of original publication of some of these articles, this was a point still in dispute.

The collection begins with a case study of the eighteenth-century idea that the world of nature provides a standard for human affairs, a concept of great importance in the Enlightenment.   Science came to acquire increasing authority in this period since its practitioners were specially concerned with the study of the natural world.   In England Joseph Priestley was an important figure, a polymath whose interests spanned religion, politics, language and science. As an influential writer who welcomed the French Revolution, he was regarded with great suspicion by the conservative Edmund Burke, who saw a link between radical politics and experimental science. The two sciences with which Priestley was most concerned were physics and chemistry. When the author was invited in 1983, the 250th anniversary of the birth of Priestley, to give a series of lectures to mark the occasion, he tried to explain, among other things, why a person who had first written books on electricity and optics should have then turned to chemistry, in which he made his main reputation. Priestley is presented primarily as an *author*, a perspective since adopted by some other scholars. Priestley's famous opponent in matters of chemical theory was Lavoisier. Lavoisier was definitely an important figure in the later Enlightenment with administrative and social concerns involving the application of his science, for example, to the construction of hospitals and prisons,[5] although that does not happen to be the particular story told here.

The article on Lavoisier probably falls into a special category. Since it was written in 1983 and looked forward to the commemoration of the Lavoisier bicentenary in 1994, there is an obvious sense in which it is dated. Great progress has been made in the last few years with the study and publication of two volumes of a definitive edition of Lavoisier's correspondence, although the tragic death of the new editor, Michelle Goupil, in 1993 constitutes a further obstacle to the early completion of the project. The year 1993-94 saw the appearance of three major scholarly treatments of Lavoisier, covering both his life and work.[6] A fourth study concentrated essentially on one aspect of Lavoisier's career.[7] The bicentenary in 1989 of the publication of Lavoisier's *Traité élémentaire de chimie* had been marked by several commemorations including a conference at the Ecole Polytechnique. One of the communications presented there focused on the question raised in the author's paper of 1983, namely to what extent Lavoisier should be seen as the leader of a school rather than a lone genius.[8] Such a question illustrates the fact that the Lavoisier paper, reprinted here in English translation, is concerned more with historiography than history. It raises in a preliminary way the question of how Lavoisier has been remembered and interpreted since his own time. It may be pointed out that it was originally commissioned for a popular science periodical, *La Recherche*, and was aimed at a French audience.

In the second section the subject of the professionalisation of science is included because it is an area where the French situation may claim special attention not unrelated to that country's early investment in institutes of advanced scientific education. There is still some disagreement as to when science became a profession but it is clear that it happened in France before Britain.[9] It also seems that chemistry was one of the first branches of science to provide a higher education followed by possibilities of a career related to that training. This was partly because of the many applications of chemistry which offered a range of employment in addition to teaching. Thus it may be claimed that the career of Gay-Lussac (1778-1850), educated at the Ecole Polytechnique, exemplifies what was effectively the first generation of professional scientists. It should be remembered that Gay-Lussac was only one of a group in the famous Society of Arcueil[10] to benefit not only from the opportunities for higher education provided by the French state but also the personal patronage of Berthollet and Laplace, both senior Academicians and formerly close associates of Napoleon Bonaparte. In France members of the Academy of Sciences saw themselves as professionals and resented the royal decree of 1816 which reintroduced a special category for honorary members, who were viewed as amateurs.[11]

This brings us to a particular field, the study of the French Academy of Sciences in the nineteenth century, where the author has long found himself ploughing a lonely furrow. Various reasons could be advanced why this subject had not at the time of writing attracted other scholars, lack of knowledge of the sources being one possibility. More important probably is the question of motivation for research. The conventional wisdom used to be that the Academy became moribund in the nineteenth century. It seemed to some scholars to be no more than an honorific institution concerned exclusively with the more superficial trappings of science. Yet the presentation of a scientist's research to his peers is surely of more than secondary importance. The rapid publication of research by the Academy in the weekly *Comptes rendus* from 1835 was a great landmark in the history of scientific communication, creating a precedent that was to be followed some thirty years later in the English-speaking world by the foundation of the journal *Nature*.

Yet the various articles on the Academy reprinted here are concerned less with scientific communication than with the judgement of the work of individual scientists, including many of the first rank. Peer review has become an important process in modern science. Its antecedents can be seen in the nineteenth-century Academy of Sciences. The evidence presented draws on extensive documentation uncovered in the 1980s. Some of these documents were printed in the nineteenth century for limited distribution on behalf of candidates for election to the Academy. It was the French who first insisted on the importance of research publications as the credentials of the scientist. There is the popular image of the nineteenth-century German university professor with his heavy tomes representing almost a caricature of scholarship. But, if at the beginning of the nineteenth century the German university professor was expected to have published a book, it was the sort of book that might establish a literary reputation in the public eye. According to Stephen Turner, who has made a detailed and prolonged study of nineteenth-century German universities, it was not until the 1840s that the publications of a university professor were expected to represent the results of original research addressed to specialists.[12]   More than a generation before, on the other side of the Rhine and in the highly competitive atmosphere of the French capital, scientists who aspired to election to the Academy (and later to university posts) carefully listed their publications as credentials. Complementing the printed lists of publications, and no less interesting, are the surviving manuscript reports on elections, drawn up by a senior specialist in the candidate's field, evaluating his work and sometimes comparing it with that of other researchers.   Here is an invaluable confidential record of how top-ranking scientists judged the work of aspiring Academicians. Such rich documentation will undoubtedly inspire further research.

The comparative method often adds to our understanding of a subject and we can make some interesting comparisons and find some startling contrasts between the Paris Academy of Sciences and the Royal Society of London. Each in its own country was arguably the single most important institution concerned with the whole range of science, yet their respective organisations could hardly have been more different. Whereas the former was always a relatively small elite supported by government funds, the latter was allowed in the eighteenth century to grow to several hundred with the prospect that the increased income from subscriptions would solve the financial problems of the Society which lacked government subsidy. To gain membership of the Royal Society it was necessary to surmount only the lowest of hurdles. Thus in the eighteenth century only a small proportion of Fellows could be described in any way as what would later be called 'scientists'. The paper included in this collection shows how standards were raised in later years. There is an interesting comparison to be made between qualifications deemed relevant and sufficient in the nineteenth century for respective entry to the Royal Society and to the French Academy.

Scientific research has increasingly required financial support. The final paper in the second section shows how in the nineteenth century the traditional Academy procedure of awarding honorific prizes for outstanding work was supplemented by the emergence of a system of grants in response to a growing need for money in the early stages of research.

In the final section the question is raised of the existence of national styles in science. Probably the best case for a specifically 'French' science in the nineteenth century would be one that related it to the French institutional framework. But scientists have fortunately also been prompted to collaborate across national boundaries, which is the subject of the last paper.

Canterbury,   MAURICE CROSLAND
January 1995

NOTES

[1] A recent author has remarked that 'the literature on nationalism and internationalism in nineteenth-century science is sparse.' Alan J. Rocke, *The Quiet Revolution. Hermann Kolbe and the Science of Organic Chemistry*, Berkeley, 1993, p.455.

[2] Thus, for example, Californian vineyards might have saved millions of dollars if their owners had studied the nineteenth-century outbreak of phylloxera in France and other European countries (*New Scientist*, No.1869, 17 April 1993, pp.27-31).

[3] This was remarked on in my Presidential Address to the British Society for the History of Science in 1977 (see X 110), and is equally true today.

[4] E. Crawford, Essay Review, *Studies in the History and Philosophy of Science, 24* (1993), 305-312 (309).

[5] Lavoisier, *Oeuvres* (6 vols., Paris, 1862-93), vol.3.

[6] In order of appearance they were:
Jean-Paul Poirier, *Antoine-Laurent Lavoisier, 1743-1794*, Paris, Pygmalion, 1993. Bernadette Bensaude-Vincent, *Lavoisier, Mémoires d'une révolution*, Paris, Flammarison, 1993. Arthur Donovan, *Antoine Lavoisier. Science, Administration and Revolution*, Oxford, Blackwell, 1993 [1994]. Donovan was the editor of a previous collection of papers: 'The chemical revolution: Essays in reinterpretation', *Osiris*, 2nd series, *4* (1988).

[7] Maurice Crosland, *In the Shadow of Lavoisier. The 'Annales de chimie' and the Establishment of a New Science*, Oxford, B.S.H.S., 1994.

[8] Maurice Crosland, 'Lavoisier, lone genius or "chef d'école"? The testimony of Fourcroy', Michelle Goupil (ed.), *Lavoisier et la révolution chimique*, Paris, Ecole Polytechnique, 1992, pp.1-12.

[9] In a famous phrase used in 1851 Charles Babbage wrote: "Science in England is not a profession, its cultivators are scarcely recognised as a class" (*The Exposition of 1851*, London, 1851, p.189). This merely took further his earlier complaint that in Britain there was no career open to anyone studying science, in contrast to the situation in Napoleonic France (*Reflections on the Decline of Science in England*, London, 1830, p.36).

[10] Maurice Crosland, *The Society of Arcueil. A View of French Science at the Time of Napoleon 1*. London, 1967.

[11] Maurice Crosland, *Science under Control. The French Academy of Sciences, 1795-1914*, Cambridge, 1992, p.406n.

[12] R. Stephen Turner, 'The growth of professorial research in Prussia, 1818 to 1848 - causes and context', *Historical Studies in the Physical Sciences, 3* (1971), 137-182 (170).

ACKNOWLEDGEMENTS

I am grateful to the following individuals and organisations for giving permission for the reproduction of articles for which they hold the copyright: Professor Haydn Mason as General Editor of *Studies on Voltaire and the eighteenth century*, Taylor Institution, Oxford (for Study I); the Council of the British Society for the History of Science (II, III and X); the editors of *La Recherche*, Revue Mensuelle, 57, rue de Seine, 75280, Paris (IV); Ms Helen Gardner, on behalf of the Royal Society, 6 Carlton House Terrace, London SW1Y 5AG (V); Mrs Gillian Anderson, Managing Editor of *Minerva*, 19 Nottingham Road, London SW17 7EA (VI, VII, VIII); Ms Vivienne Dunlop on behalf of Sage Publications Ltd., 6 Bonhill Street, London EC2A 4PU (IX and XI) and Ms Maureen Prior on behalf of Edinburgh University Press, 22 George Square, Edinburgh EH8 9LF (XII).

PUBLISHER'S NOTE

The articles in this volume, as in all others in the Collected Studies Series, have not been given a new, continuous pagination. In order to avoid confusion, and to facilitate their use where these same studies have been referred to elsewhere, the original pagination has been maintained wherever possible.

Each article has been given a Roman number in order of appearance, as listed in the Contents. This number is repeated on each page and quoted in the index entries.

PRINCIPAL PUBLICATIONS

BOOKS

Historical Studies in the Language of Chemistry. Pp.xvii + 406.
London: Heinemann, & Cambridge, Mass.: Harvard University Press, 1962.
Second Edition, New York: Dover Books, 1978.
Spanish translation: *Estudios históricos en el Lenguaje de la Química.*
Universidad Nacional Autónoma de México, 1988.

*The Society of Arcueil. A View of French Science at the time of
Napoleon I.* Pp.xx + 514. London: Heinemann & Cambridge, Mass.:
Harvard University Press, 1967.

(Ed.) *Science in France in the Revolutionary Era described by Thomas
Bugge.* Pp.xiv + 239. Cambridge, Mass. & London: M.I.T. Press, 1969.

(Ed.) *The Science of Matter. Selected Readings.* Pp.440. London:
Penguin Books, 1971.
Reprinted in series: 'Classics in the History and Philosophy of Science',
Philadelphia: Gordon & Breach, 1992.

(Ed.) *The Emergence of Science in Western Europe.* Pp.201. London:
Macmillan, 1975.
Italian translation: *L'Affermazione della Scienza moderna in Europa*, Bologna:
Societa Editrice Il Mulino, 1979.

Gay-Lussac, Scientist and Bourgeois. Pp.xvi + 333. Cambridge
University Press, 1978.
French translation: *Gay-Lussac, savant et bourgeois.* Paris: Belin, 1992.

Science under Control. The French Academy of Sciences, 1795-1914.
Pp.xix + 454. Tables, Cambridge University Press, 1992.

In the Shadow of Lavoisier: The Annales de Chimie *and the
Establishment of a New Science.* Pp.xii + 354. Tables and illustrations,
Oxford: British Society for the History of Science, 1994.

PAPERS AND CONTRIBUTIONS TO BOOKS

(* indicates inclusion in this book)

'The Use of Diagrams as Chemical "Equations" in the Lecture Notes of William Cullen and Joseph Black', *Annals of Science*, *15* (1959), 75-90.

'The Development of the Concept of the Gaseous State as a Third State of Matter', *Proceedings of the Tenth International Congress of the History of Science*, Ithaca, N.Y., 1962, vol. 2, 851-4. (1964)

'The Development of Chemistry in the Eighteenth Century', *Studies on Voltaire and the Eighteenth Century*, (Transactions of the First International Congress on the Enlightenment), *24*, (1963), 369-441.

'The Origins of Gay-Lussac's Law of Combining Volumes of Gases', *Annals of Science*, *17*, (1963), 1-26.

Les Héritiers de Lavoisier, Lecture delivered at Palais de la Découverte, University of Paris, April 1967, Pp.45, Paris, 1968.

Introduction to: *Mémoires de physique et de chimie de la Société d'Arcueil* (3 vols., 1807-17). New York: Johnson Reprint Corporation, 1967, vol. i, pp.i-xlvi.

'The First Reception of Dalton's Atomic Theory in France', in *John Dalton and the Progress of Science* (Papers presented to a conference held to mark the bicentenary of Dalton's birth), ed. D. Cardwell, Manchester University Press, 1968, pp.274-287.

'Comte and Berthollet: A Philosopher's View of Chemistry', *Proceedings of the Twelfth International Congress of the History of Science*, Paris, 1968, Fasc. VI, pp.23-27.

'The Congress on Definitive Metric Standards, 1798-99: The First International Scientific Conference?', *Isis, International Review devoted to the History of Science and its Cultural Influences*, *60* (1969), 226-231.

Introduction to: C.L. Berthollet, *Essai de statique chimique*, 2 vols. 1803, New York, Johnson Reprint Corporation, 1972, vol.i. pp.v-xxiii.

'Humphry Davy - An Alleged Case of Suppressed Publication', *British Journal for the History of Science, 6* (1972-73), 304-310.

* 'Nature and Measurement in Eighteenth-Century France', *Studies on Voltaire and the Eighteenth Century, 87* (1972), 277-309.

'The History of French Science: Recent Publications and Perspectives', *French Historical Studies, 8* (1973), 157-171.

'Lavoisier's Theory of Acidity', *Isis, 64* (1973), 306-325.

* 'The Development of a Professional Career in Science in France', *Minerva, A Review of Science, Learning and Policy, 13* (1975), 38-57.

* 'Science and the Franco-Prussian War', *Social Studies of Science, 6* (1976), 185-214.

'La Science et le Pouvoir: de Bonaparte à Napoléon III', *La Recherche, 7,* No. 71 (1976), 842-850.

* 'History of Science in a National Context', Presidential Address, *British Journal for the History of Science, 10* (1977), 95-113.

* 'Aspects of International Scientific Collaboration and Organisation before 1900' in E.G. Forbes (ed.), *Human Implications of Scientific Advance*, Edinburgh University Press, 1978, pp.114-125.

'Gay-Lussac (1778-1850): A View of Chemistry, Industry and Society in post-revolutionary France', *Endeavour*, New Series, *2* (1978), 52-56.

'Gay-Lussac: Une Etape dans la Professionalisation de la Science', *La Recherche, 9,* No.91 (1978), 625-633; reprinted in Michel Biezunski (ed.), *La Recherche en Histoire des Sciences*, Société d'Editions Scientifiques, Paris, 1983.

[With C.W. Smith] 'The Transmission of Physics from France to Britain, 1800-1840', *Historical Studies in the Physical Sciences, 9* (1978), 1-61.

'The French Academy of Sciences in the Nineteenth Century', *Minerva, 16,* No. 1 (1978), 73-102.

'The History of the International Organisation of Science: Some Preliminary Sketches', in F.R. Pfetsch (ed.), *Internationale Dimensionen in der Wissenschaft*, Erlangen: Institut für Gesellschaft und Wissenschaft, 1979, pp.37-60.

'From Prizes to Grants in the Support of Scientific Research in France in the Nineteenth Century', *Minerva, 17*, No. 3 (1980), 355-380.

'Davy and Gay-Lussac: Competition and Contrast', in Sophie Forgan (ed.), *Science and the Sons of Genius: Studies on Humphry Davy*, London, Science Reviews Ltd., 1980, pp.95-120.

'Chemistry and the Chemical Revolution', in G.S. Rousseau and Roy Porter (eds.), *The Ferment of Knowledge. Studies in the Historiography of Eighteenth-Century Science*, Cambridge University Press, 1981, pp.389-416.

'Gay-Lussac: La Biographie et l'Oeuvre d'un Savant Français', *Actes du Colloque Gay-Lussac, Ecole Polytechnique, Décembre 1978*, Paris, 1980, pp.1-11.

*   'Scientific Credentials; Record of Publications in the Assessment of Qualifications for Election to the French Académie des Sciences', *Minerva, 19* (1981), 605-631.

'The Library of Gay-Lussac', *Ambix, Journal of the Society for the History of Alchemy and Chemistry, 28* (1981), 158-170.

*   'Explicit Qualifications as a Criterion for Membership of the Royal Society: A Historical Review', *Notes and Records of the Royal Society of London, 37* (1983), 167-187.

*   'Lavoisier, le "mal-aimé"', *La Recherche, 14*, No.145 (1983), 785-791.

*   'Priestley Memorial Lecture: A Practical Perspective on Joseph Priestley as a Pneumatic Chemist', *British Journal for the History of Science, 16*, No.3 (1983), 223-238.

'Le Développement historique des Musées de la Science et le Problème de la Visibilité de la Science', in C. Beaugrand and A. Charre (eds.), *La Réinvention du Musée*, Association pour l'Institut Internationale de Recherche, Art et Science, Lyons, 1986, pp.153-159.

'The Academy of Sciences' in Edgar L. Newman and R.L. Simpson (eds.), *Historical Dictionary of France from the 1815 Restoration to the Second Empire*, Greenwood Press, New York, 1987, pp.7-10.

* 'Assessment by Peers in Nineteenth-Century France: The Manuscript Reports on Candidates for Election to the Academy of Sciences', *Minerva, 24* (1987), 413-432.

* 'The Image of Science as a Threat: Edmund Burke versus Joseph Priestley and the "Philosophic Revolution"', *British Journal for the History of Science, 20*, No.3, (1987), 277-307.

'What is History of Science?', in Juliet Gardiner (ed.) *What is History Today?*, Macmillan Education, London, 1988, pp.78-80.

'Antoine-Laurent Lavoisier: The Chemical Revolution', in Roy Porter (ed.), *Man Masters Nature. 25 Centuries of Science*, BBC Books, London, 1987, pp.101-113.

* [With A. Galvez] 'The Emergence of Research Grants within the Prize System of the French Academy of Sciences, 1795-1914', *Social Studies of Science, 19* (1989), 71-100.

'The Influence of the French Revolution on Science in Britain', *Actes du 114e Congrès National des Sociétés savantes, 1989, Section Histoire des sciences et techniques. Echanges d'influences scientifiques et techniques entre pays européens de 1780 à 1830*, Paris, 1990, pp.9-20.

'The Successors of Lavoisier', *Revue des Questions Scientifiques, 160*, No.2, (1989), 205-219.

'The Chemical Revolution of the Eighteenth Century and the Eclipse of Alchemy in the "Age of Enlightenment"' in Z.R.W.M. von Martels (ed.), *Alchemy Revisited: Proceedings of the International Conference on the History of Alchemy* (University of Gröningen, April 1989), E.J. Brill, Leiden, 1990, pp.67-78.

'The History of Chemistry seen in a broader context' (contribution to 'Science and Technology Today: The Historical Viewpoint'), *Impact of Science on Society, 159* (1991), 227-236. Reprinted in French translation in *Chimie nouvelle, 9*, no. 36 (December 1991), 1059-1064.

'Lavoisier - Lone Genius or Chef d'Ecole? The Testimony of Fourcroy.' *Lavoisier et la Révolution chimique. Actes du colloque tenu à l'occasion du bicentenaire de la publication du 'Traité élémentaire de chimie', 1789,* ed. Michelle Goupil. Paris, 1992, pp.1-12.

[With M.Y. Bektas] 'The Copley Medal: The Establishment of a Reward System in the Royal Society, 1731-1839', *Notes and Records of the Royal Society of London, 46* (1992), 43-76.

[The French Revolutionary Calendar], *Times Higher Educational Supplement,* No.1038, (25 Sept.1992), p.16.

'Anglo-Continental Scientific Relations, c.1780-c.1820, with special reference to the Correspondence of Sir Joseph Banks, P.R.S.' in R.E.R. Banks *et.al.* (eds.), *Sir Joseph Banks. A Global Perspective,* Kew, 1994, pp.13-22.

'Lavoisier et les "Annales de chimie". Un moyen de propager la nouvelle chimie au delà du dix-huitième siècle'. *Il y a 200 ans Lavoisier* (Actes du Colloque organisé à l'occasion du bicentenaire de la mort d'Antoine-Laurent Lavoisier le 8 mai 1794), ed. C. Demeulenaere-Douyère, Paris, 1995, pp.191-200.

'Lavoisier, the Two French Revolutions and "the Imperial Despotism of Oxygen"', *Ambix. Journal of the Society for the History of Alchemy and Chemistry, 42 (*1995), in press.

I

'Nature' and measurement in eighteenth-century France

Introduction

Not only the pursuit of science but also the ordinary conduct of human affairs and, in particular, commerce, requires a system of measurement. Any scheme of measurement is based on certain units and early systems often used the human body as a basis for units: the foot, the hand, the finger, etc. These units had to be defined in some way and standards of measurement were therefore introduced. It was frequently the case, however, that these standards were only accepted locally. The diversity of measurements within France was exceptionally acute. At the time of the Revolution in the north of France there were 18 kinds of the unit of length the *aune*, varying from 0.620m. to 0.845m. and in the department of Maine et Loire there were 110 different measures for grain.

The chaos was all the more painful in a country which claimed to be at the heart of civilization and which was to be a focus of the Enlightenment. One reason why France became the centre of reform was, therefore, one which might in the broader usage of eighteenth century studies be called philosophical, the desire for a uniform and rational system of measurement. But if reason was one of the key slogans of the Enlightenment surely nature was the other and so a system of measurement was sought which would be in conformity with nature. But this story is not confined to the world of ideas. The Revolution provided an ideal occasion for translating into action ideas and aspirations of previous generations. The fact that the great eighteenth-century Revolution occurred in France provides a further reason why this discussion of measurement in the eighteenth century is centred on that country. The old system of measurement was not confined to commerce but extended to the legal tenure of land. Only a fundamental political and social upheaval like the Revolution could alter customs and practices so deeply ingrained in the structure of society.

Yet the views expressed in the 1790s were in some ways no more than a recapitulation of the ideals of those who had lived through the earlier phases of the Enlightenment. The establishment of the metric system in France also served the additional purposes of marking a break with the past and demonstrating national unity.

This paper will be divided thematically rather than chronologically and it may be as well before going any further to summarize the succession of events[1]. Attempts to improve uniform standards in France date back to Charlemagne but not much tangible progress was made until the eighteenth century. One of the principal proponents of a natural system of measurement was La Condamine, who in an expedition to Peru in 1735 organised by the Académie des sciences, had carried out observations of a pendulum and made a laborious but precise measurement of an arc of longitude of about 3 degrees near the equator. He presented the standard of length he used, the *toise* of Peru, to the Académie des sciences. In 1766 Trudaine de Montigny as Intendant des finances asked Tillet, a member of the Académie des sciences, to make 80 copies of the *toise* of Peru. These standard *toises* were

---

[1] a fundamental source for the history of the establishment of the metric system is: Méchain and Delambre, *Base du système métrique décimal* (Paris 1806-1810). See also A. Favre, *Les Origines du système métrique* (Paris 1931). A good chronological account of the establishment of the metric system is given in G. Bigourdan, *Le Système métrique des poids et mesures* (Paris 1901), although the emphasis of the book is on the later history and particularly the nineteenth century.

distributed to provincial parliaments but this attempt at centralisation was stillborn and the provincial standards continued to be used as before.

The period of office of Turgot on the accession to the throne of Louis XVI provides historians with a standard example of the application of some of the principles of the Enlightenment to government under the *ancien régime*. After consultation with Condorcet, Turgot wrote to the astronomer Messier on 3 October 1775 asking him to go to Bordeaux and make an exact determination of the length of a seconds pendulum with the idea of making this a standard with which other measures could be compared. Steps were taken to allow Messier to set out immediately but unfortunately a particularly accurate clock which it was intended that he should use was found to be broken and it took six months to construct another of equal precision. By then Turgot was no longer in office. Necker in 1778 considered a reform of weights and measures impracticable and it was only in a revolutionary context that the French government finally took action.

We can see from the two abortive attempts I have mentioned that the reform of weights and measures required action on two main fronts. There were first the scientific and technical aspects which involved the definition of a unit, the composition of a scale of measurement and the construction of accurate standards. Secondly there was the political and legislative aspect. A remarkable feature of the revolutionary period is the extent of collaboration between these two sides. The initiative was taken by the Académie des sciences when on 27 June 1789 it appointed a commission on the problem of the uniformity of weights and measures. There followed two years of discussion in scientific, political and diplomatic circles which will be mentioned later. Finally after much discussion at a series of meetings in March 1791 the Académie des sciences recommended a unit of measurement, the metre, which was to be one ten millionth part of the distance along the Paris meridian from the equator to the north pole. The National assembly accepted this provisional unit and the next troubled years required an accurate re-determination of this distance by triangulation from a measured base line and determination of latitude by astronomical observation. Because of the complex nature of the operations involved, the definitive metre was not established until 1799.

France and Britain

Most of this paper will be concerned with France rather than Britain and some preliminary consideration is necessary of the respective roles played by these two countries. There had been some discussion in the Royal society in 1661 and 1662 about a universal standard of length but this idea was no more than one among a thousand thrown out in that new Society. Bishop Wilkins's idea for a universal language published in 1668 included a hasty proposal for a universal measure but this was not taken more seriously by his contemporaries than his other Utopian schemes. In France Gabriel Mouton, an astronomer from Lyons, had sent a copy of his book proposing a universal measure to Huygens in 1670. Huygens, the star of the Académie des sciences, proposed the adoption of a seconds pendulum as the unit of length[2] and the astronomer Picard[3] made a similar suggestion in 1671. After this it was France which became the centre of discussion on measures. It was also France which became preoccupied with the question of measurement of the earth. Of the many expeditions organised by the Académie des sciences in Paris probably the most famous were those sent out in 1735 to measure a degree of latitude in Lapland (Maupertuis) and Peru (Bouger and La Condamine). The immediate impact of these measurements was to vindicate Newton's ideas but in the long run they also contributed to the data of meridian and pendulum

[2] *Horologium oscillatorium* (1673), prop.xxv; *Œuvres*, xviii.348-350.

[3] *Ouvrages de mathématiques* (The Hague 1731); *Mesure de la terre*, p. 9.

I

'NATURE' AND MEASUREMENT IN FRANCE

measurement. The fact that these measurements became a French tradition with little aid or competition from Britain is largely explained by the different organisation of science in the two countries. The Académie des sciences had an advantage over the Royal society in being able to draw regularly on government funds for foreign expeditions. The Royal society usually had to rely on subscriptions from members to pay all its expenses.

Nevertheless it must not be thought that voices were still in eighteenth-century Britain on the subject of a reform of weights and measures. Yet the usual tone was notably conservative. John Whitehurst in 1787 opposed taking the length of a pendulum beating seconds at London as the basis of measurement on the grounds that this figure of 39.2 ins. conflicted with the yard of 36 ins. This would mean alteration of the 'customary measures of England and consequently create general confusion and inconveniency to the public'[4]. Sir James Steuart's plan of about 1760 counselled patience. His plan, limited to the British empire, combines British pride with ideals of the Enlightenment: 'It will appear to posterity a noble effort of British policy to have been able to break the fetters of old custom, *when contrary to reason*'[5].

Among those who interested themselves in the standardisation of weights and measures in England the prize for support of the *status quo* might be awarded to sir George Shackburgh, F.R.S., who referred to the highflown ideas of those who wished to relate units of measurement to the dimensions of the earth: 'But in truth, with much inconvenience, I see no possible good in changing the quantities, the divisions or the names of things of such constant recurrence in common life. I should therefore harshly submit it to the good sense of those kingdoms at least to preserve, with the measures, the language of their forefathers. I would call a yard a yard and a pound a pound without any other alternative than that the precision of our own artists may obtain

[4] *An Attempt towards obtaining invariable measures of length* (1787), p.vii.

[5] *Works* (London 1805), v.392 (my italics).

281

I

for us'[6]. Of course England did not have a monopoly of conservative thought. In 1792 after the Académie des sciences had decided on the basis of the new metric system the watchmaker and physicist Ferdinand Berthoud published a work in which he argued that the best way to unify measures in France was to make the provinces adopt Paris units[7].

In England some of the most radical proposals for reform came from sir John Riggs Miller who began investigating the problem of the diversity of weights and measures in England and Wales in about 1788. In a speech in the House of commons on 6 February 1790 he considered the possibility of a 'general standard from which all weights and measures might in future be raised, being itself derived from something in Nature that was invariable and immutable'[8]. Talleyrand the French minister for foreign affairs was informed of this speech and wrote to Miller on 12 May 1790 expressing his interest in a joint reform of weights and measures by France and England: 'Trop long temps les deux Nations se sont divisées par de vains prétensions ou de coupables interêts, il est temps que deux peuples libres associent leurs efforts et leurs travaux pour une recherche utile au Genre Humain'. Talleyrand's proposals were contained in a pamphlet distributed in the National assembly. He suggested collaboration between the Académie des sciences and the Royal society. Unfortunately the political climate between the countries rapidly deteriorated as the Revolution followed its course and the French decided to proceed unilaterally. The report of the commission of the Académie des sciences of 19 March 1791 reads: 'Nous n'avons pas cru qu'il fût

[6] 'An Account of some endeavours to ascertain a standard of weight and measure', *Phil. trans.* (1798), pp.133-182 (165-166).

[7] *Essai sur les poids et les mesures ou méthode simple de conserver les mesures et les poids actuellement en usage et d'établir une mesure universelle, perpétuelle et invariable* (Paris 1792).

[8] sir John Riggs Miller, *Speeches in the House of commons upon the equalisation of the weights and measures of Great Britain* (London 1790). This pamphlet includes Miller's correspondence with Talleyrand.

nécessaire d'attendre le concours des autres nations, ni pour se décider sur le choix de l'unité de mesure, ni pour les opérations'[9].

The choice of a 'natural' unit

Among the so-called 'natural' standards considered as the basis of a new system of measurement, we will consider the two main contendants, the length of a seconds pendulum and a unit of length related to the dimensions of the earth[10]. We have seen that the idea of taking the length of a pendulum beating seconds as a natural unit had been advanced in France in the 1670s. The subsequent discovery that the length was not constant but depended on the latitude undermined this idea[11]. When, however, it was taken up again by La Condamine[12] in the mid-eighteenth century, he rapturously described this measure as 'fixe, invariable, reçue des mains mêmes de la nature', although fully recognising that the problem remained of specifying the place. The poles were rejected as being inaccessible. The latitude of 45°N. was particularly convenient because it passed through France but for that very reason was suspect as an objective choice. He therefore selected the equator — 'milieu de la terre habitable, terme extrême d'où l'on commence à compter les latitudes, terme de la moindre pesanteur, terme d'ailleurs unique, le seul enfin sur lequel les différens peuples puissent vraisemblablement s'accorder' (p.503).

[9] 'Rapport fait à l'Académie des sciences sur le choix d'une unité de mesures par mm. Borda, Lagrange, Laplace & Condorcet', *Mémoires de l'Académie Royale des Sciences* (1788 [1791]); *Histoire*, pp.7-16 (15-16). Hereafter referred to as: *Rapport*.

[10] a third idea, that of taking the height of a mercury barometer registering 'normal' atmospheric pressure, was immediately rejected as impracticable.

[11] thus Fontenelle remarked in 1733: 'Ainsi il faut renoncer à l'idée flatteuse d'une mesure universelle'—quoted in *Collection des mémoires relatifs à la physique publiés par la Société française de physique* (1889), iv, p.xxxiii n.

[12] 'Nouveau projet d'une mesure invariable propre à servir de mesure commune à toutes les nations', *Mémoires de l'Académie royale des sciences* (Paris 1747), pp.489-514 (511).

There were, however, many arguments against adopting the seconds pendulum as a standard. In 1790 Tillet and Abeille, representing the Société royale d'agriculture, argued that it was unsuitable because of its variability since the length depended on temperature, arc of swing, height above sea level, air resistance and the nature of the terrain where the experiment was carried out[13]. These were all practical reasons for rejecting the pendulum but there were also idealogical reasons. It made length depend on time and what is more on an arbitrary fraction (since the second was 1/86,400 part of a day), not even a decimal fraction (*Rapport*, p.9). The difficulty of the fraction could be overcome by reference to a hypothetical pendulum with a period of one day and dividing this by 100,000 but one would still be left with the problem of basing a fundamental unit on another kind of unit. A second was itself dependent on astronomical and mechanical considerations and a geometrical unit should be logically prior to these (Méchain-Delambre iii.560). It was this basic theoretical objection which decided the commission of the Académie des sciences on 19 March 1791 to adopt a unit of length based on the dimensions of the earth. This had the advantage of being analogous to measures then in use. Finally the scientists suggested: 'Il est bien plus naturel en effet de rapporter la distance d'un lieu à un autre, au quart d'un des cercles terrestres, que de la rapporter à la longueur d'un pendule'[14].

Thus after a history of 120 years the pendulum as a basis for a unit of measurement was defeated not on grounds of practicability or of nationalism but because of a theoretical problem and because it was considered less 'natural'. As a concession, however, it was agreed that the pendulum might still be used as a ready means of finding the new unit in nature.

The seconds pendulum had been given long and serious consideration and it had even been suggested that the reason why it

---

[13] *Observation de la Société Royale d'Agriculture sur l'uniformité des poids et des mesures*, Paris, 1790, p.123.

[14] *Mémoires* (1788); *Histoire*, p.10.

was not accepted was because one of the Academy's committee, the veteran Borda, had a vested interest in a method dependent on earth measurement. Earth measurement did have a certain agricultural justification which would have been lacking if the length of the seconds pendulum had been adopted. From this point of view the unit based on the earth's dimensions would have the sympathy of the new small *propriétaires* whereas the pendulum unit would have been merely the unit of the scientists. This aspect was well expressed under the Directory: 'Il y a quelque plaisir pour un père de famille à pouvoir se dire: "Le champ qui fait subsister mes enfants est une telle portion du globe. Je suis dans cette proportion co-propriétaire du monde".'[15] Thus the metre was in theory the unit of an emergent property-owning demo-cracy. In practice there was widespread conservatism about the abandonment of the old units in ordinary life so that it was not until 1837 that the metric system became mandatory by law.

The choice of the earth as the basis for measurement had other things to recommend it. There was some precedent for it, since it had been proposed as early as 1670 (by Gabriel Mouton) and successive expeditions organised by the Académie des sciences had built up invaluable data. The earth was international and Lalande[16] could argue that it was something directly accessible to experience. 'On ne pouvait trouver dans la nature un fondement plus direct que celui de la terre même que nous habitons, ni une donnée qui convient mieux à toutes les nations'. This may be so in general but it was rather hypocritical for an astronomer to suggest that its measurement was direct. It was also a misrepresen-tation to claim that it was the earth or the circumference of the earth which was to be measured. The southern hemisphere was hardly known. Conclusions about the shape of the earth were based on the comparison of the degree of latitude near the north pole and near the equator. It would therefore be true to say that

---

[15] *Discours prononcé à la barre des deux conseils ... séance du 4 messidor an 7.*

[16] *Bibliographie astronomique avec l'histoire de l'astronomie depuis 1781 jusqu'à 1802* (Paris 1803), p.703.

the eighteenth century witnessed a detailed study of part of the northern hemisphere and in particular western Europe. It was no accident therefore that when the metre was defined it was not defined in relation to the circumference of the earth but to a quarter of that circumference on the meridian through Paris. This led one idealist, Charles François Groult[17], to claim that it was a betrayal. The unit chosen should have been a decimal fraction of the whole circumference. To take a decimal fraction of a *quarter* of the circumference was merely to pay lip service to the principles of the metric system. There was, however, an argument against such decimal idealism[18]. If the earth was considered in cross-section a quarter was 90 degrees; but this is to speak according to the degrees of the ancien régime. The new revolutionary right angle contained 100 degrees. Thus the fraction, a quarter, could be expressed in decimal terms and honour satisfied.

Nationalism

There is in the period surrounding the Revolution an interesting tension between nationalism and internationalism. Thus Laplace in 1787 made the claim that the first person to make an exact measurement of the earth was a Frenchman who lived in the sixteenth century, Jean Fernel. Condorcet writing in 1791 about the metric system could not restrain himself from a few comments on the nation which had brought about the political Revolution: 'Cette gloire étoit réservée à la France; elle a donné la première au monde ce grand exemple et d'audace et de sagesse'[19]. This was then a period of national consciousness and the Académie des

---

[17] *Parallèle entre les mesures déduites du quart de la circumférence de la terre et celles que l'on peut déduire de la circumférence même* (Paris [n.d.]).

[18] this argument is implied in a speech by Saintin: *Discours prononcé par le citoyen Saintin, professeur près l'Administration du poids public du département de la Seine* (Paris an 7 [1799]), p.13.

[19] *Mémoires* (1788); *Histoire*, p.18.

sciences realized the psychological difficulties of another nation in accepting standard which were French. For the sake of uniformity even the great nations must be prepared to forego their own system. In any case the whole scheme of weights and measures was so irrational that there would be no advantage in singling out one unit for retention.

When therefore the committee of the Académie des sciences consisting of Condorcet, Borda, Lagrange, Laplace and Monge suggested in March 1791 that the standard of length might be based on the pendulum at a latitude of 45°, they anticipated the objections that this happened to pass through the middle of France. They said that the specification of 45° should be regarded not as a place (Bordeaux) nor even a latitude (45°) but rather the *mean* position for a seconds pendulum between the two extremities of the north pole and the equator (*Rapport,* pp.9, 16). This argument of a *mean* length at 45° had however been criticised by Brisson (1790) on the grounds that it assumed the regular curvature of the earth. He therefore proposed a unit based on the seconds pendulum at Paris[20]. In other words the latitude argument was not a fully honest one. Although one strived for generality by speaking of latitude, in practice the seconds pendulum would not be related to a whole circle of latitude but to a specified place.

We have seen how at the crucial meeting of 19 March 1791 the Académie des sciences decided to abandon the pendulum and select a unit based on the dimensions of the earth. It was the task of Condorcet as Secretary of the Académie to transmit to the National assembly an account of the deliberations of the scientists. A dose of flattery was necessary to explain on the political front why the scientists had not favoured the simplest and traditional approach to the problem of a universal unit: 'L'Académie n'a pas jugé pouvoir ni s'en rapporter aux mesures déjà faites, ni se contenter de la simple observation du pendule. Elle a senti que, travaillant pour une nation puissante, par les ordres d'hommes

[20] *Mémoires de l'Académie royale des sciences, 1788* [1791], p.723.

éclairés, qui savent donner au bien qu'ils font un grand caractère et embrassant dans leurs vues les hommes et tous les siècles, elle devait s'occuper moins de chercher ce qui serait facile que ce qui approcherait le plus de la perfection, et elle a cru, enfin, qu'une grande opération, qui annoncerait le zèle éclairé de l'Assemblée nationale pour l'accroissement des lumières et le progrès de la fraternité entre les peuples, ne serait pas indigne d'être accueillie par elle'[21]. It is significant that it was Talleyrand who presented this letter to the Assembly. The man who had previously urged a political collaboration now considered that it would be sufficient to present a new system free from prejudice: 'le seul moyen d'étendre cette uniformité aux nations étrangères et de les engager à convenir d'un même système de mesures est de choisir une unité qui, dans sa détermination ne renferme rien d'arbitraire ni de particulier à la situation d'aucun peuple sur le globe'. His argument therefore was not explicitly based on the concept of a natural unit but rather a unit which was absolute and non-French. We shall see in a moment how shallow his argument was. Meanwhile it would be germane to remark that this was the stage at which the argument from nature became particularly important. Originally the idea had been for some natural unit supported by the prestige of the two great powers France and Great Britain[22]. When this proposed collaboration collapsed (after the British refusal in December 1970 to continue discussions), the argument from authority was seriously weakened and the intrinsic merits of the system chosen became correspondingly more important.

It was natural for an eighteenth-century Frenchman to think of himself as the centre of the civilized world. While this has always been appreciated it was also true in a literal sense which

---

[21] Assemblée nationale, séance du 26 mars [1791]; *Réimpression de l'ancien Moniteur*, vii.723.

[22] 'Quand ces deux nations, qui n'ont de rivals qu'elles mêmes, l'auront adoptée, toute l'Europe ne manquera pas de l'adopter aussi', D. Bonnay, Assemblée nationale, 8 May 1790, *Moniteur*, iv.323.

has not been examined. For having renounced a unit based on the parallel of *latitude* which went through the centre of France, the committee of the Academy promptly reverted to the line of *longitude* which traversed France in the other direction. They proposed that the new unit of length should be related to the meridian to be measured between Dunkirk and Barcelona. Such a meridian lay approximately equally north and south of the 45° parallel and therefore should correspond to the mean curvature of the earth. They further justified their choice by saying it would be easier to check errors since there already existed a measure of the arc from Dunkirk to Perpignan. The French commissioners would have proposed a measurement stopping short at the Pyrenees but they expressly desired the two extremes of their measurement to be at sea-level, 'qui donne l'avantage d'avoir des points de niveau invariables et déterminés par la nature' (*Rapport*, p.13).

The Committee concluded by a final rebuttal of the charge of nationalism. They were conscious that the project had begun on the basis of collaboration proposed by Talleyrand between the two great powers of France and Great Britain. The direction of Revolutionary events had made any real political co-operation unlikely and thus France was acting unilaterally. The Committee therefore insisted that there was nothing peculiarly French about the proposed system: 'si la mémoire de ces travaux venoit à s'effacer, si les résultats seuls étoient conservés, ils offriront rien qui pût servir à faire connoitre quelle nation en a conçu l'idée, en a suivi l'exécution' (p.16).

Nature and measurement

As monsieur Ehrard[23] has reminded us, the concept of Nature was not invented in the xviiith century. Yet Hazard was hardly

---

[23] *L'Idée de nature en France dans la première moitié du XVIIIe siècle* (Paris 1963).

guilty of exaggeration when he described Nature as 'l'idée maî-
tresse' of that century. It would be possible to undertake an
extensive analysis of the contemporary usage of the term. The
main conclusion likely to be drawn from such an analysis, however, would be that the same word could mean quite different
things or at least could be given contrasting emphases. One man
would use the term 'natural' to mean normal or typical, another
to express the antithesis of artificial. Something described as
'natural' might mean that it was independent of revealed religion
or of any national or political consideration, etc. Yet although
different people meant different things by 'nature' the term did
provide a theme and a slogan to which most men could subscribe
even if they did so from rather different points of view.

It is a truism that 'reason' and 'nature' were slogans of the
Enlightenment, but all too often they represented opposing views,
the one of Voltaire and the other of Rousseau. To speak therefore
of a rational system based on nature is to suggest a conflict or
tension. The romantic view of the Enlightenment as represented
by Diderot and Rousseau reacted against the analytic view that
measurement was an essential part of knowledge. However, the
possible romantic objection that to bring a human scale alongside
nature is to disfigure it or debase it must be assuaged if the scale
used is modelled on nature. It is then as if nature were to hold a
looking glass up to herself. This is not to do violence to nature
but rather to seek natural harmonies. The idea of nature as a model
was thus sufficiently broad as to be acceptable to most of the
philosophes. Rousseau himself writes: 'Je resterois toujours aussi
près de la nature qu'il seroit possible . . . Dans le choix des objets
d'imitation, je la prendrois toujours pour modèle' (*Emile*,
Pléiade iv.678-679).

It is time to turn from the general to the particular and to
consider those precise aspects of nature which made it the focus
for all discussion of the reform of measurement. The main
features of nature were its supposed constancy, security and universality. We will examine these three aspects in turn.

'NATURE' AND MEASUREMENT IN FRANCE

One of the major justifications for turning to nature as the source of a standard of measurement was the view that nature was constant and eternal. One influential writer who saw nature as continuing unchanged throughout time was Buffon: 'La Nature . . . se maintient et se maintiendra comme elle s'est maintenue; un jour, un siècle, un âge, toutes les portions du temps ne font pas partie de sa durée'[24]. This idea of a permanent substratum of nature was contrasted with man, his institutions and his monuments which were variable, temporary and transitory: 'La Nature . . . efface les ouvrages de l'homme, couvre de poussière et de mousse ses plus fastueux monumens, les détruit avec le temps'[25].

This idea of the eternity of nature was that of the marquis de Bonnay[26] in his seminal report of 1790 on the adoption of new standards for weights and measures: 'Mais comment les definir? comment les fixer? comment les preserver de cette variation inévitable que le temps amène dans tout ce qui n'est que l'ouvrage des hommes, si l'on ne détermine pas avec précision leur rapport avec ces Mesures éternelles que donne la Nature et qui ne périssent qu'avec elle?'

This view marks an interesting development from Aristotelian philosophy, which distinguished between the transient world of generation and decay which was characteristic of the earth and the perfect unchanging world of the heavens. In the seventeenth century Galileo's telescopic observations had helped to overthrow the Aristotelian view of the perfection of the heavenly bodies. In the eighteenth century the concept of heavenly perfection was transferred to life on the earth. Divine perfection was to be sought in nature. In Buffon's phrase: 'La Nature est le trône extérieur de la magnificence Divine'[27]. This was indeed the deification of nature.

---

[24] 'De la nature, seconde vue' *Histoire naturelle* (1765), xiii, p.i.
[25] 'De la nature, première vue', *Histoire naturelle* (1764), xii, p.xv.

[26] *Rapport fait au nom du comité d'agriculture et de commerce* (Paris 1790), p.7.
[27] *Histoire naturelle* (1766), xii, p.xi.

The Revolution also brought home the full realisation that even long established institutions if man-made could be overthrown. A standard based on nature was therefore, in the words of the committee of the Académie des sciences: 'le seul moyen d'exclure tout arbitraire du système des mesures et d'être sûr de le conserver toujours le même sans qu'aucun autre évènement, qu'aucune révolution dans l'ordre du monde pût y jeter de l'incertitude' (*Rapport*, p.7).

Even before the Revolution, however, there had always been a feeling of insecurity about standards. What would happen if a country defined measures in terms of a certain standard which was then lost or stolen or which was altered in some way? Earlier civilizations had guarded their standards in temples and churches. They could, however, still be broken or wear away. Some thought had been given in the eighteenth century to the material of which standards should be made. La Condamine had used an iron bar. Godin in 1756 had proposed glass as a substance not susceptible to variation. There was a fairly general feeling that gold and silver were suitable substances for standards (*Mémoires*, 1747, p.87). Finally when platinum was discovered in the mid-eighteenth century and, equally important, a knowledge of how to work this metal was developed, it became a favourite material for the new metric standards. Yet there was still a feeling that any standards of human construction could not be the ultimate arbiter. Such a standard should reflect a deeper and more permanent reality, that of nature itself.

Nature provided an ultimate arbiter, a secure custodian against storm, fire and flood. The point was made repeatedly that in any dispute nature was always available, 'on tap' so to speak, to settle the matter. This might have been a valid point if only length was concerned and as long as there was an idea of basing the standard on a seconds pendulum at that latitude. However, once the natural unit was related to the dimensions of the earth any reference back to nature involved a complex scientific expedition and a period of several years.

'NATURE' AND MEASUREMENT IN FRANCE

The main justification for seeking a 'natural' unit as universal was that it would not be linked to any one country. It was not so much that it would be *inter*national as *non*-national or rather *supra*-national. Once it was appreciated that the length of a seconds pendulum varied according to latitude, those favouring this as the basis of a universal unit of length had to decide on the latitude to be chosen. Paris, although convenient for many reasons was rejected because of its national implications. The poles were inaccessible and as a reasonable compromise the latitude of 45° was suggested, the mean between the pole and the equator. Unfortunately this line of latitude passed through the French port of Bordeaux and was therefore suspect as having a special connection with France. La Condamine also remarked that there were two latitudes of 45°, one in the southern hemisphere and it would be pure speculation to suppose that they were equivalent. He therefore supported the equator as a unique latitude appropriate for all nations and for all time: 'Enfin la convention du Pendule du Parallèle de 45 degrés, si elle pouvoit avoir lieu, ne seroit fondée que sur une sorte de convenance, et sur l'accord de quelques nations de l'Europe que nous regardons dans le moment présent comme les seules dépositaires des sciences; au lieu que la préférence donnée au Pendule équinoxial convient à tous les lieux et à tous les temps. Aucune nation, aucun siècle à venir ne pourra protester contre ce choix. Un François, il est vrai, préféroit le Pendule du Parallèle de Paris, un Européen en général pourroit, si l'on veut, opter pour celui du Parallèle de 45 degrés. Le Philosophe, le citoyen du monde choisira sans contredit, le Pendule équinoxial' (p.507). The seconds pendulum based on the equator was therefore 'natural' in this third sense of universal.

A new principle was involved in the idea of metric standards, based on nature. Although in origin many units could be said to have some 'natural' foundation in the broadest sense, as far as any government was concerned there was no question of an appeal to the origin of the unit. Units were defined by statutes and/or by standards. In other words it was enough if a ruler decreed that such was the unit. Decrees had the advantage of providing much needed guidance but obviously they could be, and often were, quite arbitrary. The cry of a new natural standard therefore implied a break with the old tradition of legal standards. It was not, of course, the intention that new standards should not also be supported by the force of law but that they should not be *merely* supported by law. In this we see a parallel to the principle of government which would, it was hoped, depend on universally acceptable principles rather than arbitrary authority or tradition. Sir John Riggs Miller, who argued in 1790 that a new system of measurement should be based on 'simple and self-evident principles'[28] had a precedent in the similar claim of the American Declaration of Independence. Miller is quite explicit in his feeling that nature could be quoted as a higher authority than the arbitrary decree of a ruler: 'The essential qualities which every Standard should possess are, that it should be taken from Nature, or connected with something in Nature and not from any work of Art, which must necessarily decay, nor from anything that is merely arbitrary and which has no other right to be a standard than that it is kept in a house which is called the Exchequer or Guildhall and which has certain marks upon it and a certain name given to it' (pp.37-38).

The 'philosophes' and the metric system

We may ask the question to what extent the rationalisation of measurement which took place at the Revolution may be laid to the credit of the *philosophes*. Such credit was sometimes given in the Convention. Thus the president of the Convention, Grégoire, on 25 November 1792 replied to a report from the Académie des

[28] *Speeches*, p.16. The reader who may doubt the radical nature of Miller's thought should compare this page with Condorcet on the same subject.

'NATURE' AND MEASUREMENT IN FRANCE

sciences in these terms: 'Citoyens, la Convention nationale applaudit à l'importance et au succès de votre travail. Depuis longtemps les philosophes plaçoient au nombre de leurs vœux celui d'affranchir les hommes de cette différence des poids et des mesures qui entrave toutes les transactions sociales, et travestit la règle elle-même en un objet de commerce. Mais le gouvernement ne se prêtait pas à ces idées des philosophes; jamais il n'auroit consenti de renoncer à un moyen de désunion. Enfin le génie de la liberté a paru; et il a demandé au génie des sciences quelle est l'unité fixe et invariable, indépendante de tout arbitraire'[29]. Thus, according to this view, only a revolutionary government was prepared to put into practice the program of the *philosophes*. But who precisely were these *philosophes?* In another political speech made in the Convention this time at the session of 19 January 1794 when David was president, the claim is made that it is the authority of philosophy ('le respect qu'imprimeront toujours les lumières de la philosophie'[30]) which should give the metric system general recognition. The context shows clearly that here by 'la philosophie' is meant science.

Consulting an admittedly partial source, the *Mémoires* of the Académie des sciences, we find no support for the idea that the reform was carried out in the name of the *philosophes*. On the contrary Condorcet[31] was proud to trace the origin of the idea of an invariable unit (based on the seconds pendulum) to Huygens, who had of course been one of the members of the Paris Académie in the seventeenth century. If then we consider in turn the origin of the idea of a unit based on nature, the propaganda on its behalf, the execution of the technical work involved in defining the units and constructing standards, all this was the work of men of science. The one aspect missing is the

---

[29] *Mémoire présenté à la Convention au nom de l'Académie des sciences 25 Novembre 1792, suivi de la réponse du président*, pp.7-8.
[30] *Procès verbaux du Comité d'Instruction publique de la Convention natio-* nale, ed. J. Guillaume (Paris 1891-1957), iii.249.
[31] *Mémoires 1788* [1791]; Histoire, p.20.

political support necessary to ensure acceptance of the work of the scientists. It is the support of the legislators for the respect of nature and the rationality of science which allowed the metric system to appear first under the patronage of the French government.

Within the fields of literature and science it is not difficult to distinguish two main groups. First there were the writers usually described as 'the *philosophes*'. Their activities and interests were mainly literary but included the realm of ideas and sometimes science although they were in no sense scientists. On the other hand there were men less well defined as a group than to-day but nevertheless the recognisable ancestor of the modern scientist. They often had in common a mathematical training and usually some connection with the Académie des sciences. In such a division Voltaire belongs to the first group and Maupertuis, for example, to the second. Potentially most important of all were those who defy classification in either category since they belong to both. Alembert, originally a mathematician, became increasingly concerned with general issues so that in his later career he may be regarded primarily as a *philosophe*. In his article 'Pendule' in the *Encyclopédie* he did no more than report the idea of La Condamine that the seconds pendulum could be used as the basis of an international unit of length. Thus on an obvious occasion when Alembert could have provided an influential argument in favour of a 'Natural' unit, he showed no particular interest. Nor does the article 'Mesure' in the *Encyclopédie* advance this idea. The greatest contribution made by the *Encyclopédie* to the reform of weights and measures was the fact of publication of tables showing their diversity and thereby possibly suggesting *implicitly* their rationalisation.

Another enormously influential figure was Buffon, who by presenting science as literature defies the categorisation suggested above. Although Buffon the *philosophe* gave to the public his views on nature in general, Buffon the *savant* also considered more technical problems. In 1745 he referred to the uncertainty of measurement of the earth[32]. He was sceptical in particular about any assumption of uniformity of the earth's curvature. This was a fundamental objection to taking any supposed measurement of the meridian too seriously and the commissioners on the metric system of the 1790s would hardly have claimed Buffon as their source of inspiration.

If any of the *philosophes* are to be considered to have inspired the metric system without taking any direct part in it[33] I would like to stake a claim for Bonnot de Condillac. In more senses than one Condillac may be considered a leading source of inspiration for French men of science in the last quarter of the eighteenth century. Condillac's relation to the metric system is of a general kind. He stands above the other *philosophes* as the patron of a rational system of knowledge based on nature (*Logique*, 1780). If ideas were confused, said Condillac, it was because men had departed from nature which should be our guide. Our language must be based on nature and must follow an order in which everything has a rational and above all a *natural* connection. He only wished that men of his persuasion had been available to advise at the formation of human languages. This far-fetched wish was to be realised after the Revolution not in ordinary language but in the language of measurement. Lavoisier's reform of chemical terminology in 1787 was based explicitly on Condillac's philosophy. Although the metric reformers of the 1790s did not invoke Condillac by name I am suggesting that they were influenced and encouraged by his philosophy.

The academicians and the administrators had to pronounce primarily in terms of scientific possibilities and political principles rather than the ideology of nature. When they did talk of nature it was a reference to a common theme of the eighteenth century but when they insisted not only on a natural unit but a *rational system* of units they were putting into practice an ideology of which Condillac was the leading exponent.

---

[32] 'Preuves de la théorie de la terre', *Histoire naturelle* (1749), i.165.

[33] thus Condorcet is excluded from this generalization.

The decimal scale

Apart from the adoption of new units of measurement, the other main aspect of the French metric system was the use of a decimal scale. If we look hard enough we can find the argument put forward that the scale of ten had a justification in nature. Thus in a pamphlet concerned with the monetary system, which Mirabeau distributed to members of the National assembly in December 1790, he claimed that 'La nature semble nous avoir indiqué ce nombre décimal'[34]. One sense in which it was 'natural' was the case of communication of numbers to deaf persons; man had a ready means of communication by means of his ten fingers. His fingers also provided man with a means of reckoning so that in these two senses at least Mirabeau could claim 'nos mains sont les types de l'arithmétique naturelle'. Nevertheless Mirabeau realised that this was hardly a strong case for a natural decimal system and it is interesting that he should claim support also from the authority of antiquity. He cited the ancient civilization of China where a monetary system on a decimal basis made calculation simple. However, Mirabeau was doing no more than exploring the possibility of a decimal system, and for weights he proposed a rationalised duodecimal system. His cautious approach was soon to be buried beneath more fundamental and far-reaching reforms.

Yet the claim that ten was in some way 'natural' reappeared in the *Instruction sur les mesures,* intended for popular consumption: 'Nous observons en général que l'arithmétique décimale a été celle de toutes les nations dont l'histoire ait conservé le souvenir. Il est probable que la conformation de la main aura déterminé une sorte de prédilection en faveur du nombre dix que chacun avoit dans ses doigts, et qui se trouvoit continuellement, sous ses

---

[34] 'Observations préliminaires sur le premier rapport du comité des mon- naies', *Discours et opinions* (Paris 1820), iii.182.

'NATURE' AND MEASUREMENT IN FRANCE

yeux, et qu'ainsi ce nombre sera devenu naturellement la base de l'échelle arithmétique adoptée dans tous les pays. Or, quoique cette préférence donnée au nombre dix paroisse avoir été l'effet de l'instinct plutôt que de la réflexion, il n'en est pas moins vrai qu'elle est puisée en quelque sorte dans la nature de l'homme, et que ce concert unanime qui en est résulté entre toutes les nations, mérite d'être respecté, et ne doit pas être contrarié sans des raisons de la plus grande force'[35]. The argument is, therefore, that the decimal system has some natural justification although not on the highest plane; our preference for ten is based more on instinct than on reason. And yet there were also strongly rational grounds for a decimal scale.

The support given by scientists to the decimal scale was usually for practical rather than ideological reasons. The Committee of the Académie des sciences in March 1791 had admitted that a scale of 10 was in a sense arbitrary, yet always described it as 'the arithmetical scale' (l'échelle arithmétique) (Rapport, p.11). It could therefore be argued that what they were trying to do was to bring the man-made system of weights and measures in conformity with the—implied natural—arithmetical scale. They further wished to rationalise the relations between weights and measures of different magnitude. Finally there was the argument of simplicity. This not only had its metaphysical and aesthetic side but also its political. Condorcet[36] hoped that a situation would be created in which 'les citoyens puissent se suffire à eux-mêmes dans tous les calculs relatifs à leurs intérêts; indépendance sans

---

[35] *Instruction sur les mesures déduites de la grandeur de la terre, uniformes pour toute la République, et sur les calculs relatifs à leur division décimale par la Commission temporaire des poids et mesures républicaines* (Paris an 2 [1794]), pp.xx-xxi.

[36] *Œuvres*, xi.583. We may compare Condorcet's reasons with those given by Gouverneur Morris in 1782 for

introducing decimal coinage in the United States: 'all calculations . . . are rendered much more simple and accurate and . . . more within the power of the great mass of the people', quoted by C. Doris Hellman, 'Jefferson's efforts towards decimalisation of United States weights and measures', *Isis* (1931), xvi.266-314 (268).

laquelle ils ne peuvent être ni réellement égaux en droits . . . ni réellement libres'. The point was, of course, that the simplest system would be intelligible to workers and peasants with a minimum of formal education whereas the complicated system of measurement under the ancien regime meant that the consumer often had to take commercial and legal transactions on trust.

Looking back later Laplace saw the adoption of the decimal scale as the most important immediate gain from the introduction of the metric system and a permanent contribution to 'the progress of civilization' (Bigourdan, p.193).

The precedent of antiquity

An interesting alternative to the appeal to the authority of nature was the appeal to the authority of antiquity. This was an argument by precedent, showing respect for the past, the distant past, by those who believed that there had once been a golden age. The idea might be expected to have had the support of classical scholars and some literary figures but what is more remarkable is that it had the earnest support of many prominent scientists, including Lavoisier. Although as early as 1702 Cassini[37] had encouraged the reconstruction of Egyptian, Greek and Roman measures the serious discussion came later in the eighteenth century. It was stimulated by the anonymous publication of a large book in 1780 entitled *Métrologie ou traité des mesures, poids et monnoies des anciens peuples et des modernes*. The author, Alexis Jean Pierre Paucton, argued that such ancient monuments as the great pyramid of Egypt constituted a permanent record of their system of measurement, which was based on the dimensions of the earth. Paucton produced a great deal of evidence which suggested a fundamental similarity in the units of measurement of different classical civilizations. This similarity he explained as

[37] Mémoires (1702); *Histoire*, pp.80-81.

deriving from a common origin, the dimensions of the earth. Here was a fundamental argument favouring the choice of the degree of the meridian as the basis of an international unit rather than the length of the seconds pendulum.

Paucton's claim was accepted by the mineralogist Romé de l'Isle, who was particularly concerned to show that the modern disharmony of weights and measures was quite unnecessary[38]. He recommended their reform as a task for the newly convened states general.

The astronomer Bailly[39] also had this great respect for antiquity and was happy to think of a precedent for the idea of a unit of measurement based on nature. As he expressed it: 'Le module des mesures itinéraires a été gravé sur les fondemens de la maison commune, pour instruire les hôtes de tous les siècles'.

The vital factors in the appeal to antiquity were first a precedent for a universal unit of length and second that this unit was related to the size of the earth. Lavoisier (Œuvres, vi.699) sums up the research of Paucton and Bailly as follows: 'Il résulte de leurs recherches qu'antérieurement à tout ce que l'histoire ancienne nous a transmis, il existait un peuple d'une antiquité très reculée, qui avait imaginé et exécuté un système métrique déduit des dimensions de la terre'. Thus even Lavoisier, who elsewhere shows the influence of Condorcet's view of human progress, felt that what eighteenth century man was achieving was no more than a recapitulation of a prehistoric accomplishment.

When, therefore, Tillet and Abeille spoke on behalf of Société royale d'agriculture in 1790 on the subject of uniformity of weights and measures, the idea of a universal system of measurement in antiquity was widely accepted. They felt that the question was still an open one and pointed out that much of the evidence of agreement or uniformity of ancient standards depended on special pleading. They therefore called for further research:

[38] Métrologie ou tables pour servir à l'intelligence des poids et mesures des anciens (Paris 1789).

[39] Histoire de l'astronomie moderne (Paris 1785), i.157.

'Nous désirons pour l'honneur de l'humanité que le résultat d'un si bel ouvrage substitue aux probabilités que plusieurs savans ont déjà rassemblées des preuves claires de l'ancienne existence d'un système métrique universel. Tout nous porte à croire que ce système existe encore et qu'il suffiroit d'écarter la rouille qui en défigure les copies pour reconnoitre que les Peuples se servent de poids et de mesures dont l'étalon-matrice, pris dans la nature, a toujours été le même'[40]. However, after 1790 research was less of the philological nature envisaged by Tillet and Abeille and more on practical lines. It was the Académie des sciences rather than the Académie des belles-lettres which was entrusted with the task of constructing a metric system. I would suggest that this classical precedent was a powerful and perhaps a decisive argument which finally persuaded the committee of the Académie des sciences to change from a natural unit based on the pendulum to one based on the dimensions of the earth. I am therefore suggesting that the change depended on the literary and archeological evidence as well as on scientific considerations.

One of the most strenuous opponents of the idea of an appeal to antiquity was Condorcet. His concept of material and moral progress based on the model of the recent advance in science would not allow him to accept the idea of a primitive system of measurement which would bear comparison with the precision of the late eighteenth century measurements. Also as a mathematician he had strict ideas on accuracy. These views are brought out in his éloge of Bourguignon d'Anville, published in 1785[41]. Anville was one of the leading geographers of the eighteenth century, and had been a member of the Académie des sciences as well as the Académie des inscriptions et belles-lettres. Yet while Condorcet's task as secretary of the former Académie was to praise his one-time colleague, there is a strong note of reservation noticeable in this *éloge*. Condorcet praises Anville's industry but

---

[40] *Observation de la Société royale d'agriculture sur l'uniformité des poids et des mesures* (Paris 1790), p.119.

[41] *Mémoires* (1782); *Histoire*, pp.69-77; see especially pp.71, 74-75.

emphasises the difficulty of his researches on ancient geography, e. g. his *Traité des mesures itinéraires anciennes et modernes* (1769). Looking at ancient writings and inscriptions, it was almost impossible to state exactly what length was intended by any particular term since the same name had been used to denote different lengths in different countries and different periods. With so many uncertainties in the data any conclusion would be far from certain. Condorcet contrasts the uncertainty of the philological approach with the certainty and precision of astronomical data. In one of his books Anville had arrived at a conclusion which was contradicted by the astronomers. Condorcet's conclusion is clear: only modern science and in particular astronomy can contribute to metrology. The classical scholar can at best provide approximate evidence. His standards of precision are totally inadequate to be of use in science.

Although the committee consisting of Borda, Lagrange, Laplace, Monge and Condorcet made no mention of a classical precedent in their report of 19 March 1791 and gave other reasons for preferring a unit based on the earth, when a further committee consisting only of Borda, Lagrange and Monge reported in May 1793 it was at pains to point out classical precedents[42]. This may be because the presence of Condorcet on the previous committee had inhibited the use of a classical rather than a scientific or practical argument. Alternatively it might be thought that the classical precedent was not so much an initial reason for the choice of the earth as a later justification for it. Even in the latter case however it is interesting to see what arguments were thought relevant. It adds to the interest to remember that this was a report by scientists to the official body of science. It had not therefore been dressed up to look respectable on the political plane.

Indeed when on 1 March 1795 Prieur de la Côte d'Or, that vigorous advocate of the metric system, presented a report to the Convention on behalf of the Committee of Public Instruction,

[42] *Mém. Acad. Sci.*, 1789 (1793), *Histoire*, 4-5.

he made a special point of disassociating the new system from the scales existing in ancient civilizations[43]. He, like Condorcet, felt that it should be justified on scientific, social and political grounds. Yet he did not deny the classical precedent and may well have felt like many of his contemporaries that it conferred on the system a higher degree of moral authority.

Conclusion: the metric system as a 'natural' system

I should like now to answer the question how successful the metric system was as a *natural* system. An obvious objection is the limitation of accuracy with which it was possible to measure the earth. Although technique of measurement had improved by the 1790s so that the measurements made then were superior to those of the early eighteenth century, the degree of error was still considerable, as some of the foreign scientists associated with the scheme pointed out[44]. If some of the more idealistic felt that they had achieved the millenium, the more hard-headed of their contemporaries could see that at best they could only provide an approximation. A dispute about a natural unit then became an argument about successive decimal figures. This may seem a trivial objection but it does detract from the dignity of a supposedly absolute system. It reduces it to a level of operational competence. It is natural only in so far as man's technique of measurement of comparatively large distances allows him to assess them in a particular period.

Another aspect which might be overlooked in the theoretical discussion of a 'natural' system is the practical point of the size

[43] Prieur Duvernois, *Rapport fait au nom du comité d'Instruction publique sur la nécessité et les moyens d'introduire dans toute la république les nouveaux poids et mesures précédemment décrétés* (Paris Ventose an 3 [1795]).

[44] *e. g.* the Danish representative at the congress on definitive metric standards held in Paris in 1798-1799. Maurice Crosland, ed. *Science in France in the Revolutionary era described by Thomas Bugge* (Cambridge, Mass. 1969), pp.205-211.

'NATURE' AND MEASUREMENT IN FRANCE

of the new units. A factor in favour of the adoption of the seconds pendulum had been that its length was approximately equal to three feet or half a *toise*. When the pendulum was rejected, the fraction chosen of the earth's circumference was selected to be of the same order of magnitude as the old unit, and conveniently differed only slightly from it. Thus neither of these proposals involved such a radical departure from the old units as might be thought from some of the abstract discussion.

After the adoption of a unit of length the second fundamental aspect of the metric system was its division and multiplication to form lesser and greater units. It was a major advance to abandon a confused and largely arbitrary system of numbers and substitute in all cases the number ten. The advantage of uniformity, however, should not obscure the departure from the natural ideal. Although pragmatically in the pre-computer age 10 can be justified in terms of convenience, the anthropomorphic origins —ten fingers, ten toes—are as clear as in the choice of the foot or finger as a unit of measurement. Once again the natural ideal had been compromised. There was also a suppressed premiss in the choice of the ten-millionth part of the meridian quadrant as the unit of length since this assumed that ten or a multiple of ten was a 'natural' fraction.

A much more fundamental criticism, however, was that the eighteenth century assumption of the equality of meridians was not justified. The meridian through Paris was merely one of an indefinite number of meridians in the northern hemisphere which could have been measured. Having rejected Bordeaux on the grounds that it would not be accepted internationally, the savants and politicians were guilty of as great a deviation from nature in selecting the Paris meridian.

So far nearly all the discussion had been concerned with the measurement of length. However, a further difficulty arises when we consider the unit of mass[45]. This was rightly related to the unit

[45] physicists consider mass rather than weight since mass may be con- sidered in absolute terms whereas the weight of a body (*e. g.* as measured by

of length and understandably to the decimal principle, but this was not enough. It had to be tied to a particular material substance. The choice of earth, one of the Aristotelian elements still accepted by some chemists in the mid-eighteenth century, might have had a certain poetic justification in so far as the metre was based on the planet Earth. Even the crudest observation would, however, reveal the impracticality of selecting a sample which could in any way be considered representative and no one seriously suggested it. The substance chosen had to be a fluid in order to ensure homogeneity. The choice fell on water, not this time in the Aristotelian sense as denoting matter in the liquid state, but the substance which Lavoisier had recently shown to be a compound of hydrogen and oxygen gases. There was, therefore, no question of water being an element; its claim to being natural rested on its common occurrence and possibly to a lesser extent on its universality on the globe.

It was in the choice of the unit of mass that one of the weakest points in the metric system is to be found. The committee appointed by the Académie des sciences reporting on 19 March 1791 and consisting of Borda, Lagrange, Laplace, Monge and Condorcet announced that for the basic material they had chosen water as 'a homogeneous substance easy to obtain at all times in the same degree of purity and density' (Méchain-Delambre, i.16). It is when this was put into practice that departures from the natural ideal were forced on the scientists. Firstly and most obviously the water used had to be pure, hence, abandoning the poetic possibilities of taking water from the oceans or from a mineral spring, it was taken from the chemists' laboratory—distilled water. Secondly it was important to specify the temperature. What was more natural than the committee's suggestion that the temperature at which water 'passes from the solid state to the liquid state', in other words the freezing point of water, 0°C. Notice however that the committee did not have to specify any

a spring balance) depends on the place
where the measurement is made.

temperature scale. As the physicist Lefèvre-Gineau said: 'il étoit essentiel d'éviter l'établissement d'une échelle de thermomètre avant celui des unités primitives du système métrique, afin de ne pas y faire entrer les quantités qui doivent y être étrangères. On a en conséquence choisi une température très constante à laquelle on suppose que l'unité de mesures d'étendue soit exposée pour la donner telle qu'elle doit véritablement être. C'est la température qui elle-même est la base de nos échelles' (iii.563).

This then was a 'natural' temperature as opposed, for example, to a lower temperature produced by a mixture of ice and salt which could justifiably be described as 'artificial cold'. Unfortunately if water was a normal substance in the sense of its common occurrence, it was not in its properties. Water was found to have an anomalous behaviour at the bottom of the ordinary temperature scale. Instead of expanding uniformly with rise of temperature like the majority of solids, liquids and gases it showed a contraction between $0°C$ and about $4°C$ and then began to expand. There was therefore at about $4°C$ a temperature of maximum density. This was less convenient than the freezing point but could be described simply as the temperature of maximum density without reference to a temperature scale. In a sense it was a 'natural' temperature as far as water was concerned but of course the original harmony had been destroyed. If 'natural' is to be understood in the sense of 'normal' or 'representative' water really was a most unfortunate choice.

There was a third factor which tended to emphasise the artificial nature of the exercise in practice. If the cube of water was weighed in air in the normal way, because of Archimedes's principle the weight registered would depend on the difference in density of the water and the materials of which the weights were made. Thus in order to obtain an absolute rather than a relative value the weighing had to be done in a vacuum. If one reads the detailed account[46] in which Lefèvre-Gineau described the

[46] *ibid.*, iii.558-580; Crosland, *Science in France*, pp.208-210.

ingenious but painstaking methods with which he devised the conditions of vacuum weighing, together with various sophisticated corrections one has difficulty in remembering that it was done originally in the name of nature.

Our final judgment on the idea of a 'natural' unit in the eighteenth century must be that it was premature. The optimistic *savants* of 1791 underestimated the problems of obtaining an exact measurement of the earth. In practice the new standard was not the earth at all but a metal bar—just as it had been in the seventeenth century. The main advancement was in the construction and maintenance of the standard and the methods used for comparison. Also the new unit was not a local one but a national one with the prospect of international recognition; and it was incorporated into an integrated system. Yet despite these incidental benefits the goal of a natural standard accessible in the laboratory lay far in the future.

For most of the nineteenth century the metre was again defined in terms of an increasingly sophisticated legal standard. In 1889 the metre was the distance between two marks on a specially constructed platinum-iridium bar at the temperature of melting ice at the international bureau of weights and measures at Sèvres. Yet the hope had not vanished of finding something in nature which is constant and could serve as the basis of fundamental units. The wave theory of light and spectroscopy, both developments of nineteenth-century science, presented the possibility of a natural standard accessible in the laboratory. In 1893 Michelson[47] was able to measure a standard metre in terms of one of the characteristic wavelengths of the red line of the spectrum of the element cadmium. Further investigation revealed that this definition was not as specific as had been thought and an isotope of the inert gas krypton was substituted. In 1960 the metre was redefined as 1,650,763.73 wavelengths of the orange-red line of krypton 86.

[47] 'Light waves and their application to metrology', *Nature* (1893-1894), xlix.56-60.

'NATURE' AND MEASUREMENT IN FRANCE

The sophistication of modern science takes us far from the eighteenth century yet the principle of natural standards, whether of length or mass or time, would seem a permanent feature of science.

II

The Image of Science as a Threat: Burke versus Priestley and the 'Philosophic Revolution'

So much of the history of science has been written from the point of view of the scientist or the proto-scientist that it may be salutary for the modern reader occasionally to consider how science and its early practitioners were viewed from the outside. We must not be too surprised if a pioneering activity performed by controversial agents was misunderstood or misrepresented and if what emerges is, therefore, sometimes less of a portrait than a caricature. We are concerned here much less with what natural philosophers actually did than what they were thought to have done, or what they were thought to stand for. The image is sometimes more influential than the reality. Considering that the period to be studied is one of major political and social unrest and that the principal spokesman, Edmund Burke (1729–1797), had made his reputation mainly in the arena of parliamentary politics, we can anticipate rather more polemic than dispassionate argument. In the formation of public opinion a colourful exaggeration or even an occasional sneer are often more effective than the objective exposition of a case. The spectacles through which Burke looked at his world sometimes magnified and often distorted, but they produced a view of knowledge and society shared by many of his contemporaries and of considerable subsequent influence.

Edmund Burke, statesman, political theorist and writer, has never attracted much attention from historians of science. He has traditionally been studied in university departments of history and politics and, although numerous books and articles have been written on these aspects of Burke, there is no published treatment of Burke's views on science.[1] As a great writer ('the supreme writer of his century' according to De

---

1 A recent bibliography, listing more than 1600 secondary works on Burke, gets no nearer to the subject of the present paper than a handful of works which relate him to the Enlightenment. See Clara I. Gandy and Peter J. Stanlis, *Edmund Burke. A bibliography of Secondary Studies to 1982,* New York and London, 1983. P. Stanlis, *Edmund Burke and the Natural Law,* Ann Arbor, Michigan, 1958, is a valuable book on its subject but the author specifically excludes physical science from his discussion (p. 5). He declines to consider the relation of his subject to science, of which he is wholly critical (p. 23). Many authors, if they mention Burke's discussion of science, imply that this was no more than the use of a metaphor, e.g. Seamus F. Deane, 'Burke and the French philosophes', *Studies in Burke and His Time,* (1968/69), **10**, pp. 1113–1137 (p. 1117). Gerald W. Chapman, *Edmund Burke, the Practical Imagination,* Cambridge, Mass., 1967, p. 278 devotes three lines to Burke's use of scientific terms. A book with considerable authority is J. J. Boulton, *The Language of Politics in the Age of Wilkes and Burke,* London, 1963, especially chapter 7, although again the emphasis is on metaphor. Michael Freeman's useful book, *Edmund Burke and the Critique of Political Radicalism* (Oxford, 1980, p. 33) has a comment to make on Burke's view of the less than human characteristics of 'scientised' men, of whom Priestley would be an obvious example.

*Unit for the History of Science, Physics Building, University of Kent, Canterbury, Kent CT2 7NR, U.K.

I should like to express my thanks to colleagues at the University of Kent for comments on different aspects of this paper, especially to Grayson Ditchfield, who generously allowed me to benefit from his expertise on the eighteenth-century English political scene and the role of the Dissenters; also to Alec Dolby and Crosbie Smith, with whom I discussed the conclusions. In the University of Kent library I have been able to draw on the valuable resources of the Maddison Collection, which includes strong Priestley holdings.

Quincey²), he has also been studied for his prose style.

Yet in the 1790s, Burke's reaction to the French Revolution made an important contribution to intellectual history. He had many things to say about the threat to the established order from new ideas and science. He also went out of his way to attack Joseph Priestley (1733–1804), who combined the roles of Dissenting minister and man of science. We shall see how Priestley's science came under attack from Burke's pen. But, quite apart from Priestley, Burke felt the need to warn his contemporaries about what would later be called scientism, represented for him by a number of *philosophes,* and particularly by Condorcet.

Burke is, of course, by no means the first to criticize early modern science. Earlier in the eighteenth century, Pope, Swift and Rousseau all commented adversely on it. But Burke, writing in an age of revolution, adds a special intensity and gives a new direction to his critique. The harm done by science to the individual through pride, as suggested for example by Pope,[3] is now less important than the harm done to society by the ambition of social engineering.

Unlike the poet Blake, who can see nothing but evil and atheism in Newtonian science,[4] Burke represents a succession of increasingly hostile attitudes towards science. There is a historical progression which parallels the political development of the French Revolution and its increasing threat to the three pillars of the social order in Britain: the monarchy, the aristocracy and the Church of England.[5] Burke had always been suspicious of abstract reasoning and theory[6] and his instinctive reaction to the claims of science is one of caution. This develops in the 1790s through distrust to outright hostility and finally to extreme hatred. Science and philosophy are seen as a threat as never before. Burke, therefore, is writing in a period of intense political, social and cultural change. He is an exceptionally articulate chronicler of the fears of the English upper classes, faced with a world in which the Anglican-aristocratic hegemony seemed to be threatened by atheism and egalitarianism, and appeals to order and tradition were replaced by appeals to Nature and reason.

The eighteenth century was a period of spectacular growth in several branches of science. This growth was represented as progress. In other words, science was rapidly creating a new world. Of course, there had also been important scientific movements in the seventeenth century, but only in the eighteenth century are there widespread claims about the relevance of science *as applied to society.* This is brought to a head by the French Revolution. To express this in slightly different terms, the *philosophes* in their glorification of science had invested it with greater social space. It was not only that there

2 *Blackwood's Magazine,* December 1828, extracted by A. M. D. Hughes (ed.), *Edmund Burke. Selections,* Oxford, 1930, p. 22.

3 Alexander Pope's famous injunction: 'presume not God to scan' (*Essay on man,* Epistle II, line 1) is supplemented by, for example, 'Trace Science then, with Modesty thy guide' (*Ibid.,* line 43). See also Marjorie Nicolson and G. S. Rousseau, *'This Long Disease, My Life'. Alexander Pope and the Sciences,* Princeton, 1968.

4 See, for example, William Blake's association of Newton with materialism, and the *philosophes* with the Industrial Revolution in Geoffrey Keynes (ed.), *Poetry and Prose of William Blake,* London, 1939, p. 107 (miscellaneous poems) and p. 449 (Jerusalem, 1).

5 J. C. D. Clarke, *English Society, 1688–1832,* Cambridge, 1985.

6 J. MacCunn, *The Political Philosophy of Burke,* London, 1913; reissued New York, 1965.

was new knowledge. The boundaries of scientific knowledge had been vastly extended, making it a potential threat to the established order.[7] With the Revolution, science and its devotees could be seen as an actual threat.

It is therefore a matter of interest to know how society responded. Burke articulates the ideas, prejudices and misconceptions of his time. He can be seen as a spokesman for the conservative response to the claims of the French Revolution. If he misunderstands science, there is a lesson to be learned. Although later adopted as a philosopher of conservatism, Burke was not a Tory but a Whig, and one who had earlier found common cause with Priestley in supporting the claims of the American colonists. The French Revolution, however, forced him to rethink his position and his party loyalties. A man who, in his earlier career had fought to limit the prerogatives of the King, now devoted himself to the task of limiting the influence of the new ideas associated with the Revolution.

The French Revolution was the political revolution which, before the twentieth century, had the greatest impact on society. It sent waves throughout Europe but nowhere, perhaps, were hopes and fears raised to a greater height than in Britain, the traditional enemy of France and rival for 'great power' status. In no other country did the French Revolution produce such a reaction and a political polarization. But although the reaction is usually described in purely political terms, Burke appreciated that it was also a social revolution. He blamed the revolution on writers, intellectuals and men of science, whom he described collectively as 'philosophers'. It was their ambitions, their ideas and their claims to new knowledge which were at the heart of the revolution.

It was the events of 1789 which prompted Burke's most famous and influential work: *Reflections on the Revolution in France*, first published in November 1790. A literary *tour de force*, a further reason for the success of the book was that it was soon vindicated as a work of prophecy. In portraying the humiliation of the French king and queen, it set the scene for the violence to come. In one of the more famous passages of the book, Burke describes the seizure of Marie Antoinette, and declares that:

> the age of chivalry has gone—that of sophisters, oeconomists and calculators has succeeded; and the glory of Europe is extinguished for ever.[8]

Opinions are divided about the real message of the *Reflections*. One historian writes dismissively:

> Burke's pamphlet (sic) was an *œuvre de circonstance*, reflecting his fears of the immediate impact of the French revolution on English society and politics rather than its ulterior threat to the fabric of European civilization.[9]

7 The threat of the new knowledge to Burke's world is explored in a second article by the present author. The metaphor of intellectual space has most recently been applied to a seventeenth-century context by Stephen Shapin and Simon Schaffer, *Leviathan and the Air Pump. Hobbes, Boyle and the Experimental Life*, Princeton, N.J., 1985, pp. 332ff.

8 Edmund Burke, *Reflections on the Revolution in France* (1790), edited and with an introduction by Conor Cruise O'Brien, Harmondsworth, Middx., 1981, p. 170. All quotations from the *Reflections* will be referred to this Penguin edition, being the one most widely available. It is to be regretted that, although there is an excellent scholarly edition of Burke's *Correspondence*, there is not yet a standard scholarly edition of his works. The plethora of editions of his works, although indicative of his great influence, serves merely to increase the difficulties of the modern commentator.

9 Albert Goodwin, *The Friends of Liberty. The English Democratic Movement in the Age of the French Revolution*, London, 1979, p. 99.

But I believe that Burke was deeply concerned with both of these aspects, and I prefer the assessment of Cobban:

> For Burke the Revolution was not a mere change of government by force, as revolutions had been in the past, but a turning point in the history of western civilization.[10]

The French Revolution was one with universal ramifications as opposed to the so-called 'Glorious Revolution' of 1688, which was, as Englishmen proudly proclaimed, peculiar and local.[11] The Whigs were committed to the 1688 settlement and it was one of Burke's major tasks to convince his fellow countrymen that the Revolution of 1789 was of a totally different order. Whatever the deficiencies of Louis XVI, they did not justify a revolution so fundamental that not only France but the whole of the Western world would never be the same again. Burke claimed that he was taking up the cause of humanity.[12] This canvas is much broader than the context of parliamentary debates; it is broader than any national politics. There is a vast literature on Burke's politics and political thought, on some of which I have thankfully drawn.[13] The purpose of this paper is not to dispute the political importance of Burke. Rather it is to suggest that concentration on Burke's politics has to a certain extent detracted from the understanding of his importance in social and intellectual history. Burke himself in a private letter refers to 1789 as 'the Great Revolution in Human affairs which has begun in France'.[14] It has been perceptively remarked that there is a growing universality in Burke's concerns in the last years of his life.[15] Now at the end of his parliamentary career, Burke was able to transcend party politics and devote his final years to warning his contemporaries of the dangers of the new age.

Burke appreciated that in France a number of revolutions were taking place simultaneously. Not only was there a political revolution but also a social one. Traditional social distinctions were disappearing. The moral revolution was not the least important

10 *The Correspondence of Edmund Burke*, (ed. T. W. Copeland *et al.*), 10 vols, Cambridge, 1958–1978, **vi** (ed. Alfred Cobban and Robert A. Smith) p. xx. Subsequently this will be referred to simply as *Correspondence*. See also Robert A. Smith, 'Burke's crusade against the French Revolution: principles and prejudices', *Burke Newsletter*, (1966), 7, pp. 552–569.

11 J. M. Roberts, *The Mythology of Secret Societies*, St Albans, 1974, pp. 162–163.

12 *Reflections*, e.g. p. 297. Mary Wollstonecraft, however, pointed out that Burke's 'humanity' excluded the vast majority of ordinary people.

13 There are useful comments on Burke's *Reflections* in George Fasel, *Edmund Burke*, Boston, Mass., 1983 and, most recently, in Marilyn Butler, *Burke, Paine, Godwin and the Revolution Controversy*, Cambridge, 1984. On Burke's reaction to the French Revolution see Philip Anthony Brown, *The French Revolution in English History*, London, 1965, chapter IV. C. P. Courtney, *Montesquieu and Burke*, Oxford, 1963, (especially chapter VIII), is more relevant than the title of the book might suggest. Isaac Kramnick, *The Rage of Edmund Burke*, New York, 1977, has relevant material, whether or not one agrees with the author's psychoanalytic thesis. Frank O'Gorman, *Edmund Burke, his Political Philosophy*, London, 1973, in emphasizing Burke the parliamentarian, seems to deny the possibility of a philosophical basis for Burke's political thought. David Cameron (*The Social Thought of Rousseau and Burke. A Comparative Study*, London, 1973, pp. 80–81), while reminding us that Burke was not a philosopher, is more positive. The most recent writer to argue that Burke was not a philosopher is F. P. Lock, (*Burke's Reflections on the Revolution in France*, London, 1985), since he wished to focus on Burke's use of rhetoric to persuade his contemporaries. Frequent use of metaphor does not, however, preclude the possibility of a system of ideas behind the words.

14 Letter from Burke to Lord Fitzwilliam, 21 November 1791, *Correspondence*, op. cit. (10), **vi**, p. 450.

15 Frank O'Gorman, op. cit. (13), p. 14.

aspect. Indeed, Burke spoke of the change as a 'revolution in sentiments, manners and moral opinions'.[16] The French had also brought about a revolution in finance by their reckless introduction of paper money.

But above all it was a revolution in ideas.[17] Ideas embraced all aspects of human life. These were the intangibles which could pass over national frontiers and affect public opinion. When war came, one of the tasks Burke set himself was—as the editors of his *Correspondence* put it—to persuade the Pitt Administration 'to regard the conflict as a war against the ideas of the Revolution'.[18] The new ideas were being propagated with all the fervour of a religious movement:

> The present revolution in France . . . is a revolution of doctrine and theoretic dogma.[19]

It is a 'philosophic revolution',[20] i.e., a revolution in ideas brought about by 'philosophers'. Burke writes of the revolutionaries in France as having 'inverted order in all things'.[21] The French, he says, are

> a people who attempt to reverse the very nature of things,[22]

a comment one must interpret in a social and economic context. The most wide-ranging comment from Burke's pen, written after the execution of Louis XVI, is that the Jacobin Republic has

> made a schism with the whole universe, and that schism extended to almost everything great and small.[23]

Indeed, it is Burke's frequent comments on the relation of the French Revolution to 'the universe' which reminds us of the close parallel he sees between society and the natural world.

Priestley and chemistry

The satire on science contained in Swift's *Gullivers Travels* (1726) was directed at the Royal Society, but the undistinguished history of that body in the period of the Enlightenment produced a situation in which few people in the late eighteenth century could take it as representative of English science. In the absence of competing national institutions, one turns to individuals. If any one person were to be chosen as the exemplar of experimental science in the decade before the French Revolution, the most likely

16 *Reflections,* op. cit. (8), p. 175.
17 It is worth noting that when Paine claimed that every man had 'natural rights', he explained such rights as being in the first place 'intellectual rights or rights of the mind', Thomas Paine, *The Rights of Man,* Part I 1791), Everyman edn., London, 1966, p. 44.
18 *Correspondence,* op. cit. (10), vii, p. xi (my italics).
19 *Thoughts on French Affairs* (1791), *Reflections,* Everyman edn., London, 1967, p. 288.
20 *Reflections,* p. 237.
21 Ibid., p. 161. There is an obvious parallel with the situation arising from the civil war in seventeenth-century England. See, e.g., Christopher Hill, *The World Turned Upside Down; Radical Ideas during the English Revolution,* London, 1972.
22 Ibid., p. 359.
23 *Letters on a Regicide Peace* (1795–1797), *Works,* Bohn's Standard Library, London, 1903, iii, 215.

candidate would be Joseph Priestley. Priestley was known primarily to Burke as a troublesome, outspoken Dissenter, a man who, as a Unitarian, could hardly be considered as a Christian at all,[24] and someone who was to become increasingly involved in radical politics. In 1782, Priestley had written to him about his latest research on gases, a subject he considered Burke might find of some interest.[25] Priestley would have been pained if he had foreseen Burke's later jocular response to his prolonged and serious studies of different kinds of air. Perhaps if Priestley had done research in a different field, say in agricultural chemistry in the way Davy was later to do, Burke might have been more impressed, in view of the potential economic utility of the enterprise. Yet in England the late eighteenth century was rather a lean period for science as opposed to technology.[26] It is understandable that Burke should think of France as the leading scientific nation. In England, scientific innovation was to be found less in the capital than in the Midlands and the North, where industry was fast developing. Priestley, as one of the most eminent members of the Lunar Society of Birmingham, was particularly visible in the 1780s, and could understandably be taken as a representative of physical science in England at the time. Priestley had begun his scientific career in natural philosophy (or 'physics') but it was as a chemist that he was able to make his greatest reputation. Chemistry was the most innovative of all the sciences in the last quarter of the eighteenth century.

The fact that Priestley wrote variously on theology, science and politics helps us to understand Burke's conception of science as having strong ideological implications. Nor should we, from our twentieth-century perspective, assume that Priestley kept science separate from theology and politics. It was, after all, in a late edition of one of his scientific works, the *Experiments and Observations on Air*, that Priestley saw fit to comment on the established church. He said that the growth of knowledge

> will ... be the means under God of extirpating all error and prejudice and of putting an end to all undue and usurped authority in the business of religion, as well as of science.[27]

It was in the same work that he made the bold claim that:

> the English hierarchy ... has ... reason to tremble even at an air pump or an electrical machine.[28]

Sometimes, and particularly in the period 1790–1794, Burke's attacks on science were indirect attacks on Priestley. As Coleridge writes of Burke: 'Are his opponents calculators? *Then* calculation itself is represented as a sort of crime'.[29] Priestley's science

24 In a Trinitarian age religious radicalism was strongly associated with anti-Trinitarianism.

25 Priestley writes: 'you seemed to give some attention to the object of my experiments', 11 December 1782. R. E. Schofield (ed.), *A Scientific Autobiography of Joseph Priestley, 1733–1804*, Cambridge, Mass., 1963, p. 216. For Burke's awareness of Priestley's scientific work as early as 1772 see footnote 213.

26 A situation often explained by 'the shadow of the great Isaac Newton'.

27 *Experiments and Observations on Air*, Birmingham, 1790, vol. i, preface. J. T. Rutt (ed.), *The Theological and Miscellaneous Works of Joseph Priestley*, 26 vols, London, 1817–1832, xxv, p. 375, referred to hereafter as Priestley, *Works*.

28 Ibid.

29 Coleridge, 'Essay on the grounds of government' (first published in *The Friend*, 1809). *The Collected Works of Samuel Taylor Coleridge* (ed. Kathleen Coburn) vol. iv (*The Friend*, vol. i, ed. Barbara E. Rooke), London and Princeton, 1969, p. 188.

of chemistry provided a convenient butt for ridicule, partly through its historical association with fraudulent alchemy and partly through its current emphasis on gases. These invisible substances which potentially could make all the difference between life and death, which could produce explosions, understandably provided reactions varying from scorn to alarm.

Priestley is the person most relevant to the attack on science. The Burke-Priestley conflict had been partly disguised by the fact that in the *Reflections* of 1790, although Burke had now come to fear all the Dissenters,[30] his principal subject of attack is another Dissenting minister, Richard Price, who in 1789 had used the anniversary of 4 November of the 'Glorious Revolution' to preach a sermon in praise of the French Revolution. Burke mentioned Price repeatedly by name but he did not name Priestley. Anxious not to engage such a powerful adversary[31] too directly, Burke referred to Priestley indirectly and politely as a distinguished author, 'a man . . . of great authority and certainly of great talents'.[32] Although commentators have instantly picked up this thinly disguised reference to Priestley, they have failed to understand properly half a dozen references, mostly critical of chemistry, in the *Reflections,* which I interpret as veiled attacks on Priestley, or rather gentle derision of Priestley's science. Even as Burke wrote, his enemy Price was a sick man who soon died (March 1791). It was Priestley who now became Burke's principal opponent and not simply by default. Priestley had dared to reply to the *Reflections* in an open letter to Burke.[33] In this Priestley antagonized Burke, first by his praise of the French Revolution, second by advocating a philosophy of the rights of man and the principle of individual liberty, an 'abstract principle' which Burke had attacked, and third by his bold attack on the system of the establishment of the Church of England. After this it is not surprising if Burke, instead of subtly mocking chemistry in elegant phrases, should have turned on Priestley, and on any person or activity associated with him, the full fury of his purple prose. We should remember too that public emotions were further aroused by France declaring war on Britain on 1 February 1793. In May 1794, at the height of the Terror in France, the British government suspended Habeas Corpus and arrested leaders of radical societies. The time for irony had long passed. Burke's prediction of the extremes to which the Revolution would go were by now largely confirmed.

The chemical references in the *Reflections* can be divided into two main categories, the alchemical and the gaseous. It seems that as late as 1790 it was still possible for a

---

30 While writing the *Reflections,* Burke told a correspondent that he hoped in the book to expose the Dissenters 'to the hatred, ridicule and contempt of the whole world', Burke to Philip Francis, 20 February 1790, *Correspondence,* op. cit. (10), vi, p. 92.

31 In a letter to Fox of 9 September 1789, Burke had described Priestley as 'a very considerable Leader among a Set of Men powerful enough in many things, but most of all in Elections', *Correspondence,* op. cit. (10), vi, 15. At this point Burke was actually urging Fox to *cultivate* the Dissenters for the benefit of the Whigs! This illustrates the powerful influence of the 'October days' on Burke's thinking.

32 *Reflections,* p. 148.

33 *Letter to the Rt. Hon. Edmund Burke,* Birmingham, 1791. See also R. R. Fennessy, *Burke, Paine and the Rights of Man,* London, 1963, p. 209.

popular audience to associate chemistry with alchemy.[34] Burke makes the assertion that every French revolutionary is acting in defiance of the process of nature, like a 'projector and adventurer', like an 'alchymist and empiric'.[35] The republicans are characterized as 'metaphysical and alchemistical legislators'.[36] They are fools who have been misled 'just as the dream of the philosopher's stone induces dupes, under the most plausible delusion of the hermetic art to neglect all rational means of improving their fortune'.[37] In the sale of confiscated lands, the revolutionaries are 'carrying on a process of continual transmutation of paper into land and land into paper'.[38] This he describes as the grand *Arcanum*.[39] Finally, Burke mischievously describes as 'alchymy'[40] the melting down of the metal from church bells, which was actually one of the more sophisticated examples of applications of the new chemistry.[41] Thus, science and its practitioners are condemned by association with vain hopes, fraud and deceit.

The second kind of chemical reference in the *Reflections* relates to the discoveries of the 1770s and 1780s. Priestley and the French chemists had written much about gases, which allows Burke to remark that the metaphysicians were sometimes forced to 'descend from their airy speculations'.[42] Probably the best known of Burke's scientific references is where he compares the spirit of liberty to a gas given off in a chemical reaction. The light wit may be contrasted with Burke's later fury:

> The wild *gas*, the fixed air is plainly broken loose—but we ought to suspend our judgement until the first effervescence is a little subsided . . .[43]

Helmont's wild gas had been characterized as breaking out of vessels, so Burke refers to 'the expansive power of fixed air in nitre'.[44] It cannot be mere chance that immediately after referring to Priestley, Burke speaks of 'a mine that will blow up at one grand explosion all examples of antiquity'.[45]

34 Although a great time lag often exists between distinctions made by men of science and their understanding by the general public, we may recall that as late as 1782 James Price, F.R.S., (not to be confused with the Dissenting minister) had claimed to have transmuted mercury into gold. The University of Oxford presented Price with an honorary degree, but his dishonesty was soon exposed and he committed suicide. See James Price (né Higginbotham), *An Account of Some Experiments on Mercury, Silver and Gold, made at Guildford in 1782*, Oxford, 1782, and Hector C. Cameron, *Sir Joseph Banks*, London, 1952, pp. 151–157.

35 *Reflections*, p. 282.

36 Ibid., p. 300.

37 Ibid., p. 359.

38 Ibid., p. 308.

39 Ibid., p. 226.

40 'This is their alchymy', Ibid. p. 369.

41 See W. A. Smeaton, *Fourcroy, Chemist and Revolutionary, 1755–1809*, Cambridge, 1962, pp. 120–121.

42 *Reflections*, p. 370.

43 Ibid., p. 90. c.f. G. Canning and H. Frere, 'The Loves of the Triangles', *The Anti-Jacobin*, (1798), 2, pp. 164–165.

44 Ibid., p. 268.

45 Ibid., p. 148.

Gunpowder and insurrection

Among the most powerful images conjured up by Burke was that of the explosion of gunpowder. Gunpowder rivalled gases or airs as a symbol for the danger posed to the British government by the writings of Priestley. It was a particularly appropriate symbol for a man who combined strong scientific and political interests. It was a reference which could be understood by all levels of society and cause the maximum alarm. Priestley was foolish enough to speak of gunpowder in a political context, and by so doing he played right into Burke's hands. He was soon to be known as 'gunpowder Joe'. It was in vain that Priestley's supporters could claim in 1790 that there was no need for alarm because, when he had spoken of gunpowder, it had been purely as a figurative expression.[46] Gunpowder in the 1790s was much more than a metaphor, it was a symbol derived from Priestley's own writings, which he could not disavow. It came from the pen of England's leading man of science, for whom explosions were part of the stock in trade, or at least for whom that was the popular image. Priestley's friends could later insist that his studies had tended to preserve the health and life of man,[47] but Burke was only too conscious of the destructive potential of science. In Burke's eyes, Priestley's mention of gunpowder linked science and sedition. The association of gunpowder with science was later strengthened by the contributions of several leading French chemists to a crash programme of saltpetre production. Burke discussed at some length the involvement of Guyton de Morveau in the manufacture of gunpowder.

Priestley's first allusion to gunpowder, other than in a purely scientific context, came in some remarks which he had annexed in 1787 to a sermon entitled 'Reflections on the present state of free enquiry in this country'. Speaking of the growing power of the Unitarians, he first used a traditional organic metaphor, referring to sowing seeds which would eventually shoot up. Warming to his theme, he continued imprudently:

> We are, as it were, laying gunpowder, grain by grain, under the old building of error and superstition, which a single spark may hereafter inflame so as to produce an instantaneous explosion.

Alluding to the Church of England, he warned that:

> that edifice, the erection of which has been the work of ages, may be overturned in a moment.[48]

But, he concluded,

> till things are properly ripe for such a revolution, it would be absurd to expect it, and in vain to attempt it.[49]

The reference to gunpowder did not go unnoticed in March 1787 when the House of Commons was debating the repeal of the Test and Corporation Acts.[50] As evidence of the

---

46 Debate on Repeal of Test and Corporation Acts, 2 March 1790, Speech by Mr W. Smith, *Cobbett's Parliamentary History of England*, vol. xxviii, cols. 443–444.

47 Thomas George Street, *A Reply to a Letter from the Rt. Hon Edmund Burke to a Noble Lord. Being a Vindication of the Duke of Bedford's Attack upon Mr Burke's Pension*, London, 1796, p. 61.

48 Priestley, *Works*, op. cit. (27), **xviii**, 544.

49 Ibid., p. 545.

50 See G. M. Ditchfield, 'The parliamentary struggle over the repeal of the Test and Corporation Acts', *English Historical Review*, (1974), **89**, pp. 551–577.

potential danger of giving more power to Dissenters, Sir William Dolben quoted one sentence from the passage reproduced above, giving a particular emphasis and solemnity to the words: '. . . gunpowder, grain by grain . . .'[51]

Perhaps all this would have been forgotten if Priestley had not spoken of gunpowder for a second time. In his published *Letters to the Rev. Edward Burn* (1790) he referred mischievously to:

> those grains of gunpowder . . . at which they have taken so great an alarm and which will certainly blow up at length and perhaps as suddenly, as unexpectedly and as completely as the overthrow of the late arbitrary government of France.[52]

This is a reminder that the situation had changed considerably since Priestley's first reference to gunpowder. Whereas in 1787 Priestley could dampen the inflammatory effect of his words by saying that the time for revolution had not yet come, after July 1789 the revolution had arrived, so that Priestley was almost referring to the events in France as a threat. Priestley added a provocative warning, attacking the clergy of the Church of England for their conservatism and complacency:

> If I be laying gunpowder, they are providing the match.

These words in print were almost too good to be true to the opponents of reform in a further Commons debate of 1790 on the repeal of the Test and Corporation Acts. Burke, who had previously been sympathetic to the granting of concessions to Dissenters, now opposed them. He effectively demolished the opposition by quoting from the published works of the leading Dissenters to show how dangerous they had become. In the case of Richard Price he was able to quote from the sermon of 1789, in which he had glorified the cause of the French Revolution. Turning to Priestley, he referred to the mention of gunpowder,

> which he considered as a serious indication on the part of Dr Priestley, at least, of a determination to proceed step by step till the whole of the church establishment was levelled to the foundations.[53]

Gunpowder, therefore, was seen as destructive to the established Church and thus to the state. A later speaker in the debate, William Smith, defended Priestley, saying that he had only mentioned gunpowder a second time because he had previously been misunderstood. The historian, however, might conclude that Priestley had been trying to score additional points without recanting anything.

As an aspect of Priestley widely reported, the gunpowder reference was incorporated into a popular tract published in Birmingham and alleged to be the work of one John Nott, button burnisher. He addresses Priestley:

> If you ben't (sic) melancholy mad, as I guess you to be, what makes you rave so much about *gunpowder*. You never wrote but you tell us church people, that you're laying it *grain by grain*

51 *Parliamentary History*, 28 March 1787, 22, col. 831. See also Priestley, *Works*, op. cit. (27), **xviii**, p. 544n.

52 *Parliamentary History*, 2 March 1790, 28, col. 438n. Ronald E. Cook, *A bibliography of Joseph Priestley, 1733–1804*, London, 1966, p. 45.

53 *Parliamentary History*, 2 March 1790, 28, col. 438–439.

under the churches and mean to blow 'em all up together very soon. And yet you take upon you mightily because they tell you of it again. Now, prithee, Mr Priestley, how would you like it yourself, if they were to send you word that they had laid trains of gunpowder under your house or meeting house? Why you would be frightened out of your seven senses.[54]

Perhaps, with knowledge of the Birmingham riots of 1791, this passage might be viewed as a provocation to retaliation. Priestley, who could be interpreted as having raised the discussion to an inflammatory level, would find his own house in Birmingham burned by the mob.

In the *Reflections*, Burke had written of 'the expansive power of fixed air in nitre'[55] and after the declaration of war he was aware that the manufacture of gunpowder was one of the tasks undertaken by the French chemists. It is time to examine the French dimension. The loss of India to the British in the Seven Years War deprived France of (Indian) saltpetre (potassium nitrate), the principal constituent of gunpowder. Thus, the French were compelled to make the most of indigenous sources of saltpetre as an excrescence on the outer walls of buildings. Such places as cellars and outbuildings might be particularly rich in saltpetre. Under the ministry of Turgot (1775–1776) the Régie des Poudres was established and Lavoisier was brought in to advise on the extraction of saltpetre and the manufacture of gunpowder. With the revolutionary war, production had to be increased, and the work of (Guyton de) Morveau and Fourcroy was particularly important.[56] These two chemists played a prominent role in politics and were elected to the National Convention in September 1792 and July 1793, respectively. Burke knew of Morveau's work in connection with the crash programme to extract saltpetre, possibly because he had lent his name to a booklet with a title beginning with the motto: 'Death to the Tyrants',[57] an unusually dramatic and even political beginning to a work on applied science. Politics and applied chemistry overlapped so much at this time that there was even a patriotic French song composed about 'Republican saltpetre' and liberty.[58]

In his *Letter to a Noble Lord* (1796), addressed to the Duke of Bedford, Burke fancifully describes in some detail how the French chemists could constitute a threat to the buildings on his estate:

They have calculated what quantity of matter convertible into nitre is to be found in Bedford House, in Woburn Abbey . . .[59]

---

54 John Nott (pseudonym), *Very Familiar Letters Addressed to Dr Priestley*, Birmingham, [1790], p. 13. Burke is referred to humorously (p. 21) as using 'plaguy (sic) hard words'.

55 *Reflections*, p. 268.

56 For a general account of saltpetre production in France, see Robert P. Multhauf, 'The French crash program for saltpetre production, 1776–94', *Technology and Culture*, (1971), **12**, pp. 163–181 (p. 174).

57 *Mort aux Tyrans. Programme des Cours Révolutionnaires sur la Fabrication des Salpêtres, des Poudres et des Canons*, Paris, An 2 [1793–1794].

58 *Le Salpêtre Républicain.*

59 *Letter to a Noble Lord* (1795), *Works*, v, p. 143. Burke was incensed by the criticism of his recent government pension by the Duke of Bedford, who failed to appreciate the dangers of the Revolution.

These buildings, together with

> Churches, play-houses, coffee-houses, all alike, are destined to be mingled, and equalized, and blended into one common rubbish; and, well sifted, and lixiviated, to crystallize into true, democratic, explosive, insurrectionary nitre.

Burke went on to explain how greatly the leaders of the French Republic valued science and 'chemical operations' which enabled revolutionary gunpowder to be extracted from the chateaux and feudal fortresses of the nobility. Burke was therefore making a much more interesting and original claim than about the explosive properties of gunpowder. To speak of a revolution as an explosion is a commonplace metaphor. But it was not necessary to explode the gunpowder; the very process of *manufacture* began with destruction. Burke saw the revolution as using the extraction of saltpetre for military purposes as an excuse for destroying the proud property of the nobility. The extraction of saltpetre is a metaphor for the undermining of society by dismantling its very buildings— 'They consider mortar[60] as a very anti-revolutionary invention',[61] he quipped. The image of crumbling buildings was a very powerful one,[62] all the more so as it had some relation to fact. But at the same time as Burke was condemning the revolutionaries, he was condemning the French chemists, who associated themselves with such destructive tasks.

The new saltpetre process had provided Burke with additional evidence, in rhetoric if not in logic, that chemistry was associated with revolution. His *Letter to a Noble Lord* received a reply from Thomas George Street, who wrote as follows:

> Those wonderful chemical operations, of which all France partook in 1794 . . . operations that forced admiration even from the enemies of France, extort from him nothing but a sarcasm and a sneer. The mention of chymical operations naturally connects with it, in Mr Burke's, as well as in every other person's mind, the name of Priestley . . .[63]

In England in the early 1790s, for anyone to refer to chemistry—even French applied chemistry—was, therefore, to bring Priestley to mind.

'Philosophers' and Frenchmen

Burke speaks of 'this philosophic revolution'[64] and he often uses the term 'philosopher' as a term of reproach. Who were these philosophers and why was Burke so opposed to them? For Burke, the word 'philosopher' was a portmanteau word used sometimes polemically and loosely with atheistic associations. Only a minority of the persons described by Burke as 'philosophers' were academic philosophers. A much larger proportion would be more justly called *philosophes*, that is to say that they were notable

---

60 Burke had used the same metaphor in 1791, speaking of the 'untempered mortar' holding together the revolutionary building (*Speeches*, London, 1816, vol. iv, p. 26).

61 *Letter to a Noble Lord* (1795), *Works*, v, p. 143.

62 In a parliamentary debate Burke had scathingly described the French revolutionaries as 'the architects of ruin'; they had 'completely pulled down to the ground their monarchy, their church, their nobility, their law, their revenue, their army, their navy, their commerce, their arts and their manufactures', *Parliamentary History*, 9 February 1790, **28**, col. 354.

63 Street, op. cit. (47), p. 23.

64 *Reflections*, p. 237.

writers who concerned themselves with political and religious affairs, like Rousseau. They often, like Voltaire, displayed knowledge of science in their writings. Thus Burke writes:

> We are not the converts of Rousseau; we are not the disciples of Voltaire.[65]

But there are also writers who had made major contributions to science, like the mathematician Condorcet and the chemist Priestley. These were *natural* philosophers. When Burke accused Richard Price of being 'much connected with . . . intriguing philosophers [and] with political theologians . . .'[66] he might well have meant in the first place Priestley the man of science, and in the second place Priestley as a clergyman involved in politics.[67] Priestley's many roles were a source of some perplexity to Burke as they may be for the historian. It is the interaction between science and politics which is particularly striking in this period. It was not only that political theorists often made appeals to the natural world and science; it also happened that many prominent radicals themselves contributed to science. There was a most noticeable overlap in France between the small scientific community and men involved in the Revolution. Apart from Condorcet, Burke often mentions the astronomer Bailly, president of the first National Assembly.[68] He also mentions the lawyer-chemist Morveau and the Jacobin mining engineer and scientist, Hassenfratz.[69]

In defiance of the evidence, Burke even blames the storming of the Bastille on the philosophers:

> They are modern philosophers; which when you say of them, you express everything that is ignoble, savage and hard-hearted.[70]

Burke was not the first to attribute sinister political power to 'philosophers'. The counter attack in France on the *philosophes* goes back several decades.[71] In 1789, the abbé Barruel had published a tract, blaming the revolution on a conspiracy of philosophers and Freemasons, an argument he was later to develop into an influential four-

---

65 Ibid., p. 181.
66 Ibid., p. 93.
67 Given the especial political disabilities of the Unitarians, it is hardly surprising that Priestley should have become an advocate of political change. The majority of Dissenters, however, were not Unitarians and many had little interest in politics.
68 *Reflections*, pp. 370, 371. *Thoughts on French Affairs* in Everyman edition of *Reflections*, London, 1967, p. 293. Interestingly, it was Bailly who introduced the modern concept of a scientific revolution (e.g. 'the Copernican revolution') in 1785, I. Bernard Cohen, *Revolutions in Science*, Cambridge, Mass., 1985, pp. 221–223.
69 *Letter to a Noble Lord*, Works, v, p. 144.
70 *Letter to a Member of the National Assembly* (1791), in Everyman edition of *Reflections*, London, 1967, p. 261. Burke was referring to 'those who have made the exhibition of 14 July', the date of the fall of the Bastille.
71 Especially after the Jesuits had been expelled from France. John M. Roberts, *The Mythology of Secret Societies*, London, 1972, pp. 183, 211 and chapter VI. See also R. J. White, *The Anti-philosophers*, London, 1970, pp. 115–117.

volume work.[72] Although Barruel was soon (September 1792) to take refuge in England, where he was welcomed by Burke, the latter scrupulously avoided the grosser exaggerations of the French cleric. For Burke, the revolution was essentially a conspiracy between politicans and philosophers. The politicians might receive more publicity, and they certainly gave practical direction, but

the philosophers were the active internal agitators, and supplied the spirit and principles.[73]

Burke launches a crusade against 'the philosophical fanatics'[74] and the 'new philosophic doctrines'.[75] He speaks not only of the 'shallow speculations of the . . . shortsighted coxcombs of philosophy'[76] but of 'philosophic spoilers',[77] and 'philosophic financiers'.[78] Another term of abuse was 'metaphysics', so that to describe something as 'barbarous metaphysics'[79] is to insult it twice over. The trouble with the philosophers in his opinion is that they are 'entangled in the mazes of metaphysic sophistry.'[80] Neither 'metaphysical abstractions'[81] nor 'metaphysic declarations'[82] are to be trusted.

It should be noted, however, that in attacking metaphysics Burke was hardly being original. It had been a favourite topic with Bolingbroke and other political writers who opposed the philosophical Whiggism of the school of John Locke.[83] The philosophers were to be mistrusted as 'men of speculation'.[84] Burke makes a shrewd attack on irresponsible academics who do not pay the price of their mistakes, 'never intending to go beyond speculation'[85] into 'the world in which they live'. They are 'only men of theory'[86] and 'they despise experience as the wisdom of unlettered men',[87] a comment Burke makes immediately after a reference to Priestley. He has nothing but contempt for 'the learned professors of the rights of man',[88] a phrase he uses on more than one occasion. He asks:

---

72 *Le Patriote Véridique, ou Discours sur les Vraies Causes de la Révolution Actuelle*, Paris, 1789, e.g., pp. 11, 22. *Mémoires pour Servir à l'Histoire du Jacobinisme*, 4 vols., Hamburg, 1798. A late Scottish perspective of the conspiracy thesis is provided by John Robison, *Proof of a conspiracy against all the religions and governments of Europe carried on in the secret meetings of freemasons, illuminati and reading societies*, Edinburgh, 1797. See J. B. Morrell, 'Professors Robison and Playfair, and the Theophobia Gallica: natural philosophy, religion and politics in Edinburgh, 1789–1815', *Notes and Records of the Royal Society of London*, (1971), **26**, 43–63 (47). A reasoned argument on the limited power of philosophers had to wait until 1801, J. J. Mounier, *On the Influence Attributed to Philosophers, Freemasons and to the Illuminati on the Revolution in France*, London, 1801. Mounier points out (p. 113) that some philosophers like Condorcet had no influence at the beginning of the revolution and themselves became victims of it.

73 *Regicide Peace*, II, *Works*, v, p. 246. For a later statement of the harmful effects of philosophy on politics, brought to my attention by David Knight, see the 1817 letter of Coleridge to Lord Liverpool: *Letters of S.T. Coleridge*, (ed. E. L. Griggs), Oxford, 1959, vol. 4, pp. 757–763.

74 *Reflections*, p. 256.
75 Ibid., p. 347.
76 Ibid., p. 141.
77 Ibid., p. 273.
78 Ibid., p. 359.
79 Ibid., p. 338.
80 Ibid., p. 105.
81 Ibid., p. 90.
82 Ibid., p. 345.
83 E. J. Payne (ed.), *Select Works of E. Burke*, Oxford, 1898, vol. ii, p. 307, c.f. p. 332.

84 *Reflections*, p. 183.
85 Ibid., p. 155.
86 Ibid., p. 128.
87 Ibid., p. 148.
88 Ibid., p. 207.

What is the use of discussing a man's abstract right to food or to medicine? The question is upon the method of procuring and administering them. In that deliberation I shall always advise to call in the farmer and the physician rather than the professor of metaphysics.[89]

France, noted both for its writers and its scientists, is for Burke a 'nation of philosophers'.[90] Even a century previously Louis XIV had used 'the imposing robes of science, literature and the arts' to cover over the despotism of his government.[91] One writer accuses Burke of opposing anything on principle if it is linked with France:

> So complete is ... [his] detestation, not unmixed with dread, of everything that is French that he not only hates French politics but also French literature and French science'.[92]

Burke writes of 'the simplicity of our [British] national character and ... a sort of native plainness and directness of understanding'.[93] He stands for the natural feelings and good sense of the British as opposed to the foppish sophistication and taint of atheism of the French. Later in Parliament Burke was to indulge in a bit of play acting in order to dramatize the threat from the continent. There were rumours that French orders for arms had been placed with a Birmingham firm. In December 1792, towards the end of his speech on a Bill on the treatment of aliens, Burke drew out a dagger,[94] previously concealed and, with great vehemence, threw it on the floor. He told the House of Commons that we must:

> prevent the introduction of French principles and French daggers. When they smile, I see blood trickling down their faces ... I now warn my countrymen to beware these execrable philosophers ...[95]

Whereas in 1790 Burke had been content with gentle jibes about, for example, 'the aeronauts of France'[96] (an allusion to the fact that French savants had pioneered ballooning), by 1795 he is referring bitterly to 'the sect of the cannibal philosophers of France'.[97] With the same hyperbole he writes of 'the new school of murder and barbarism set up at Paris'.[98] We should beware, however, of reading nineteenth-century ideas of nationalism into this eighteenth-century writer. Burke is quite explicit on how he views France:

> My ideas and my principles led me in this context to encounter France not as a state, but as a faction.[99]

89 Ibid., pp. 151–215.

90 Ibid., p. 239. In 1795 Wilberforce complained in the House of Commons that it was not only French politics but 'French philosophy' that was being imported into Britain (*Parliamentary History*, vol. 32, col. 292). Political cartoonists also associated the French Revolution with 'philosophers', e.g., the allegory 'Philosophy run mad' (1793) and an unsigned print by Gillray portraying Fox as 'A democrat—or reason and philosophy', Mary Dorothy George, *English Political Caricature, 1793–1832*, Oxford, 1959, vol. ii, pp. 1, 3.

91 *Parliamentary History*, (9 February 1790), **28**, col. 354.

92 Street, op. cit. (47), p. 60.

93 *Reflections*, p. 186.

94 The dagger is obviously a symbol of armed aggression, see *Correspondence*, vii, p. 328.

95 *Speeches*, iv., p. 99.

96 *Reflections*, p. 376.

97 *Letter to a Noble Lord, Works*, v, p. 139.

98 *Letter to a Member of the National Assembly*, op. cit. (70), p. 269.

99 *Regicide Peace*, II, *Works*, v, p. 231.

A few pages further on comes the most significant statement:

> It is not France extending a foreign empire over other nations, it is *a sect aiming at universal empire*, and beginning with the conquest of France.[100]

Thus, although Burke is to some extent contrasting the situations in two different states, Britain and France, what really concerns him is the transmission of ideas beyond their country of origin. He was particularly worried about subversive ideas from France undermining the British government and way of life and, from his viewpoint, scientists like Priestley, political writers like Paine and Godwin and the French *philosophes* were all all in the same category. The British radicals, described by other writers as English Jacobins, are alluded to by Burke as 'the Frenchified faction'.[101]

Burke claimed of France that:

> there was perhaps no country in the universe in which . . . men of letters . . . were so highly esteemed, courted, caressed and even feared.[102]

We may recall that in the eighteenth century the expression 'men of letters' was applied to men of science as much as to essayists, so it is no surprise to find the name of the astronomer Bailly immediately after this remark. For Burke, the dominant place of writers and intellectuals in France would have almost sufficed as a total explanation of why the Revolution had taken place in that country rather than in a neighbouring one. For Burke:

> men of letters (hitherto thought the peaceable and even timid part of society) are the chief actors in the French revolution.[103]

Many people thought of the *philosophes* as individuals, writing independently, but Burke insists that there was collusion, with writers forming a cabal:

> The literary cabal had some years ago formed something like a regular plan for the destruction of the Christian religion. This object they pursued with a degree of zeal which hitherto had been discovered only in the propagators of some system of piety. They were possessed with a spirit of proselytism in the most fanatical degree.[104]

These authors, many of whom 'stood high in the ranks of literature and science',[105] Burke describes as 'atheistical fathers'[106] with 'a bigotry of their own':

> they have learned to talk against monks with the spirit of a monk. But in some things they are men of the world. The resources of intrigue are called in to supply the defects of argument and wit . . .[107]

100 Ibid., p. 234 (my italics).
101 *Letter to a Noble Lord, Works*, v, p. 141. In a letter to Lord Grenville in 1792 Burke does speak of the English Jacobins. He mentions several French revolutionaries including Condorcet 'and their Brethren, the Priestleys, the Coopers and the Watts', *Correspondence*, vii, p. 177. James Watt (1769–1848), son of the engineer, had presented an address to the Jacobin club in Paris on 13 April 1792.
102 *Thoughts on French Affairs*, loc. cit. (68), p. 293.
103 Ibid., p. 292.
104 *Reflections*, p. 211.
105 Ibid., p. 212. It should be noted that literature is associated with science. The antithesis between literature and science that is taken for granted in the twentieth century had not yet arisen.
106 Burke is comparing them to the early church fathers.
107 Ibid., p. 212.

So, whereas Burke was quite prepared to argue with any individual writer, he says that 'writers, especially when they act in a body'[108] are to be feared because of their influence on public opinion.

To command that opinion, the first step is to establish dominion over those who direct it,[109]

which is why radical writers have formed a cabal. Although this paper is concerned with perceptions rather than realities, one might remark that Burke was not without evidence in claiming that writers were organized. For example, the London publisher Joseph Johnson, who regularly published Unitarian tracts, also specialized in the publication of scientific works, not only of Priestley but also of Benjamin Franklin and Erasmus Darwin. By 1791, Johnson's bookshop in St Paul's churchyard had become a meeting place for radicals.[110]

Burke writes of 'those cabals of literary men, called academies'[111] and specifically relates these 'political Men of Letters' to the Académie Française and the Académie des Sciences.[112] Swift had satirized the members of the Royal Society in *Gulliver's Travels*[113] and Burke does not miss the opportunity in his *Reflections* to pour ridicule on the idea of a country governed by philosophers when he refers ironically to 'the learned academicians of Laputa and Balnibarbi'.[114] A few years later, Burke abandons gentle satire for direct attack:

> Never before did a den of bravoes and banditti assume the garb and tone of an academy of philosophers.[115]

In other words, to introduce a Biblical metaphor, the philosophers are really wolves in sheep's clothing.[116] In December 1791, Burke writes:

> We have seen all the academicians at Paris with Condorcet, the friend and correspondent of Priestley at their head, the most furious of the extravagant republicans . . .[117]

In Condorcet, mathematician, permanent secretary of the French Academy of Sciences, and politician, we see the best example of the overlap between science and politics in France. In so far as the secretary held an even more important position in the Academy than even the president, we can see some foundation for Burke's mistrust of the Academy.[118] Earlier that same year (1791) Condorcet had, on behalf of the Academy of

---

108 Ibid., p. 213.
109 Ibid., p. 212.
110 Gerland P. Tyson, *Joseph Johnson: A Liberal Publisher*, Iowa City, 1979, p. 121.
111 *Thoughts on French Affairs*, loc. cit. (68), p. 291.
112 *Reflections*, p. 211. Burke refers to 'the two academies of France'.
113 Jonathan Swift, *Gulliver's Travels* (1726). See especially Part III, *A voyage to Laputa*, etc. A referee has suggested that Swift's critical portrayal of science may have had more than a superficial influence on Burke. Burke, however, was much more sympathetic to a science-based technology than Swift (see footnote 215).
114 *Reflections*, p. 238.
115 *Letter to a Noble Lord*, *Works*, v, p. 139.
116 Priestley is portrayed in one contemporary caricature as a wolf in sheep's clothing. George, op. cit. (90), i, p. 213.
117 *Thoughts on French Affairs*, loc. cit. (68), p. 315.
118 See Roger Hahn, *The Anatomy of a Scientific Institution. The Paris Academy of Sciences, 1666–1803*, Berkeley, 1971.

Sciences, written a letter of sympathy to Priestley[119] after the destruction of his house in the Birmingham riots. Condorcet was a particular bête noir for Burke, who referred scathingly to him in February 1793 as 'the most humane of all murderers'.[120] Unfortunately for Condorcet, as a Girondin, he was swept up and perished in the Jacobin tide of that same year. But it did not suit Burke to distinguish Girondins from the more radical Jacobins; for him they were equally dangerous revolutionaries.

Reductionism

One of the most serious accusations of Burke against the new science was that it was reductionist. We should note that he uses the term 'philosopher' where we would say 'scientist':

> These philosophers consider man in their experiments no more than they do mice in an air pump, or in a recipient of mephitic gas.[121]

This is surely an allusion to Priestley's use of mice to test the 'goodness' of different kinds of air.[122] In a mephitic gas a mouse would, of course, die by suffocation. Chemistry is here under attack not only because it was the science in which Priestley had made an international reputation for himself, but also because for Burke it provided an ideal target as representing new and dangerous science. Were these experimenters not reckless adventurers who would treat men at large in the same way as they treated mice in their laboratories? Burke claimed that 'philosophers . . . would sacrifice the whole human race to the slightest of their experiments'.[123] Priestley had shown that he was prepared to sacrifice a few mice, hoping no doubt in the end to make some contribution to the welfare and greater happiness of mankind. But for Burke, cruel experimental science could hardly be justified in terms of the ultimate benefit to man. Burke accuses 'philosophers' of complete lack of scruples: 'to cut up the infant for the sake of an experiment'.[124] They are heartless: 'their imagination is not fatigued by the contemplation of human suffering',[125] and again:

> Nothing can be conceived more hard than the heart of a thoroughbred metaphysician.[126]

119 Priestley was proud of the support he received from France. He translated Condorcet's letter and had it published in his *Appeal to the Public on the Subject of the Riots of Birmingham*, 2nd edn., Birmingham, 1792, pp. 154–156. For the text of another similar letter from 'the chemists of Paris' see R. E. Schofield, op. cit. (25), pp. 257–258. They claim to defend 'the cause of tolerance, of liberty and of philosophy'.

120 *Speeches*, iv, p. 119.

121 *Letter to a Noble Lord, Works*, v, p. 142. In modern times mice are still a favourite animal for medical experiments. In 1984 54.4% of such experiments were carried out on mice compared with only 0.6% on cats and dogs (*New Scientist*, No. 1505, 24 April 1986, p. 27).

122 Priestley, in his paper 'Observations on different kinds of air' (*Phil. Trans.* (1772), 62, pp. 147–267 (p. 214)), describes how he has substituted a purely chemical test using 'nitrous air' for his original test using mice. Henceforth, he no longer had occasion to keep 'so large a stock of mice'.

123 *Letter to a Noble Lord, Works*, v, p. 141.

124 *Reflections*, p. 277.

125 *Letter to a Noble Lord, Works*, v, p. 142.

126 Ibid., p. 141.

The bleak new world of radical philosophy and science is not simply amoral, by the end of his life Burke had concluded that it is intensely evil.[127]

A particularly famous passage in the *Reflections* relates to the treatment of the French King, and especially the Queen,[128] who in October 1789 had been forced by a hungry and armed mob to abandon their palace at Versailles to live in Paris. Burke's source for this news was greatly distorted, but it was this event rather than the more famous earlier storming of the Bastille in July 1789 which was decisive in converting him from spectator to arch anti-revolutionary. In fact, the treatment of the Royal family in 1789 was probably understandable in the circumstances. It was only later, with the attempted flight of the Royal family, their capture, imprisonment and final execution that the violence became extreme. Burke's passionate defence of the monarchy may be contrasted with the reductionist attitude of Tom Paine who, in his reply to Burke, can see in the crown no more than a metaphor, even a conspiracy.[129]

Already in his youth Burke had written of the common decencies of civilization and the evils of new reductionist literature and philosophy:

> What shall we say to that philosophy which would strip [human nature] naked? Of such sort is the wisdom of those who talk of the love, the sentiment and the thousand little dalliances that pass between the sexes, in the gross way of mere procreation. They value themselves as having made a mighty discovery; and turn all pretences to delicacy into ridicule.[130]

Burke, however, is no puritan. Sex for him is not dishonourable but essentially mysterious.[131]

In the *Reflections* Burke clearly represents a tradition of chivalry in the treatment of the female sex, which he contrasts with the reductionism of the revolutionaries:

> On this scheme of things, a king is but a man; a queen is but a woman; a woman is but an animal; and an animal not of the highest order . . .[132]

In a later work Burke accuses the revolutionaries of 'endeavouring to persuade the people that they are no better than beasts'.[133]

In the logical extremes of revolutionary reductionism there is nothing sacred about a church—it is simply a building; thus churches, playhouses and coffeehouses all have the same status.[134] There is no special horror in the crimes of regicide or parricide or sacrilege. The murder of a king or queen, of a father, of a bishop, is not distinguished from ordinary homicide. In revolutionary egalitarianism quantity is more important

---

127 'It is like that of the principle of evil himself: incorporeal, pure, unmixed, dephlegmated, defecated evil'. Ibid. Notice the choice of adjectives taken from science.

128 A modern account is given in Alfred Cobban, *A History of Modern France*, 3rd edn., Harmondsworth, Middx., 1963, i, p. 161. Burke nevertheless harboured a deep distrust of Marie Antoinette. *Correspondence*, vi, p. xvii.

129 *Rights of Man*, Part II (1792), loc. cit. (17), pp. 178–180. See also David K. Weiser, 'The imagery of Burke's *Reflections*', *Studies on Burke and His Time*, (1974–1975), 16, pp. 312–329 (p. 312n).

130 H. V. F. Somerset (ed.), *A Notebook on Edmund Burke*, Cambridge, 1957, p. 91.

131 Generation 'is hid . . . not because it is dishonourable but because it is mysterious', Ibid., p. 92.

132 *Reflections*, p. 171.

133 *Letters on a Regicide Peace*, *Works*, v, p. 212.

134 *Letter to a Noble Lord*, *Works*, v, p. 143.

than quality. Indeed, as previous social distinctions are abolished, mathematics and particularly arithmetic, acquires a new importance in politics.

In another part of the *Reflections* Burke takes a homely example and points out that:

> the coarse husbandman should well know how to assort and use his sheep, horses and oxen and should have enough common sense not to abstract and equalise them all into animals without providing for each kind an appropriate food, care and employment.[135]

Burke says that the French legislators 'have attempted to compound all sorts of citizens . . . into one homogenous mass'.[136] Finally, he makes the most telling accusation, that the revolutionaries 'reduce men to loose counters'.[137] Abstraction for Burke was the beginning of fanaticism. If a person thinks persistently in terms of categories or numbers, he will finally forget what the numbers stand for. The French philosophers, said Burke, love mankind dearly but could not abide men.[138]

We must go on from Burke's deeply felt feelings that human beings are more than numbers to his general hostility to mathematics. First, there was the romantic objection that mathematics was an arid and abstract field of study, far removed from the main concerns of humanity.[139] But it was because of its application to politics that Burke felt it necessary to make several criticisms of mathematics, and here there is not only a feeling of abhorrence but something of an argument. Burke insists that 'the constitution of a country' is not simply 'a problem of arithmetic'.[140] He argues that the new philosophers, by emphasizing a purely quantitative approach, are lacking in traditional learning.[141] He points out that Aristotle's logic required analysis of problems under many different headings, of which *quantity* was only one, *quality* and *relation* being equally valid headings for analysis.

In the *Reflections* in characteristic style, Burke prefers scorn to detailed argument and speaks dismissively of 'a government of five hundred country attornies and obscure curates' making decisions on behalf of a population of 24 million.[142] He suggests that 'the most considerable of their acts have not been done by great majorities',[143] but only in his *Appeal from the New to the Old Whigs* (1791) does he challenge the very concept of a majority,[144] possibly a small one, making decisions on behalf of an entire population. It is necessary first to have unanimous agreement about the powers of majorities. He also points to a long-standing tradition, exemplified by English juries, that absolute

---

135 *Reflections*, p. 300. c.f. New Testament parallel in which sheep are separated out from goats, Matthew, xxv, 32.

136 *Reflections*, p. 300.

137 Ibid.

138 Louis I. Bredvold and Ralph G. Ross (eds), *The Philosophy of Edmund Burke: a Selection of his Speeches and Writings*, Ann Arbor, Michigan, 1960, p. 5.

139 Burke speaks of 'the severity of geometry'. *Letter to a Member of the National Assembly*, loc. cit. (68), p. 248. Several previous eighteenth-century writers had expressed reservations about the excessive claims of mathematics, e.g., Diderot, *De l'interpretation de la Nature* (1754), opening paragraphs.

140 *Reflections*, p. 141.

141 Ibid., p. 301.

142 Ibid., p. 141.

143 Ibid., p. 276.

144 *Appeal from the New to the Old Whigs*, *Works*, iii, p. 82ff.

unanimity is required, for example, before any major new course of action is adopted. If there is 'adding, subtracting, multiplying and dividing' in politics, then it is on the moral plane, not the mathematical.[145] If I understand this correctly, it means that in society and politics two plus two does not necessarily always equal four.

We can now appreciate why Burke has no confidence in 'an arithmetical constitution', but the phrase Burke actually uses to condemn the French innovators is 'a geometrical and arithmetical constitution'.[146] The reference here to geometry is partly explained by the fact that since the time of Descartes, geometry has been held up as a model of certainty, whereas, said Burke, 'In politics, the most fallacious of all things . . . [is] geometrical demonstration'.[147] As Burke says elsewhere: the rights of man may be 'metaphysically true [but] they are morally and politically false'.[148] Such principles are based on a number of questionable axioms, of which one is that men always act rationally.[149] Those who have founded the new French constitution 'have much but bad metaphysics; much but bad geometry; much but false proportionate arithmetic'.[150] Unfortunately, the political world is not 'all as exact as metaphysics, geometry and arithmetic ought to be' since it includes, for example, a moral dimension which they have omitted. The very abstraction of mathematics makes it an unsuitable model for politics:

'The lines of morality are not like ideal lines in mathematics. They are broad and deep as well as long. They admit of exception; they admit of modifications . . .'[151]

The critical reference to geometry, however, goes beyond the model of geometry as a deductive system of reasoning. When Burke later in the *Reflections* refers to 'a geometrical constitution' and a 'geometrical policy',[152] he is talking about applied geometry, land measurement and the administrative division of France. Burke alleges that 'the French builders' are constructing a geometrical system of local and national administration 'like their ornamental gardeners',[153] that is to say in geometrical patterns like the formal gardens of Versailles, contrasting with the naturalistic English garden. Burke then proceeds to explain how the French have divided their country into 83 exact squares called Departments. This was not historically accurate, Burke having confused one of several projects with reality.[154] In fact, the division between Departments followed natural boundaries. Once again Burke is presenting a caricature, but not one to be ignored since it was widely believed and it represents the evil triumph of mathematics. This geometrical option, he says, requires 'nothing more than an accurate land surveyor, with his chain, sight and theodolite'.[155] This, therefore, was a sorry sort of society, run by

145 *Reflections*, p. 153. For an exegesis of this difficult passage, see Burleigh T. Wilkins, *The Problem of Burke's Political Philosophy*, Oxford, 1967, p. 178.
146 *Reflections*, p. 144.
147 Ibid., p. 286.
148 Ibid., p. 153.
149 E. J. Payne, op. cit. (83), p. 332.
150 *Reflections*, p. 296.
151 *Appeal from the New to the Old Whigs, Works*, iii, p. 16.
152 *Reflections*, p. 314.
153 Ibid., p. 285.
154 Thomas Paine seems to have been the source of Burke's misinformation. *Correspondence*, vi, pp. 74–75.

technicians. The historic provinces were ignored, the old regional royalties were forgotten and France was reduced to a 'new pavement of square within square', a reference to the Communes, into which the Departments were divided, thus annexing hateful arithmetic[156] to vile geometry.

There was a further area related to mathematics which became the subject of Burke's hostile comments, that of economics and accounting. One of the best known passages of the *Reflections* is a romantic allusion, quoted near the beginning of this paper, to the passing of the age of chivalry and its replacement by a new era under the control of 'sophisters, economists[157] and calculators'.[158] On the cultural level, therefore, Burke's principal motive for deprecating economics was similar to that for his criticism of mathematics. We have, for example, a reference to 'the mathematics and arithmetic of the exciseman'.[159] Economics, like mathematics, becomes a perjorative term and we have 'oeconomical politicians',[160] whose gods are commerce and trade. In similar vein there is a sneer about 'the book-keepers of politics'.[161] These references have a direct social implication. Economics and accounting are associated with trade, which is socially degrading for those involved, notably the Dissenters, who thus compare unfavourably with those whose wealth comes from the ownership of the land. Thus, he later chides the landowning Duke of Bedford for his supposed 'readiness in all the calculations of *vulgar* arithmetic',[162] rather than in understanding 'moral proportions', a higher activity on a Platonic scale of values.

But there is a final context which is not the least important. In the analysis of Richard Price's sermon, Burke uses the phrase: 'the calculating divine computes',[163] which seems distinct from the ordinary duties of a man of religion. In contrast to some of Burke's jibes, this was a remarkably apposite comment about Price, who was deeply involved in some aspects of social arithmetic. It was said that on one occasion a member of Price's congregation went to him hoping for words of Christian consolation, but instead was treated to a lecture on annuity tables. So, just as an attack on chemistry in the early 1790s rebounded on Priestley, the criticism of mathematics and economics can be seen as relating principally to Price.

---

155 *Reflections*, p. 286.

156 Burke discusses in some detail the system of elections and assemblies which he portrays as abstract and mathematical. Ibid., pp. 287–289.

157 C.f. letter from Burke to Claude François de Rivarol, 1 June 1791: 'It is better to forget once for all the Encyclopédie and the whole body of Economists . . .'. (*Correspondence*, vi, p. 267.) J. J. Mounier pointed out that most economists, far from supporting the revolution, taught respect for property (*On the Influence Attributed to Philosophers . . .*), London, 1801, pp. 24–25.

158 *Reflections*, p. 170. The attack on economics did not prevent Burke himself writing a small work on economics, posthumously published, Donal Barrington, 'Edmund Burke as an economist', *Economica*, (1954), n.s. 21, pp. 252–258.

159 Ibid., p. 299. We may recall that Tom Paine had been an exciseman.

160 Ibid., p. 174.

161 Ibid., p. 176.

162 *Letter to a Noble Lord*, *Works*, v, p. 114 (my italics).

163 *Reflections*, p. 95.

The natural and the political worlds

It is necessary to ask why Burke should have apparently gone out of his way from time to time, particularly after the French Revolution, to speak about the science of his time. First, we must remember that science or natural philosophy had a significant place in the culture of Western Europe in the eighteenth century, and nowhere more than in Britain and France. There was no artificial barrier between arts and science. Also, Burke liked to show in his writings that he was a well-educated person. There may even sometimes be an affectation of learning. Secondly, he discussed science in a deliberate attempt to comment on the territory of his opponents. He could support the status quo more effectively if he could cast doubts on the new premises of the revolutionaries. He was particularly concerned to set up barriers against the rising tide of rationalism. Thirdly, association with science or popular images of science could sometimes be used to discredit his opponents. If they dabbled in science, they could be portrayed as alchemists. If they were chemists they could be associated with explosions. Finally, there is the assumption of a correspondence between the world of man and that of nature, an idea that goes back to the Greeks but had a particular political slant in early modern European support of the monarchy.[164] But even if the rational grounds for such a relationship are disputed, the very derivation of *images* from the natural world tended to lend them objectivity.[165]

The parallel Burke draws between the visible and the invisible worlds, that is between the physical world on the one hand and the moral and political on the other, is very relevant to the argument of this paper.[166] Sometimes Burke argues simply by analogy. Thus, he says in a most unconservative way that a

> system of change . . . is perhaps as necessary to the moral as it is found to be in the natural world.[167]

There are several instances of Burke explaining a point by drawing an analogy between the principles of mechanics and politics,[168] but at other times Burke goes much further than simple analogy: the political world *corresponds* to the natural world:

> Our political system is placed in a just correspondence and symmetry with the order of the world.[169]

In other words, the order of the world (presumably as discovered by science) gives authority to a political system. Civilization requires man to 'move with the order of the universe'.[170] It is important to note that the natural world is introduced to support an

164 Thus the sun at the centre of the universe was compared to the King. William Harvey, court physician to Charles I, used this analogy as well as that between the heart and the monarchy, *The Circulation of the Blood*, (tr. Kenneth J. Franklin), London, 1963, pp. 3, 108.

165 David K. Weiser, op. cit. (129), p. 324.

166 Burke speaks of 'the great primeval contract of eternal society . . . connecting the visible and invisible world according to a fixed compact sanctioned by the inviolable oath which holds all physical and moral natures, each in their appointed plan'. *Reflections*, p. 195.

167 *Abridgement of English History, Works*, vi, p. 236.

168 E.g., *Reflections*, pp. 122, 153. See also *Letter to William Elliott* (1975), 'The momentum is increased by an extraneous weight. It is true in moral as in political science'. *Works*, v, p. 80.

169 *Reflections*, p. 120.

170 Ibid., p. 196.

argument, but only as additional support, not as its main justification. Burke discusses the natural world partly because he believes that there should be harmony between it and society. Action in both the natural and the political world, he says, 'draws out the harmony of the universe'.[171]

A further reason, however, for the association is that Burke's enemies 'cloaked their judgement in the idiom of scientific explanation'.[172] This would be true, for example, of Helvetius. Since there were people like Priestley in England and a whole host in France known to Burke, including Bailly, Condorcet and Guyton, who were as much political figures as men of science, it seemed very natural to draw parallels between the world of politics and science. Both Bailly and Priestley draw analogies between revolutions in society and in science.[173] Considering that Condorcet and Burke were ideologically poles apart, it is interesting that in some cases they could come to similar conclusions, that one could argue from physical science to the problems of society. But whereas for Condorcet there was a continuum between physical and social science, for Burke the connection is obviously more tenuous.

In the last years of his life Burke reflected on the use of analogies between the physical and political worlds and untypically gives advice on the validity of such reasoning:

> Parallels of this sort rather furnish similitudes to illustrate or to adorn, than supply analogies from which to reason.[174]

Again, after using the standard metaphor of 'the body politic', Burke writes:

> These analogies between bodies natural and politic, though they may sometimes illustrate arguments, furnish no arguments of themselves.[175]

But even if Burke is here admitting that the parallel may sometimes be no more than useful rhetoric for the politician, what matters for our purposes is less the validity of the argument than whether it was widely believed.

Nature was an important concept for Burke, as for many eighteenth-century *philosophes*. But whereas Rousseau saw nature as the antithesis of Western civilization, Burke follows Montesquieu[176] in seeing nature as supporting eighteenth-century society. Thus, both considered the idea of a hierarchical society as 'natural'.[177] Although he never defines what he means by nature, Burke feels strongly that political and social institutions should be in 'conformity to nature'.[178] But the *philosophes* too appealed to the authority of nature and claimed that mankind in the eighteenth century had reached a new peak. Burke scorned such claims by reminding his audience that they had not suddenly become supermen:

> But no name, no power, no function, no artificial institution whatsoever, can make the men of whom any system of authority is composed any other than God, and nature, and education and

171 Ibid., p. 122.
172 D. Cameron, op. cit. (13).
173 I. Bernard Cohen, op. cit. (68), pp. 473–447.
174 *Letters on a Regicide Peace*, I, *Works*, v, p. 153.
175 *Letter to William Elliott*, *Works*, v, p. 78.
176 See C. P. Courtney, op. cit. (13).
177 Burke speaks of 'the principles of natural subordination'. *Reflections*, p. 372.
178 Ibid., p. 121.

habits of life have made them. Capacities beyond these the people have not to give . . . They have not the engagement of nature, they have not the promise of revelation, for any such powers'.[179]

Thus, basically man cannot rise above himself unless he has the help of either God or nature. But unlike certain revolutionary movements in the seventeenth century, the new intellectuals in France were hardly concerned with God. Hence, in their purely secular ethic, the understanding of nature became crucial.

Burke, in his various statements about nature and natural law, provides one of those subject areas where it might be quite misleading to construct a fully articulated theory, and claim that it represented a completely consistent philosophy.[180] Nevertheless, Burke made use of such ideas on several occasions and it is desirable to consider some of the issues involved, particularly in so far as they involve science. There is considerable special pleading. Thus, Burke falls into the trap of tending to equate his own beliefs with the laws of nature. He confuses 'Nature' with 'the nature of things' in eighteenth-century England, i.e., the system of property ownership and constitutional government. Thus, in opposition to the revolutionaries, he claims that the existing order has the authority of nature. We must 'preserve the method of nature in the conduct of the state'.[181] It is as a child of the eighteenth century that he appeals to nature in the first place. But in appealing to nature he combines two traditions. There is the old idea of natural laws as in Aquinas, where it is necessarily linked with God. There is also in the eighteenth century a purely secular concept of natural law which was related to impersonal natural forces. We cannot blame Burke too much for drawing on both concepts, since there was considerable confusion in the eighteenth-century usage, which resulted from the fact that the two pioneering seventeenth-century philosophers, Hobbes and Locke, had continued to use the old terminology to express new ideas which maintained a concept of 'natural law' but without the Deity.[182]

In his *Reflections* Burke makes repeated appeals to 'nature'. The revolutionaries are 'at war with nature'.[183] Otherwise expressed they seem to want 'to wage war with heaven itself'.[184] The revolutionaries have thought out 'grand theories', to which they 'would move heaven and earth to bend'.[185] Burke sees the French Revolution as 'an usurpation on the prerogatives of nature'.[186] 'The levellers . . . change and pervert the natural order of things';[187] constitutions should be 'after [i.e. in accordance with] the pattern of nature'.[188] We must 'preserve the method of nature in the conduct of the state'.[189] There are certain basic laws of nature. In the political world the right of self-preservation or self defence is

179 Ibid., p. 128.
180 The best attempt to explain Burke's ideas on natural law is probably P. Stanlis, op. cit. (1).
181 *Reflections*, p. 120.
182 Thus Hobbes relates 'natural law' to man's self-preservation rather than to God. F. S. McNeilly, *The anatomy of Leviathan*, London, 1968, p. 183.
183 *Reflections*, p. 138.
184 Ibid. p. 92.
185 Ibid., p. 323.
186 Ibid., p. 138.
187 Ibid.
188 Ibid., p. 120.
189 Ibid.

one of the most basic laws.[190] In the scientific world the refraction of light as it passes from one medium to another is a 'law of nature'.[191] And, as I suggested earlier, Burke spells out a correspondence between the political world and the world of nature. There is a principle of action and reaction ('counteraction') both 'in the natural and in the political world'.[192] There is a 'great ruling principle of [both] the moral and the natural world'.[193] There is a natural law which governs 'the moral and physical disposition of things, to which man must be obedient'.[194] But if he is not, and 'the law is broken, nature is disobeyed', then we leave the world of reason and order and risk madness.[195] In other words, in the moral and political realms laws may be broken, but only at man's peril. Natural laws cannot be broken, but they may sometimes be circumvented. Although atheists 'cannot strike the sun out of heaven',[196] they can obscure it with smoke. Burke would have agreed that another difference between the laws of society and the laws of nature is that only the latter can provide accurate predictions of the behaviour of physical bodies.[197] Society is not a physical but a 'moral essence'.[198]

For Burke, it is the fundamental place of God which links laws of nature and laws of society. The basis of the legal system has to be seen in absolute terms. Everyone in English eighteenth-century society lived

'in subjection to one great immutable pre-existent law'.[199]

It is this theocentric view which allows Burke to claim that:

we are knit and connected in the eternal frame of the universe,[200]

a statement which out of context might be thought to be concerned with physics or astronomy but is actually about principles of government and the legal system.

In his *Reflections* Burke devoted considerable attention to the proliferation of paper money, the famous *assignats*. Such 'paper currency', he says, is merely 'fictitious wealth'.[201] In polemical vein he is scathing about speculators. The ambiguity of the term is germane to Burke's purpose since it describes both financial investors and philosophers.[202] In their financial affairs the French have become 'a people who attempt to reverse the very nature of things'.[203] In a pamphlet, *Thoughts and Details on Scarcity*, written in 1795 in an attempt to influence governmental policy,[204] Burke claims that the

190 Ibid., p. 150.
191 Ibid., p. 152.
192 Ibid., p. 122.
193 Ibid., p. 200.
194 Ibid., p. 195.
195 Burke speaks of exile into 'the antagonistic world of madness, discord, vice, confusion and unavailing sorrow'.
196 *Regicide Peace*, II, *Works*, v, p. 245.
197 C. P. Courtney, op. cit. (13), p. 152.
198 *Regicide Peace*, I, *Works*, v, p. 153.
199 *Speech on the Impeachment of Warren Hastings* (1788), *Works*, vii, p. 99.
200 Ibid.
201 *Reflections*, p. 224.
202 E.g. Ibid., pp. 360ff. Burke associates 'the most desperate adventurers in philosophy and finance'.
203 Ibid., p. 359.
204 F. P. Lock, op. cit. (13), pp. 18–19.

'laws of commerce' are also 'the laws of Nature and consequently the laws of God'.[205] He felt that the attempt in France to control food prices went against one of the basic laws of economics, that of supply and demand.

Burke is here attributing to economics the status of a science. On a spectrum of knowledge economics would be placed between politics[206] and physical science. Thus, although Burke often argues by simple analogy between the moral and the natural world, the case of economics seems to suggest that the connection does not depend solely on this analogy; economics helps to provide a continuum between human concerns and the natural world.

Burke believes that classification is important in any well ordered society and that:

every such classification, if properly ordered, is good in all forms of government.[207]

Indeed, he insists that it is just as important in a republic as in a monarchy. Unfortunately, Burke does not discuss differences between these two types of society, of which the second is presumably more hierarchical than the first. He is in a much stronger position in claiming that there are

many diversities amongst men, according to their birth, their education, their professions . . .[208] etc.

It was therefore a great mistake for the revolutionaries to

have attempted to confound all sorts of citizens, as well as they could, into one homogeneous mass.[209]

But the importance of classification goes beyond the world of politics. It also relates to literature, art and science, and here collection and classification go together. Burke believes that a civilized society should spend some of its surplus wealth on libraries, archives, paintings and monuments; also on

collections of the specimens of nature, which become a representative assembly of all classes and families of the world, that by disposing facilitate, and, by exciting curiosity, open the avenues to science.[210]

Thus, for Burke not all science is bad. As someone who believes in differentiation in society, he is interested in differentiation in the natural world. He is therefore interested in classification and the classificatory approach to nature.[211] He thus, like Rousseau, approves of natural history and supports the philosophy of the collector.[212] The method

---

205 *Thoughts and Details on Scarcity* (1795), *Works*, v, p. 100.

206 The comment has recently been made that 'one of the large unanswered questions is how Burke's economic theory is related to his political theory', Gandy and Stanlis, op. cit. (1), p. 213.

207 *Reflections*, p. 301.

208 Ibid., p. 299.

209 Ibid., p. 300.

210 Ibid., p. 272.

211 For an introductory bibliography to the growing literature on classification see David Knight, *Ordering the World*, London, 1981, pp. 207–209.

212 Elsewhere in the *Reflections* he uses the metaphor of an 'ample collection of known classes, genera and species, which at present beautify the hortus siccus [i.e. collection of dried plants]', ibid., pp. 95–96.

of collection was gradual, it was modest and it did not expose men to the same dangers as experimental science. Indeed, as early as 1772 the *Annual Register,* edited by Burke, had devoted several pages to a favourable review of Priestley's *History . . . of . . . Vision, Light and Colours,* saying that:

> Nothing can be more agreeable than a view of the *gradual progression* of human industry and the *gradual unfolding* of knowledge.[213]

In this view of knowledge by accretion rather than by sudden leaps there is plenty of opportunity for mature reflection. New ideas emerge by evolution rather than revolution. Certainly, government should be based on experience rather than experiment.

Finally, I should like to quote another passage from the *Reflections,* which is partly critical and partly supportive of science. In so far as it is based on the application of power, it is more concerned with technology than pure science. It illustrates that Burke could lend support to the ultimate utilitarian justification of science. He draws a parallel between politics and mechanics; both being concerned with power. To try to destroy any power in society

> would be like the attempt to destroy . . . the expansive force of fixed air in nitre, or the power of steam, or of electricity, or of magnetism. These energies always existed in nature, and they were always discernible.[214] They seemed, some of them unserviceable, some noxious, some no better than a sport to children, until contemplative ability, combining with practical skill, tamed their wild nature, subdued them to use, and rendered them at once the most powerful and the most tractable agents, in subservience to the great views and designs of men.[215]

From this extremely important passage I would like to select two major points. First, Burke is happy to dismiss a part of eighteenth-century science as no more than children's games. Yet some science cannot be so easily dismissed because it is actually dangerous. Recognizing some parts of science as endangering mankind, man must not allow these powers to overcome society but must himself become the master. Although apparently the enemy of experimental science, Burke reveals himself here as the potential friend of controlled technology.

CONCLUSION

Burke was not a philosopher, but he did have a philosophy. The philosophy was a basically conservative one but with ramifications much wider than politics, and this was particularly true in the 1790s, when the French Revolution forced on Burke a reappraisal of the values of western civilization. He had much to say about religion, law and many

213 *Annual Register,* 1772, pp. 232–235 (232, my italics). For a view which contrasts the early and acceptable writing of Histories by Priestley with his later experimental career in pneumatic chemistry, see Maurice Crosland, 'Priestley Memorial Lecture: A practical perspective on Joseph Priestley as a pneumatic chemist', *B.J.H.S.* (1983), **16**, pp. 223–238 (231–232).

214 Although these powers may have been *capable* of being discovered for a long time, it was only in the eighteenth century that they began to be developed.

215 *Reflections,* p. 268.

other aspects of society, but this paper has deliberately focused on his repeated attacks on science and its practitioners and on new ideas in general. Burke believed firmly that it was 'pernicious to disturb the natural course of things'.[216]

In eighteenth-century England, Newtonian science had traditionally been seen as broadly supportive of the established religion and government.[217] Before the French Revolution, therefore, there was no good reason why Burke should have spoken out against the ideas of natural philosophy, although he may have been sceptical about its ultimate authority. But with the Revolution, he began to see dangers on all sides and these included science and particularly experimental science. The revolutionaries seemed to have applied the experimental method to society. Burke speaks with trepidation of 'this new experimental government'[218] set up by the revolutionaries in France, while Priestley in England argued that government could be brought to perfection by experiment.[219]

Nor did Burke like the analytical method of science which, when studying a community, might try 'to dissolve it into an unsocial, uncivil, unconnected chaos of elementary principles',[220] a method which, incidentally, characterized chemistry. It is the romantic in Burke which objects to the disposition 'to pull every thing in pieces'.[221] This is a fundamental aspect of his organic philosophy and provides additional evidence which leads us to reject the extreme view that, when Burke speaks of science, it is simply as a metaphor.

Burke does not, of course, speak of 'scientists', a word which did not exist before the 1830s; also, science itself in the eighteenth century in Britain (if not in France[222]) was largely 'natural philosophy' or 'natural history'. In the days before professional science, there was not even an occupational group corresponding to 'scientists'. Most people who contributed to science did so on a part-time basis, earning their living in some other capacity, often in one of the traditional professions. England's most famous chemist was by no means unique in the diversity of his interests, only in his zeal and fame. It was only in the nineteenth century that there was a growing differentiation of social roles which led to the recognition of science as a profession. What Burke was attacking, therefore, was not a clearly defined homogeneous group but a few notable individuals and a number of less clearly identified people and attitudes. On the European front he was attacking above all a mentality, one which had come to the fore in the revolutionary period and which believed in reductionism and social engineering. In so far as the French *philosophes* had been prominent in advocating new ideas of society based on science, Burke's description of the enemy as 'philosophers' has some justification. He totally

216 Ibid., p. 271.
217 See, e.g., Margaret Jacobs, *The Newtonians and the English Revolution, 1689–1720*, Hassocks, 1976. See also a critique of Jacobs in Colin Russell, *Science and Social Change, 1700–1900*, London, 1983, pp. 52ff.
218 *Reflections*, p. 276.
219 Isaac Kramnick, 'Eighteenth-century science and radical social theory: the case of Joseph Priestley's scientific liberalism', *Journal of British Studies*, (1986), 25, pp. 1–30 (24).
220 *Reflections*, p. 195.
221 Ibid., p. 283.
222 Different specialisms were identified in the regulations of the Paris Academy of Sciences, dating back to 1699 and were well accepted in eighteenth-century France.

opposed Priestley, who advocated carrying 'the same spirit into the study of history and of human nature that [philosophers] do in their laboratories'.[223] In so far as he perceived before many of his contemporaries the growth of a political and social movement which claimed the authority of science, we can claim Burke as a prophet of scientism, or rather anti-scientism. Condorcet had arrived. Saint-Simon and Comte were in the future.

It is clear that much of Burke's criticism of science and mathematics was directed towards the application of these disciplines to society. It was social engineering as much as science which was under attack. Yet we must insist that Burke was fundamentally opposed to a great deal of science and for three main reasons.

In the first place, Burke distrusted excessive reliance on the powers of reason:

> we are afraid to put men to live and trade each on his own private stock of reason; because we suspect that this stock in each man is small.[224]

He had more faith in the collective wisdom of established society.

Secondly, Burke felt that the universe was essentially a divine mystery which should not be studied too closely by man.[225] Thus he writes:

> Dark and inscrutable are the ways by which we come into the world. The instincts that give rise to this mysterious process of nature are not of our making.[226]

In the same vein he goes on to speak of 'physical causes unknown to us, perhaps unknowable'.

This brings us to Burke's scepticism of the authority of men and of science to make pronouncements about the natural world. He accepted the Scriptures and traditional authority, but when men claimed by observation and deductive reasoning to arrive at certain knowledge he felt that they were often going too far. In some earlier writings he criticized Newton's attempt to explain gravitation in terms of the hypothesis of an ether and concluded:

> That great chain of causes which, linking one to another even to the throne of God himself, can never be unravelled by any industry of ours.[227]

Burke was very influential in the nineteenth century. There were very many editions of his works[228] and he was widely read, particularly by the upper social classes, who had the greatest political power. His defence of the status quo and his attacks on the

223 *Experiments and Observations on Natural Philosophy*, Birmingham, 1779–1786, iii, pp. xvi–xvii. Priestley's claim to be applying scientific method to theology can hardly be accepted at face value. As has been pointed out, he often decided in advance what was 'true Christianity' and what were its corruptions. He then turned to history to find support. He did not consult his sources with an open mind. See Gerald R. Cragg, *Reason and Authority in the Eighteenth Century*, Cambridge, 1964, pp. 236–237.

224 *Reflections*, p. 183.

225 Basil Willey, *The Eighteenth-Century Background*, London, 1940, p. 232. Ray B. Browne, *The Burke-Paine Controversy. Text and Criticism*, New York, 1963, p. 147.

226 *Appeal from the New to the Old Whigs, Works*, iii, p. 79.

227 Robert Montgomery (ed.), *Edmund Burke: Being First Principles Selected from his Writings*, London, 1853, p. 4. There is some discussion of Burke's scepticism in Francis P. Canavan, *The Political Reason of Edmund Burke*, Durham, N. C., 1960, pp. 33–34.

228 See William B. Todd, *A bibliography of Edmund Burke*, London, 1964, especially pp. 142ff.

reductionist methodology of social mathematics, and to a lesser extent on chemistry, helped confirm the prejudices of the English upper classes against rationalism and science. Much of science for Burke was either frivolous—'a sport to children'[229]—or dangerous. Even if science did not threaten the establishment, it certainly did not deserve state support.

229 *Reflections*, p. 268. A seventeenth-century parallel is provided by Henry Stubbes who, in his attack on the Royal Society, speaks of 'Toyish Experiments'—quoted by Michael Hunter, *Science and Society in Restoration England*, Cambridge, 1981, p. 151. It should be noted that no claim is made in this paper for the *originality* of Burke's ideas. He was all the more influential because in many cases he is doing no more than reminding his audience of their prejudices.

III

A PRACTICAL PERSPECTIVE ON JOSEPH PRIESTLEY AS A PNEUMATIC CHEMIST

Two major problems in understanding Joseph Priestley (1733–1804) are that he wrote so much and over such a wide area. The nineteenth-century edition of his collected works fills 25 volumes[1]—and that leaves out the science! In discussing a man like Priestley, therefore, one cannot hope in a single lecture to do justice to the wide range of his interests or even to summarise adequately his many contributions to science. Fortunately much of the scientific work is fairly well known, for example his discovery of many new gases or 'airs', as he preferred to call them.[2] It might be appropriate, therefore, to try to put Priestley's pneumatic chemistry in a wider context and in particular to relate it to his career. Priestley was not only an important man of science. He was also an outspoken theologian, a literary figure and a family man, and all of these roles (and several others, including his political role on behalf of Dissenters) will have to be taken into consideration when the definitive biography is written.

Throughout his life Priestley was a compulsive writer. In his autobiography he explained that he had early 'acquired a habit of composing with great readiness'.[3] One reason for his writing was simply as an aide-memoire in learning. But he also wanted to express his own ideas. We should remember that, as a young man, he suffered from a serious impediment of speech, without which, he said: 'I might have been disputatious in company',[4] having previously recalled that, as a student, he inclined to 'the heterodox side of almost every question'.[5] Thus a partial

* Unit for the History of Science, Physics Building, University of Kent, Canterbury, Kent CT2 7NR.
On the occasion of the 250th anniversary of the birth of Joseph Priestley in March 1983, the author was invited to deliver a series of three historical lectures as Brotherton Visiting Professor in the Department of Inorganic Chemistry at the University of Leeds. The present text is an abbreviated version of the first lecture.

On specific points in the preparation of this lecture I am grateful for comments from Ted Caldin, Geoffrey Cantor, Grayson Ditchfield, David Knight and Crosbie Smith, none of whom, however, are to be blamed for the arguments presented.

[1] *The theological and miscellaneous works of Joseph Priestley*, ed. J.T. Rutt, 25 vols., in 26, London, 1817–32.

[2] A detailed analysis of Priestley's work on gases is given in J. R. Partington, *A history of chemistry*, London, 1961–70, vol. 3, chapter VII. It is unfortunate that Partington saw 'the discovery of oxygen' as a simple event, reflecting credit on Priestley alone. Philosophical discussion of this point does not, however, detract from the major importance of Priestley in the history of the practical science of pneumatic chemistry.

[3] The most recent and, therefore, probably the most accessible edition of Priestley's Memoirs is: *Autobiography of Joseph Priestley*, ed. Jack Lindsay, Bath, Adams and Dart, 1970, p. 74.

[4] *Ibid.*, p. 78.   [5] *Ibid.*, p. 76.

explanation of his extensive writing is that, for a man with a lively, independent and original mind, it developed as a substitute for oral controversy. He was always more concerned with subject matter than with style:

> 'My object was not to acquire the character of a fine writer but of a useful one'[6]

and he, therefore, excused himself for his 'hasty performances'. Nevertheless, he was ambitious for a literary reputation of some kind. We must consider what paths to fame lay open in eighteenth-century England to a would-be author with deeply held religious views.

In this lecture I would like to attempt to answer two general but fundamental questions. First, why did Priestley want to write about the natural world? Second, why did he take up chemistry? After all, his first studies in science were in the area we would now call physics. He made a study of electricity and he was very successful—his substantial book on the subject went through five editions[7]—why did he move to something different? Priestley's chemistry tends to loom so large to-day that we could easily forget that his scientific education began with mathematics and natural philosophy.

Priestley's original scientific interests would have fitted in well with the tradition of natural theology, that the natural world provided evidence of the wisdom and beneficence of the Creator. But if we ask why he studied and wrote about science, there is evidence that what drew him further into science was religious conflict. Priestley began his scientific career as an educator and writer. And from writing as simply summarising other books, he was led actually to participate in scientific activity.

Science and Religious Conflict

A twentieth-century scientist might be tempted to ask why Priestley, the man of science, wasted so much of his time writing about religion. It would be better to turn the question round and ask how did the young minister of religion become interested in science, and particularly in chemistry? There is another intriguing question—why did Priestley move so often? His many moves almost tend to suggest that he was running away from something. There may be some connection between the answers to these two questions. Perhaps his moves can be explained both negatively and positively. He was sometimes distancing himself from religious disputes but he was also attracted by the ambition to improve his position and his income. When he went to Leeds in 1767 he became one of the best paid dissenting ministers of his day but by moving on in 1773 to a position

---

[6] *Ibid.*, p. 90.
[7] The first edition was published in 1767 and the second (enlarged) in 1769. See Ronald E. Crook, *A bibliography of Joseph Priestley, 1733–1804*, London, 1966.

with Lord Shelburne, he more than doubled his income without too much loss of independence.

I am not sure that a fully satisfactory explanation has yet been given of Priestley's changes of career and his many moves around England, from his first ministries in Suffolk and in Cheshire. After six years at Warrington Academy, he spent the next six years as Unitarian minister at Leeds and then about the same time as 'librarian' to Lord Shelburne in Wiltshire. From 1780 he was in Birmingham. After his house in Birmingham had been burned down in the notorious 'Church and King' riots of 1791, he moved to London and finally to Pennsylvania.

Some of these changes can be explained in terms of religious difficulties and/or better financial rewards. I think that these two considerations were important, but anyone who wanted to understand the whole of Priestley's career would need to add several more parameters. A third guiding principle in his career was a lifelong theological quest, a fourth was a striving for intellectual achievement, a fifth was moral and political concern and a sixth was a desire for personal independence. Some recent analyses of Priestley have been very theoretical and abstract. I think that these six practical considerations provide us with a basis for a much more realistic analysis of Priestley's life. If one were to think along these lines, one would understand rather better what made Priestley tick.

Priestley's autobiography reveals a certain deprivation in childhood. It was not only that his mother died when he was only seven but even when his mother was alive she had little time for him, the eldest, because Joseph tells us she was 'having children so fast'.[8] During his infancy he was therefore cared for mainly by his grandfather. On the death of his mother—Priestley's own poignant words are 'being without a mother'[9]—he was cared for by an aunt, a Mrs. Keighley. Perhaps this led to a feeling of insecurity in later life. Priestley's aunt was a fairly broad-minded Calvinist, who often entertained various dissenting ministers from the neighbourhood. The aunt hoped that her studious nephew might himself become a minister and the boy readily agreed. He was sent to study at the Dissenting Academy at Daventry.

For reasons of space, I must now pass over Priestley's education to his first job. This was as a minister at Needham Market in Suffolk. His salary depended on his pleasing his theologically heterogeneous congregation, something he did not find easy. When it was further discovered that he was an Arian,[10] and did not accept the divinity of Jesus Christ, his congregation

---

[8] *Autobiography, op. cit.* (3) p. 69.

[9] *Ibid.*, p. 70. His father was still alive but 'incumbered with a large family'. After the death of his wife the father married again.

[10] Although, as an Arian, Priestley believed in the pre-existence of Christ, he denied his deity, but he did not believe either that Christ was an ordinary human being. For a brief guide to Priestley's religious position see articles 'Arianism' and 'Unitarianism' in F. L. Cross (ed.) *The Oxford dictionary of the Christian Church*, 2nd ed., London, 1974 and especially Miachael R. Watts, *The dissenters*, London, 1978, pp. 471–7.

rapidly diminished and he was only able to make ends meet with the help of several local charities. He had originally hoped for some financial support from his dear aunt but this never came, partly, he says 'owing . . . to the ill offices of my orthodox relations'.[11] A neighbouring minister hinted to him that his troubles would be at an end if only he would conform to the Church of England but Priestley would not hear of this. During his three years at Needham he pursued his theological studies and had yet more doubts which further distanced him from his Calvinistic roots. His theological problems were compounded by his financial ones. He therefore considered supplementing his income by teaching. A clergyman would often teach catechism but Priestley thought this was an area better avoided. He recalled: *'because I was not orthodox* I had proposed to teach the classics, mathematics, etc. for half a guinea per quarter'[12] and also to take pupils as boarders. Here then was Priestley looking desperately for neutral ground as a basis for education and a decent livelihood. He was not yet doing science but he had made the essential transition through education, drawing on a potentially useful curriculum free from the taint of heterodoxy in religion.

In his first job Priestley had very few books of his own. In order to earn some money to buy books he planned to learn up a new subject and then charge for teaching it. He chose to give a course of 12 lectures on the use of globes, a traditional subject which might now be classified as geography. Building on this slight educational experience, when he subsequently went to Nantwich in Cheshire he decided to set up a school. The profits from the school he said

'soon enabled me to purchase a few books and some philosophical [i.e. scientific] instruments, as a small air pump, an electrical machine, etc.'[13]

Having purchased these scientific instruments and books he then incorporated natural philosophy into the curriculum for the senior pupils.

He was then offered a post as tutor at the Warrington Academy, where he was to teach languages and belles lettres. Here, he said, he was 'singularly happy' in the company of fellow tutors who shared his theological views. Priestley wrote several books relating to his duties but a visit to London and an introduction to Benjamin Franklin helped turn his mind towards the new and exciting science of electricity. He would write a book on that too, a *History of Electricity*. One of the advantages of writing a history of electricity (a good example of progress) was that 'the principal actors in the scene . . . [were] still living',[14] for example, Franklin, Watson and Canton, whom he counted among his friends and who lent him

---

[11] *Autobiography, op. cit.* (3)., p. 79.   [12] *Ibid.*, p. 84. (my italics)   [13] *Ibid.*, p. 85.
[14] Preface to 1st ed., p. xi, reprinted in *History and present state of electricity* reprinted from 3rd edn., London, 1775, Johnson Reprint, New York and London, 1966.

relevant books. But not only did he describe the achievements of others, he repeated some of their experiments and added a few of his own.

We now come to one of the most famous and fruitful periods in Priestley's life, the period in Leeds, and we know that he at first lived south of the river in Meadow Lane in a house next to a brewery. Priestley was fascinated by a by-product of the fermentation, the 'fixed air', what we now call carbon dioxide. He described his first experiments with 'fixed air' as no more than amusement. When he moved house, however, he no longer had a ready source of the gas and he had to learn to make it for himself if he was to continue with his hobby. Then as he modestly put it, one experiment just led to another.[15]

So here we have the gradual transition from theology to natural philosophy and chemistry. Some might think that Priestley revelled in theological controversy. Yet, if theological writing had become a habit and he almost saw it as his duty to put pen to paper on those frequent occasions when he thought that he had discovered some theological truth, it could, nevertheless, be a duty which was painful. Writing in August 1770 to a friend, Priestley confided that he now realised that many of his theological tracts caused considerable offence and particularly among fellow Dissenters:

> 'By one means or another I believe that I have more enemies among Dissenters than in the Church [of England]. I shall soon be obliged to court the Papists and Quakers in order to have any friends at all, *except a few philosophical people . . .*'[16]

He therefore saw science at this time as likely to provide the principal basis of friendship. In the Prospectus he issued in February 1771 for his *History of Vision* he spoke of the encouragement he hoped for 'from the liberal-minded friends of science, which is of no particular party, either in politics , or religion . . .'[17] A proposal in 1771 that he should join Captain Cook's second voyage round the world came to nothing because of clerical opposition, based, not on his scientific competence, but on his well-known theological opinions.[18] In his ministry in Leeds he said that, in general, he had 'no unreasonable prejudices to contend with'.[19] In 1772, however, he described a meeting with a number of dissenting ministers in the neighbourhood who, 'came to oppose and wrangle'[20] but he managed to achieve some agreement with them. The Methodists, 'very numerous in Leeds'[21] Priestley tells us, scandalised by his heterodox views, composed a hymn, asking God

[15] *Autobiography, op. cit.* (3), p. 94.
[16] J.T. Rutt, *Life and correspondence of Joseph Priestley*, 2 vols., 1831, Vol. 1. p. 118. (my italics) Priestley to Rev. T. Lindsey, Leeds, 30 July 1770.
[17] Robert E. Schofield, *A scientific autobiography of Joseph Priestley, (1733–1804)*, Cambridge, Mass. and London, p. 76.
[18] *Ibid.*, No. 44.   [19] *Autobiography, op. cit.* (3), p. 92.
[20] Schofield, *op. cit.* (17), No. 53.   [21] *Autobiography, op. cit.* (3), p. 94.

'The Unitarian fiend [to] expel
And chase his doctrine back to Hell'.[22]

Science could provide some relief. Thus in Leeds he often discussed his experiments on different airs with the surgeon, William Hey, whom he described as 'a zealous Methodist' and author of several theological tracts opposing his Unitarian friend. However, Priestley reported that they 'always conversed with the greatest freedom on philosophical subjects, *without mentioning anything relating to theology*'.[23]

In the turmoil of eighteenth-century theological controversy and Priestley's various attempts to find a suitable position in English society, the pursuit of natural philosophy provided a haven. Men of different theological and political persuasions could come together to study the apparently neutral world of science. This is why he felt so betrayed when, after his house had been burned down in the Birmingham riots, he came to London for refuge and found himself 'shunned' (Priestley's own term)[24] by his fellow members of the Royal Society because of his religious and political opinions rather than welcomed by them as a scientific colleague. But this says more about the social composition of the Royal Society in the eighteenth century[25] and about the climate of fear in Britain in the 1790s than it says about the nature of science.

Among the books and pamphlets he published in his London period was one on airs with a most interesting and significant dedication to his former colleagues in the Lunar Society in Birmingham. He said that his home in Birmingham had been sacked because he was considered to be 'a fomenter of sedition and an enemy to the peace and constitution of my country'. Yet, if he was really such a person, would not his colleagues have been aware of this in their frequent meetings?

'But', he wrote, 'You know that neither politics nor religion were ever the subjects of our conversation. *Philosophy engrossed us wholly*'.[26]

This passage is Priestley's alibi, although, of course, the logic is faulty, since the Lunar Society only met once a month, giving Priestley plenty of time for his other interests. He was claiming to be a natural philosopher and therefore to be neutral in any religious or political dispute:

---

[22] Quoted by F. E. Mineka, *The dissidence of Dissent*, Chapel Hill, N. Carolina, 1944, p. 19, who refers to a review of Joseph Nightingale's *Portraiture of Methodism*, in *Monthly Respository*, 1808, *3*, 103, which quotes the 431st hymn in the 'Large Hymn Book'. Mineka implies that the hymn referred to Priestley, an interpretation accepted by John A. Passmore, *Priestley's writings on philosophy, science and politics*, New York, 1965, p. 17.

[23] *Autobiography, op. cit.* (3), p. 95. (my italics)

[24] *Ibid.*, p. 130. He was particularly upset by Cavendish's indifference to him.

[25] Priestley was particularly concerned about the rejection of his friend Thomas Cooper, apparently because he was a politically active Unitarian. For a recent comment on social and intellectual considerations in applications for membership of the Royal Society in the eighteenth century, see Maurice Crosland, 'Explicit qualifications as a criterion for membership of the Royal Society: A historical review', *Notes and records of the Royal Society of London*, 1982–83, *37*, 167–87.

[26] *Experiments on the generation of air from water*, London, 1793, pp. iv–v. (my italics)

'Happy would it be for the world if their pursuits were as tranquil and their projects as innocent, and as friendly to the best interests of mankind, as ours'.

This then is Priestley's case for the neutrality of science, which I think was important in Priestley's career around 1770, although it is understandably most explicit in his writings only after his life had been in danger. In such a situation it is perhaps not for us in our more just society to accuse Priestley of some inconsistency, of contributing to political unrest by his outspoken support for the French Revolution and the rights of Dissenters, and then running to science for protection.

But leaving aside Priestley's many different roles, what justification is there for the claim that science was neutral? Even in the pursuit of natural philosophy Priestley did not forget that his first concern was theology and he pointed repeatedly to the natural world as evidence of a beneficent Creator. Here of course all Christians could agree, so long as the natural theology was intended to supplement revelation and not to replace it—as in Deism. Indeed Priestley could use natural theology to emphasise that he was after all a (sort of) Christian, since he had now become even more heterodox and was so often regarded by his more orthodox Christian contemporaries as nothing less than an atheist. There had already been precedents in the Boyle lectures for natural theology being used to support Christianity but to gloss over sectarian division.[27]

Priestley actually claimed to have not a weaker but 'a stronger bias than many other persons in favour of Christianity' and he said that this was based on his scientific research:

'I view with rapture the glorious face of nature and I admire its wonderful constitution, the laws of which are daily unfolding themselves to our view'.[28]

For him the natural philosopher was something of a missionary explorer with the task of exploring 'the hidden powers which the Deity has impressed on matter'.[29] Priestley was distressed by the fact that his friend Benjamin Franklin was an unbeliever.

'A philosopher' (i.e. a scientist), he said, 'ought to be something greater and better than another man. The contemplation of the works of God should give a sublimity to his virtue . . . A life spent in the productions of divine power, wisdom and goodness, would be a life of devotion'.[30]

The Socinian Priestley actually considered himself a better Christian than his more orthodox comtemporaries. Their Christianity he stigmatised as 'corrupt' whereas his Christianity was 'rational' and better fitted for 'philosophical and thinking persons'.[31] If he presented his theology as theology particularly acceptable to those versed in philosophy, it is not

[27] See Margaret Jacob, *The Newtonians and the English Revolution*, 1689–1720, Hassocks, Sussex, 1976, p. 144.
[28] *Experiments and observations on different kinds of air*, Birmingham, 1790, i, p. xl.
[29] *Ibid.*, i, p. x.   [30] Priestley, *op. cit.* (14), p. xxiii.   [31] *Autobiography, op. cit.* (3), p. 111.

surprising that in his writings on natural philosophy he should have stressed the theological implications. Thus, to sum up, Priestley was attracted to science for a number of reasons:

 (i) as rational recreation
 (ii) as the subject matter of education and hence a source of income
 (iii) as a haven from theological controversy
 (iv) as a means of advancing knowledge and furthering his own reputation
 (v) as a means of emphasising his Christianity
and (vi) as a basis for a system of materialism, which, he thought, would place Christianity on a sound philosophical foundation.

These very different roles for science suggest some inconsistency. But the inconsistency is partly explained by intellectual development. The materialist phase really only dates from about 1777, when Priestley published his *Disquisitions relating to matter and spirit*. This resulted in a clash with the Jesuit, Ruggiero Boscovich, whose theory of point atoms Priestley had used to draw conclusions which were the antithesis of orthodox Christianity.[32] My thesis about Priestley using the supposed neutrality of science is not invalidated at all by the fact that, after only a short interval, he launched himself back into the fray more vigorously than ever. Priestley, obsessive writer and controversialist, could not be suppressed nor seduced, not even by the wonders of pneumatic chemistry.

The Financial Aspect: From Literature to Science

But if I have explained some of the attractions of science for Priestley, I still have to explain how Priestley passed from what is now called 'physics' to chemistry. Probably the main reason for his turning to the study of 'airs' was a financial one, the problem of finding sufficient subscribers for his book, *History and present state of discoveries relating to vision, light and colours*, mostly written in 1770. By November 1771, Priestley realised that in financial terms it was going to be a failure.[33] This financial worry was increased by financial difficulties experienced by the controversial *Theological Repository*, a journal founded by Priestley in 1769 and forced to close in 1771, after only three volumes, leaving Priestley in debt.[34]

The *History . . . of . . . vision* was written as the second in a series of so-called 'histories' of all the experimental sciences. Had it succeeded, he would have written further books in this series, e.g. on magnetism, 'upon the same extensive plan'.[35] It was largely because of the failure of this book

---

[32] Schofield, *op. cit.* (17), Nos. 79, 80.
[33] *Ibid.*, No. 39.
[34] See *ibid.*, No. 31. (23 Dec, 1770).) In 1772 Priestley told Price that he had lost £50 (half his annual salary) in the *Theological repository*, and was therefore obliged to abandon it. J. T. Rutt (ed.) *op. cit.* (16) p. 184.
[35] Schofield, *op. cit.* (17), pp. 73, 76.

that he turned to chemistry. The change of subject has been noted by previous historians, but not, I think, the change of method and its significance.

In his prospectus of December 1770 to his *History . . . of . . . vision* Priestley recommended his historical method:

> 'The historical method, adopted in this work, has many obvious advantages over any other, being particularly calculated to engage the attention of the reader, and communicate useful knowledge with the greatest ease, pleasure and certainty'.[36]

In order to write his book Priestley needed to buy all the most important books previously published on the subject and by July 1770 he had already spent about £100 (i.e. a full year's salary) although there were still many important sources lacking. He became uneasy and confided to his friend Lindsey.

> 'now that I am in for it, [I] shall risk a good deal more [than £100]'[37]

and he continued to order books.

In his *Memoirs* Priestley emphasised how large his investment had been in books to enable him to write his *History . . . of . . . vision*. He spoke repeatedly of the expense.[38] Indeed he seems to have been one of those people whose life is a struggle between idealism on the one hand and financial problems on the other, not an uncommon predicament. He had hoped to make some modest profit to invest in further work, but, being unable to obtain a sufficient number of subscribers, he actually made a substantial loss.[39] Priestley had a great urge to write, but in natural philosophy his method of reviewing the literature of a subject has proved fallible. He had not been successful in his *History . . . of . . . vision* in condensing in a pleasing way the great mass of information contained in his sources. He did not have the imagination nor the style to write anything like a novel. In any case he would have regarded this as frivolous. In his youth he had tried versifying, but he was clearly no poet. Rather than continue with his 'histories', he would have to try an alternative approach to authorship which did not involve a large financial investment. Priestley described the failure of his optical work and remarked:

> 'I am obliged to abandon it and apply wholly to original experiments'.[40]

---

[36] *Ibid.*, p. 78.

[37] *Ibid.*, No. 30. In the Preface to his *History and present state of discoveries relating to vision, light and colours*, London 1772, he explained to the reader that the knowledge he was going to provide, being widely dispersed in the sources, would not only require a great deal of time to extract independently but would also cost 'several hundred pounds in any one branch of science' (*op. cit.* p. ii.).

[38] In his *Memoirs* Priestley described the writing of the book as 'an undertaking of great expense' (*op. cit.* (3), p. 95). He mentioned the expense three times in the Preface to *History . . . of . . . vision*, London, 1772, pp. ii–v. In his *Memoirs* Priestley several times related his increasing expenses in the 1770s and 1780s to the needs of his family, Autobiography, *op. cit.* (3), pp. 116, 120.

[39] Schofield, *op. cit.* (17), No. 39.   [40] *Autobiography, op. cit.* (3), p. 95.

What an admission! He was *obliged* to do original experiments. But for Priestley this was a new way of writing books, *starting with experiments* and it was also the beginning of his chemical career. In October 1771 he compared his early work on airs with his book, *History . . . of . . . vision*. He had spent more time on the experiments than on the first volume of the book 'But', he said,

> 'They have not been very expensive, whereas the other work is exceedingly so'.[41]

In a letter of September 1772 he confirmed the failure of his *History . . . of . . . vision* and said that he would now probably 'write the history of discoveries relating to air'.[42] Although he spoke here of writing yet another *History*,[43] it is significant that all his books on pneumatic chemistry make a break by abandoning this title in favour of the title: *Experiments and observations*. Priestley said that he was in a very good position to write on air

> 'in consequence of having made so many observations of my own on the subject. I mean, however to prosecute my experiments as I have opportunity, *keeping off such as would involve me in expense*'.[44]

He wrote pointedly to Richard Price that the new Pyrmont [soda] water he had learned to make artificially would cost Price five shillings but 'will not cost me a penny'.[45] Thus for an enterprising man with an interest in natural philosophy, and living next door to a brewery, an introduction to pneumatic chemistry was available as what, in the language of modern commerce, might be called a free trial offer. Priestley was to continue to find the study of airs both intellectually satisfying and inexpensive. It was to be the best investment he ever made.

Priestley's study of gases not only marks the beginning of his chemical career but the beginning of a new approach to science. His previous researches had been primarily literary and bibliographical, i.e. he had based his books on electricity and optics on previous *writings* on these subjects and his credentials were his bibliographies and footnotes.[46] This follows well-established traditions of scholarship in the arts. Priestley had, however, deviated from this tradition by supplementing his literary

[41] Schofield, *op. cit.* (17), No. 37. In his first book on pneumatic chemistry Priestley explained that he had been obliged to abandon his plan to write the history and present state of all the branches of natural philosophy 'because I see no prospect of being reasonably indemnified for so much labour and expense' *Experiments and observations on different kinds of air*, 3 vols., London, 1774, 75, 77, i, p. xix).
[42] Priestley to Dr. Price, Leeds, 27 September 1772, Rutt, *op. cit.* (16), p. 183.
[43] It is interesting that in 1776, when Priestley had finally become fully an experimentalist, he defended himself against criticism that he had published a mistaken observation in his *History . . . of . . . vision* by saying that he had then been writing 'as an historian', Priestley, *op. cit.* (41), Vol. 2, p. xvi. [44] Priestley to Dr. Price, Leeds, 22 September 1772, Rutt, *op. cit.* (16), p. 183. (my italics)
[45] Schofield, *op. cit.* (17), p. 107.
[46] In the Preface to his *History . . . of . . . vision*, Priestley recalled that he had begun with a list of books which he owned and also some which he wanted. He then produced a second and larger catalogue of books. He finally printed a third catalogue of the books and journals he had used, presenting these as his credentials. See *op. cit.*, 1772, p. vi. and Schofield, *op. cit.*, (17), p. 79.

researches and repeating some of the experiments described. This had sometimes led him to make original experiments. But these earlier books are primarily descriptions of other people's works[47] and are not based on original experiments.[48] The studies of airs in the 1770s are entirely Priestley's original experiments. Priestley was no longer dependent on books lent by friends or sent at great expense from London. By interrogating nature directly, rather than relying on books, he had at the same time reduced his costs and increased his independence, both goals close to the heart of our hero.

One might conceivably wish to claim that the change from optics to the study of gases was the result of some sudden enlightenment or the influence of some awe-inspiring figure in the history of philosophy. Instead the evidence seems to point to something more prosaic. In 1770–71 Priestley was almost overwhelmed by problems arising from the excess of his expenditure over his income. But there should be no misunderstanding: the important point is not that chemical apparatus is cheaper than physical apparatus. It is rather that simple apparatus is less expensive than a comprehensive library. This interpretation solves the puzzle of why Priestley, who knew something of chemistry since his Warrington days,[49]. did not take it up until about 1770.

Not all parts of eighteenth-century chemistry were inexpensive. For example, recently discovered minerals, being scarce, might be expensive. Certainly the platinum used for crucibles in the late eighteenth century was expensive.[50] But the chemistry of gases happens to be one of the least expensive areas in the whole of chemistry. As Priestley remarked at the end of his first publication on gases,

'The apparatus with which the principal of the preceding experiments were made is *exceedingly simple and cheap*'.[51]

Given his economic circumstances, he had made an appropriate choice. When Priestley left Leeds his friend William Hey begged that he would give him the earthen trough he had used to collect gases. This Priestley was happy to do, since he says, 'it was such a one as is there commonly used for washing linen'.[52] In other words Priestley adapted kitchen utensils as

[47] The contrast with *The History . . . of . . . vision* is all the greater because his Yorkshire friend John Michell had helped him with several technical problems which Priestley might otherwise have checked himself.
[48] Although Priestley had included a number of original experiments in his *History of electricity*, Schofield points out that he was then thought of, even by Franklin, simply as an author. It was his work on airs which revealed, both to himself and to the world, his great talents as an experimentalist, Schofield, *op. cit.* (17), p. 118.
[49] Schofield has pointed out that Priestley, when at Warrington, had attended a course of lectures on chemistry by Dr. Mathew Turner, Schofield, *op. cit.* (17), pp. 9–11.
[50] Donald McDonald and Leslie B. Hunt, *A history of platinum and its allied metals*, London, 1982.
[51] *Phil. Trans.*, 1772, 250. (my italics) In 1774 Priestley remarked 'Phosphorus is too expensive for me to have much to do with', Priestley to Rev. N. Cappe, Calne, 28 August 1774. Rutt, *op. cit.* (16), i, p. 275. Such considerations of thrift provide a marked contrast with the work of Lavoisier.
[52] *Autobiography, op. cit.* (3), p. 95.

234

chemical apparatus. His other early tools: mice, candles and green plants also cost next to nothing.[53]

But it was not enough for Priestley to have the interest in airs and the necessary apparatus. He also had to have the talent. About this time he spoke of the desirability of chemists studying charcoal, saying that 'the subject seems to be fairly within our reach'.[54] Here then is Peter Medawar's analysis of science—it is 'the art of the soluble'.[55] Science is a success (within a limited area) because scientists have chosen those problems which they can solve. The secret of Priestley's success in the history of airs is that he chose an area within his reach, an area ripe for exploitation. It required certain powers of observation and manual dexterity. He was able to adapt some of the skills he had acquired in his electrical studies. Yet pneumatic chemistry did not require advanced mathematics or very complex or expensive apparatus. He was going to build on the work of Hales with a prospect of great discoveries:

'. . . by working in a tub of water . . . we may perhaps discover principles of more extensive influence than even that of *gravity* itself . . .'[56]

In June and July 1772 he was able to inform Franklin of the success of his experiments on air. He had 'never been so busy or so successful in making experiments'.[57] He had now tried the effect of spirit of nitre on copper and even a little gold. This gave him occasion to remark that if it were not for his considerable success with pneumatic experiments,

'frugality and an attention to a growing family will, at length get the better of experimenting, and then I shall write nothing but *Politicks or Divinity* . . .'[58]

This surely confirms my analysis that Priestley was above all a *writer*. He wrote extensively on religion, politics and science. It was the failure of his purely literary efforts in science which drove him to experiments and was the basis for his subsequent reputation as a chemist. After his discovery of 'dephlogisticated air' (oxygen), he could write triumphantly:

'. . . I may flatter myself . . . that there is not in the whole compass of philosophical writing a history of experiments so truly *ingenious* as mine . . .'[59]

In pneumatic chemistry Priestley had at last found his true metier!

But chemistry represented more than career building, successful manipulation and observation or intellectual exercise. Chemistry was a subject of great potential use to mankind. Priestley claimed of the new science that he had adopted:

[53] Some of Priestley's earlier apparatus relating to his studies of natural philosophy, notably his air pump, were also put to good use in pneumatic chemistry.
[54] *The history and present state of electricity*, 1st edn., London, 1767, p. 607. Also in 3rd edn., 1775, ii, p. 199.
[55] P. B. Medawar, *The art of the soluble*, Harmondsworth, Middlesex, 1969, p. 11, 'Good scientists study the most important problems they think they can solve'.
[56] Priestley, *op. cit.* (41), ii, p. viii.   [57] Schofield, *op. cit.* (17), No. 49.
[58] *Ibid.*, No. 48.   [59] Priestley, *op. cit.* (41), ii, p. ix.

'Chemistry is perhaps of more various and extensive use than any other part of natural knowledge'.[60]

A fundamental attraction of the subject, therefore, was its utility. This is evident from his very first publication on gases: *Directions for impregnating water with fixed air* (1772), in which he presented the product as artificial spa water with beneficial medical properties. Following Hales, Priestley's scientific career was partly motivated by a spirit of pious utilitarianism.[61]

Priestley was obsessed by the 'goodness' of the airs he collected. The metaphor is a moral one and this may be significant. For Priestley, air was 'injured' by the respiration of human beings or animals and it was a matter of human concern to restore it. It was more than a coincidence that it was Priestley who discovered the function of plants in restoring air which had been spoiled by the respiration of animals. It was also a matter of great theological significance, since it showed that God would not allow mankind to be suffocated by continual exhalation but had carefully arranged for green plants to restore the balance.[62] That man and nature are the creation of a beneficent providence is an important theme in the Enlightenment, a movement to which Priestley clearly belonged.

Priestley's Refusal to Accept Lavoisier's New System of Chemistry

I should like finally to discuss briefly why Priestley did not accept the new theory of Lavoisier. There is no one simple answer but a number of considerations may be mentioned. First the phlogiston theory had proved a useful and adaptable theory in all of Priestley's researches on different types of air. One does not abandon easily a theory which has rendered good service over the years. Secondly, Priestley's chemistry was almost exclusively the chemistry of gases and it is natural that he should have thought in terms of volumes of gases. He was particularly struck by *changes* in volumes of gases, as in his famous test for the 'goodness' of air, using nitric oxide. For Priestley chemistry was the chemistry of gases collected in jars over water or, later, mercury. It was not the analysis of minerals. It was not a science depending on the balance.

It has also been pointed out that Priestley's approach to gases was remarkably mechanistic. For someone best remembered as a chemist, it is surprising to find the mechanical and the physical approach so prominent. Schofield would say the Priestley was really a physicist rather than a chemist.[63] The new chemistry of Lavoisier involved not only a new

---

[60] Priestley, *op. cit.* (28), i, p. vi.

[61] The phase 'pious utilitarianism' is used by John McEvoy, *Ambix*, 1978, 25, 93.

[62] 'The discovery of the provision in nature for restoring air, which has been injured by the respiration of animals, having long appeared to me to be one of the most important problems in natural philosophy . . .' *Phil. Trans.*, 1772, 183.

[63] Most recently in Robert E. Schofield, 'Joseph Priestley and the physicalist tradition in British chemistry', in L. Kieft and B. R. Willeford (eds.), *Joseph Priestley, Scientist, theologian and metaphysician*, Lewisburg, 1980, pp. 92–117.

interpretation of combustion and other chemical reactions but a whole new approach to the understanding of *chemical composition* based on a new logic of elements and this was just too much for Priestley.

Chemical combination and decomposition were not central to Priestley's chemistry.[64] He reveals himself, as McEvoy says, as an 'aerial philosopher',[65] that is, an aerial natural philosopher, rather than a chemist, in describing his airs as being 'extracted', 'emitted' or 'diminished'. Such terms are clearly physical and mechanical; perhaps they suggest the influence of Hales. Yet when Priestley spoke of airs being 'generated', water being 'impregnated' and the 'nascent state',[66] I detect a more organic metaphor, redolent of the Aristotelian tradition. But, however we characterise Priestley's perspective of chemistry, it was hardly commensurate with that of Lavoisier.

Because Priestley was a pioneer in the study of so many airs, there is an obvious temptation to view him as one of the first early modern chemists, but the balance of the evidence is that he belongs to an older tradition involving a more intuitive approach to the natural world. Since the time of Galileo the main stream of successful physical science had focussed on primary qualities that could be measured at the expense of the secondary. It might be attractive to see Priestley's chemistry simply as a derivative of his earlier studies of natural philosophy in the Newtonian tradition. But this simple view is weakened by the discovery of Priestley's persistent concern with colours, tastes and smells. These together with volumes were major parameters of Priestley's studies of airs. His famous test of the 'goodness' of airs involved both a colour change and a change of volume.[67] Unfortunately most of Priestley's airs were colourless, although he was delighted when the combination of 'marine acid air' and 'alkaline air' produced a 'beautiful white cloud'.[68] His detailed studies of the reactions of 'nitrous acid' concentrate on colours and mention as many as any analysis of a painting.[69] His interest in colour extended to an explanation of the similarity in the colour of mercury calx and red lead by relating them to (red) spirit of nitre.[70]

Priestley spoke of 'tasting' airs[71] and wholesome atmospheric air was described as 'sweet'.[72] Priestley also paid attention to 'odiferous sub-

---

[64] Priestley confided to Keir in 1778 that he was 'afraid of tripping on chemical ground'. He says: 'My walk is between what is called *chemistry* and other branches of *Natural Philosophy*'. Schofield, *op. cit.* (17), No. 77.

[65] John G. McEvoy, 'Joseph Priestley, "Aerial Philosopher": metaphysics and methodology in Priestley's chemical thought, from 1762 to 1781', *Ambix*, 1978, 25, 1–55, 93–116, 153–175; 1979, 26, 16–38. I have found pp. 154–155 in the third article in this series particularly instructive.

[66] '. . . a true inflammable air is first produced, and in the *nascent state*, as it may be called', Priestley, *op. cit.* (41), i, p. 187. The other terms quoted here are widely used and without apology or explanation.

[67] With the new air that he was to call 'dephlogisticated air', Priestley remarked that, on addition of 'nitrous air' (nitric oxide), 'the redness was really deeper and the diminution somewhat greater than common air would have admitted'. Priestley, *op. cit.* (41), ii, p. 41.

[68] Priestley, *op. cit.*, (41), i. p. 170.   [69] Priestley, *op. cit.* (28), i, 383–7.

[70] Priestley, *op. cit.* (41), ii, p. 61.   [71] Priestley, *op. cit.* (41), ii, 102.   [72] *Ibid.*, 98.

stances'[73] and 'putrid effluvia'.[74] On the other hand, following Hales and Cavendish, he did occasionally consider physical properties of his airs, such as density.[75] However, Priestley was far from being the simple Newtonian. In keeping with his religious convictions, Priestley considered the moral aspects of the airs he examined. He was concerned with 'restoring air which had been injured'[76] and suggested that 'the air which nature has provided for us [i.e. atmospheric air] was as good as we deserve'.[77] These moral concerns, his interest in secondary qualities, his preference for the language of ordinary discourse,[78] and—let us face it—his charming amateurism, place Priestley in a different world from the strictly quantitative and specialised science which was the new chemistry of Lavoisier.

Priestley was much more concerned to relate his bewildering variety of new airs to each other[79] than to traditional chemical substances. This provides a constrast with Black, whose genius lay in relating one particular gas, 'fixed air' (carbon dioxide) to a few substances like chalk. This relationship has an obvious parallel with Lavoisier relating the gas oxygen to the oxides of metals and non-metals. When Priestley did occasionally speculate about the relation of his airs to solids, he introduced phlogiston as an all too convenient and infinitely adaptable entity, mediating between the solid and the gaseous state.

We must also remember Priestley's whole life represented the freedom of the independent thinker. He was not someone who would bow readily to external authority. He did try to enter into a dialogue with Lavoisier and his colleagues but his protestations and the few remaining experimental anomalies which he continually stressed, were largely ignored by the French chemists. Also after his house had been burned and his papers destroyed in the Birmingham riots, Priestley no longer had favourable conditions for the pursuit of his work. He had little encouragement from others and he was increasingly isolated, never more so of course than when he went in 1794 to live out the last years of his life in a remote part of Pennsylvania.

Conclusion

Priestley was primarily a writer—a compulsive writer. It is this I think which brings together his science and his theology. But neither can be understood without the other. Priestley was one of the great polymaths in English history. In 1771, after his first experiments on gases, he wrote reprovingly to his friend Lindsey:

[73] Priestley, *op. cit.* (28), ii, 406.   [74] Priestley, *op. cit.* (41), i, pp. 80–5.
[75] Thus Priestley compared the weight of a bladder filled with different gases in turn, Priestley, *op. cit.* (41), ii. p. 94.
[76] *Phil. Trans.*, 1772, 166.
[77] Priestley, *op. cit.* (41), ii, p. 101.
[78] 'No person has ever been more temperate or more cautious than I have been in the introduction of new terms . . . ', Priestley, *op. cit.* (28), i, pp. 8–9. (cf. Priestley, *op. cit.* (41), i, pp. 23–4.)
[79] E.g. Priestley, *op. cit.* (41), i. pp. 62, 68, 77.

'I do not desire to know how I stand with the public, though I have no reason to think I should have stood amiss as a *philosopher* if theology has been out of the question'.[80]

The value of Priestley's theological writings is a matter of controversy even to-day but after 250 years his reputation as a major figure in the history of eighteenth-century science is secure. Certainly without the benefit of the work done on gases by British men of science, and notably by Cavendish and Priestley,[81] Lavoisier would not have been able to build chemistry anew.

The story of Priestley's scientific career has some general significance for the history of science. It reveals that in the first phase of his scientific work (electricity and optics), his credentials (and incidentally his financial liabilities) were principally the books he had bought, borrowed, read and listed. In the second, chemical phase, (which was possible because of low initial expense) his only credentials were his experiments. The change represents yet another way of looking at the difference between the arts and the sciences. In the humanities the scholar takes pride in the books he has read, whereas the laboratory scientist bases his reputation on the experiments he has performed. The experiments are written up and published as papers and, less commonly and directly in modern times, as books. A scientists' reputation, therefore, depends not at all on the books he has read but on the papers he has published.[82] Thus the development of Priestley's career symbolises the difference between arts and science. Despite what is sometimes claimed for the 'scientific revolution' of the seventeenth century, it shows that in a study of the eighteenth century, we may still witness the emergence of early modern science.

[80] Rutt, *op. cit.* (16), i, 129.

[81] Although I would not wish to exaggerate the achievements of Priestley, I have a higher opinion of him than T. H. Huxley, who said that Priestley could not be said to stand on the level of Black or Cavendish, 'Joseph Priestley' (1874), *Science and education essays*, London, 1895, pp. 1–37 (15).

[82] Maurice Crosland, 'Scientific credentials: Record of publications in the assessment of qualifications for election to the French Academy of Sciences', *Minerva*, 1981, *19*, 605–31 (published 1983).

IV

LAVOISIER: A NEGLECTED *SAVANT* ?

It is paradoxical that the life and work of a scientist, whose name seems as familiar as that of Lavoisier, is still relatively unknown and unexplored by historians of science. Everyone knows that the chemist met his death on the guillotine, but what exactly did he achieve during his lifetime? Compared with other major scientists like Galileo, Newton and Darwin, not much more than a beginning has been made to study the career and achievement of Lavoisier. There have been several recent detailed biographies of these other men but the only full scholarly biography of Lavoisier, one of the greatest of all French scientists, was one published in the nineteenth century! Over the past twenty years historians of science have reassessed the work of Newton and, with Westfall's biography of Newton entitled *Never at Rest*, we have a profound and definitive interpretation of the famous English scientist which presents a portrait very different from that accepted a generation ago. Similarly a large number of scholarly books on Charles Darwin have recently appeared and no-one to-day can claim he has suffered from lack of attention. Lavoisier in his turn has attracted the attention of a number of scholars, mainly outside France, but there is a long way to go before Lavoisier studies reach the level of studies of Galileo, Newton and Darwin, to mention only these same three giants in the history of modern science.

But this neglect is not new. Little public regret was shown in France in the years following the execution of Lavoisier in 1794. The Academy of Sciences, with a long tradition of delivering éloges of its deceased members, waited nearly 100 years before providing one. It is almost as if there was a conspiracy of silence. The person who did most in the nineteenth century to revive the memory of Lavoisier and make available to the public his many writings was J.B. Dumas (1836). Perhaps he exaggerated the importance of Lavoisier to compensate for 40 years of neglect. Perhaps the name of Lavoisier was cited later for purely nationalistic purposes (rather like waving a flag) without any real understanding of his contribution to science. But in the late twentieth century, with the passage of time, we should be able to view him and his achievements more dispassionately.

LAVOISIER'S LIFE AND WORK

We begin with a very brief resumé of some of the main features of the life of Lavoisier. He was born in Paris in 1743, the son of a lawyer. At the age of five his mother died and he was brought up by a maiden aunt. He had the good fortune to attend the Collège des Quatre Nations, where, after the normal classical and literary education, his final year was devoted to mathematics and physical science. Thinking of a career, however, a scientific career being hardly possible under the ancien régime, he took his license in law, but continued to study science by attending lectures at the Jardin du Roi. After writing several scientific papers, he managed to enter the Academy of Sciences in the most junior rank in 1768 at the age of 25, and the Academy henceforth became the centre of his scientific life. His livelihood was as a member of the Ferme Générale, a private company which collected taxes for the government. He was also able to do useful work for the government as a member of the Régie des Poudres, responsible for the manufacture of gunpowder. Lavoisier's great scientific output might make us forget that he was never a full-time scientist. His wife tells us that he had one full day each week in the laboratory but that normally he did scientific work in the early morning and again in the evening. But though his time was limited, his energy was boundless. His income too was ample and he was able to pay for the best apparatus. His wealth was sometimes envied by other scientists and in the Revolution he was arrested as a member of the Ferme Générale. His important contributions to science, even his work on the metric system in accordance with the new revolutionary ideology, did not save him from execution during the Terror in 1794.

As regards Lavoisier's scientific achievement, it is more difficult to speak, since it is still not fully explored. There are some recent signs of a reappraisal, and the Lavoisier of the year 2000 will certainly be very different from the Lavoisier of 1950. Lavoisier used to be presented as being preoccupied with problems of combustion, but recent examination of his oxygen theory suggests that his interests were much wider. He was concerned as much with a new theory of acidity. He gave an important place in his system to caloric, which he used to explain the three states of matter. He made good use of the principle of conservation of matter and wrote the first real chemical equation. He provided a new understanding of chemical composition and his table of elements provided the building blocks of the new chemistry. This is only a partial list of his main achievements in one science, but going beyond chemistry, Lavoisier also made a major contribution to physiology when he compared respiration to combustion. He began his scientific studies in geology but by the end of his life he had contributed to several other sciences, ranging from physics to economics. In 1791 Lavoisier laid the foundations for a modern economic study of the theory of National Revenue and its assessment, a study which was not taken up again until 1913, a situation which prompted a modern economist (Sellier) to write a paper entitled "Un précurseur sans disciples: Lavoisier". He also did important work in administration, and in applied science. To do justice to Lavoisier's contributions to chemistry alone one would need at least a whole article. Here I must be brief in order to return to the subject of the present article, which is how *other writers* have reported on his work.

AFTER THE POLITICAL REVOLUTION

After the Terror of 1793-94 and the fall of Robespierre in 'thermidor' (July 1794) France witnessed the beginning of a more constructive period. If one looks very carefully at the records one can find a few isolated references to the death of Lavoisier. The Lycée des Arts, for example, in 1796 was the scene of an éloge by Bouillon-Lagrange, who is remembered principally as the author of an elementary textbook. Fourcroy too in the safer world of the Directory penned an appreciation of Lavoisier, but, on the whole, the French scientific community wanted to forget the political turmoil of the previous years and their effects on science. It was not only that, as ordinary human beings, it was natural for them not to dwell on the misfortunes which had removed some of their colleagues, but also, in their emerging role as members of the new professionalised scientific community, they needed to emphasise the bonds which held them together rather than their difference of political opinions and personal fortunes.

There is a sharp contrast between the leading scientists in the years after 1795 and Madame Lavoisier who, far from forgetting, wanted to use the period of relative calm to repair the injustices to her husband. Thus, although Madame Lavoisier continued social contacts with members of the former Academy of Sciences, there was a deep division on the question of recalling the events of 1793-94, which had included the closing of the Academy and, as far as she was concerned, had culminated in the judicial murder of her husband. It is also possible that Lavoisier's colleagues, including Fourcroy, Guyton and Berthollet, might have done more to protect him during the Terror; they might even have managed to save him. To bring up the name of Lavoisier after 1795 was to invite an inquest that might not have totally exonerated his scientific colleagues. During his lifetime Lavoisier somewhat distanced himself from his colleagues by his superior wealth. After his death he might also be an embarrassment. But the difference extends from the economic, political and social plane to the cognitive. That is to say that, although they all followed Lavoisier in the new theory and collaborated with him in the new nomenclature, each had a slightly different conception of what chemistry was. Fourcroy tended to emphasise pharmacy and medicine, Guyton had been more interested in affinity and Berthollet became principally interested in the conditions of chemical reactions. Of Lavoisier's associates, it was probably Berthollet who published the most original new work after

4

Lavoisier's death. Chemistry was, therefore, moving on rather than staying at the point where Lavoisier had left it. Also around the year 1800 people were tempted to look forward optimistically into the new century rather than to look back to the troubles and divisions of the old century.

MADAME LAVOISIER

After the death of Lavoisier it was his widow who, at a personal level, did most to keep alive the memory of the chemist. First she reclaimed from the authorities the apparatus and papers which had been confiscated in 1793. With the order for restitution came an official acknowledgement that her husband had been unjustly condemned. For the remainder of her lifetime, which extended for another half century, she was the guardian of his papers, which then passed to the Chazelles family.

In the last years of his life Lavoisier had been planning to publish a collection of his memoirs. Madame Lavoisier took up this task, seeking first the co-operation of Séguin, but finally proceeding on her own. She explained:

> "ces fragments n'auraient point paru, s'ils ne contenaient (page 78 du second volume) un Mémoire de M. Lavoisier qui réclame, d'après les faits qu'il y expose, la nouvelle théorie chimique comme lui appartenant."

This referred to Lavoisier's claim that the new theory of chemistry was not "the theory of the French chemists", as it was sometimes called, but essentially his own work. "Elle est la mienne!" he cried from the grave. As befitted a work of piety, the *Mémoires* were not sold but given to a number of institutions and influential scientists in the hope that they might have a maximum impact. In practice the volumes distributed were mostly put on library shelves and their contents have still to be studied by historians of science. The vast majority of the stock (some 1100 copies) remained with Madame Lavoisier. It was in any case an incomplete work.

Of her married name Madame Lavoisier said "J'ai regardé comme un devoir, comme une réligion de ne point quitter le nom de Lavoisier". Indeed her only action reflecting adversely on her devotion of the memory of her husband was her acceptance in 1805 of Count Rumford as a second husband, but even then she insisted on being known as Madame Lavoisier de Rumford. The marriage was short-lived and by the time of his first wedding anniversary, the eccentric Count Rumford was unkind enough to describe his wife as a "female Dragon".

Twenty five years after Lavoisier's death his widow was asked to provide some details of his life for an article in the *Biographie Michaud*. She obliged, but at the same time registered a complaint about the treatment of the great chemist by the Academy of Sciences:

"L'on demandera toujours comment après 25 ans l'académie qu'il a servi avec tant de zèle, qu'il a tant illustré, n'a point encore versé des larmes sur sa tombe et ne lui a point décerné les honneurs de l'éloge."

The éloges delivered by the secretaries of the Academy of Sciences of their deceased members are well known, and certainly the honour had been accorded to many lesser figures as a matter of routine. Madame Lavoisier had a right to feel bitter.

J.B. DUMAS AS AN ADVOCATE OF LAVOISIER

In the Lavoisier story his great protagonist, Jean-Baptiste Dumas (1800-1884), does not come on the scene until 1836, and it seems more than a coincidence that this was the year of the death of Madame Lavoisier. Whatever his feelings (and they were strong), Dumas did not feel it proper, as long as the widow was alive, to take on himself the responsibility for proclaiming that Lavoisier had been unjustly neglected. Marie Anne Lavoisier died suddenly on 10 February 1836 at the age of 78. Three months later Dumas was delivering a famous series of lectures at the Collège de France on the history of chemistry. In these lectures he made strong claims for Lavoisier and immediately became the leading advocate in a crusade for doing belated justice to the memory of the famous French chemist.

Dumas' historic tribute to Lavoisier was thoughtful but it was emotionally charged, doubly so, since he deliberately chose to deliver his lecture on 7 May, the eve of the anniversary of the death of Lavoisier. Emotion was understandable because of the long neglect of Lavoisier. Dumas also associated himself personally with his fellow countryman, both as a chemist and as a Frenchman. He almost blamed Lavoisier's colleagues for seeming to take collective credit for a theory, which both Lavoisier and Dumas insisted was essentially his alone. But the emotion was increased by a further injustice: Lavoisier's unfortunate death by the guillotine. Dumas only stopped short of speaking of the martyrdom of Lavoisier. But he hinted at this in his use of religious language: No *sacrifice* was too great for Lavoisier in pursuit of his scientific research. In the end he sacrificed his life. Dumas proposed as a monument to the eighteenth-century chemist, an edition of his works: "je doterai les chimistes de leur *évangile*".

Dumas defended Lavoisier against the accusation that he had allowed himself to be diverted from the pure path of science by collecting taxes for the Ferme Générale. Lavoisier, he said, needed the money for costly experiments, and anyway he was able to do good in the world of economics and administration. His other career was justified by his exceptional industry: "Pendant quatorze ans, sa pensée toujours féconde et sa main toujours infatigable n'ont pas un seul instant connu le repos". It was quite natural for Dumas not to mention 'mistakes', for example, Lavoisier's erroneous theory of oxygen as the universal principle of acidity. Instead he emphasised that the recent liquefaction of gases by Faraday and Thilorier served to confirm a prediction of Lavoisier. His life was described as: "si belle, ...si honorable, si pure". He even characterised Lavoisier with the words "modestie et simplicité", an assessment which few modern historians of science would accept without qualification. He concluded:

> "Après une vie si honorable, après une mort si cruelle, qu'avons nous fait pour Lavoisier? Ou trouver un monument qui rappelle sa mémoire, un simple buste qui lui soit consacré? La France, helas! semble l'avoir oublié".

When Dumas was elected President of the Academy of Sciences in January 1843 he immediately took advantage of his position to write to the Minister of Education, asking for government money to pay for a complete edition of Lavoisier's *Oeuvres*. The year 1843 happened to be the centenary of Lavoisier's birth, a further justification for urging action. After careful examination and costing, publication was agreed in principle, although another twenty years elapsed before the first volume was published (1864). Dumas himself was only able to supervise the publication of two-thirds of the Lavoisier material available, that is four out of six volumes. After the dislocation caused by the war of 1870 and the Commune, Dumas abandoned the publication but asked his student Henri Debray (1827-1888) to take over the task. Debray himself died within a few years of Dumas and the Lavoisier torch was taken up by Edouard Grimaux (1835-1900).

GRIMAUX, BERTHELOT AND WURTZ

Grimaux not only published two more volumes of Lavoisier papers, making up the final part of Dumas' plan, he also undertook a full biography of the eighteenth-century Frenchman. The full title of his work is worth quoting: *Lavoisier, 1743-1794, d'après sa Correspondance, ses Manuscrits, ses Papiers de Famille et d'autres documents inédits*. The author obviously believed in presenting his credentials. This was no quickly penned overview of Lavoisier but a carefully researched biography, the first to be based on an extensive study of the Lavoisier papers, then in the possession of the Chazelles family but later dispersed. The main biography is complemented by supporting documentation and a number of appendices, giving invaluable information, for example, about Madame Lavoisier and about the *Annales de chimie*, the journal founded by Lavoisier and his colleagues to propagate the new science.

By a coincidence Grimaux published his work in 1888, one year before the centenary of the French Revolution. This gave another chemist, Marcellin Berthelot, a wonderful opportunity. He would celebrate the centenary by pointing to the existence of another revolution - "la révolution chimique", a phrase which was given currency by his book of that title published in 1890.

CHRONOLOGICAL SUMMARY OF LAVOISIER STUDIES

	1743	Birth of Lavoisier
	1772	Sealed note for Academy on combustion
1789	1789	*Traité élémentaire de chimie*
Beginning of French Revolution	1794	Execution of Lavoisier
	1803	Posthumous publication of Lavoisier's *Mémoires de chimie*
1815 (Second) Restoration of Bourbons		
	1836	Death of Madame Lavoisier
	1836	J.B. Dumas: *Leçons sur la philosophie chimique*
1848 Revolution 1852 Second Empire (Napoleon III)		
	1864- 1868	Publication of vols. 1-4 of *Oeuvres de Lavoisier*
1870-1871 Franco-Prussian War, Third Republic		
	1884	Death of Dumas
	1888	Grimaux's biography of Lavoisier
	1890	Berthelot: *La Révolution chimique. Lavoisier.*
	1900	
1914-1918 First World War	to	Another period of neglect
1939-1945	1930	
Second World War	1943	Bicentenary of birth of Lavoisier
	1950s and	Growth of History of Science in American and British universities
	1960s	
	1980s /90s.?	

But first he adroitly covered the embarrassment that the Academy of Sciences had never provided an éloge for one of its most illustrious members.

Berthelot had just been elected as one of the permanent secretaries of the Academy of Sciences and, at the public meeting of the Academy for 1889, he siezed the opportunity to present an éloge of Lavoisier. The éloge itself was very derivative, relying heavily on Grimaux's book but Berthelot was also able to do something more original. Pointing out that Grimaux had concentrated on the *life* of the famous chemist, Berthelot wrote a book about his *work*. At the same time he published a summary of the contents of Lavoisier's laboratory note books. This is an invaluable guide to the historian of chemistry and enables him to date more accurately the major steps in Lavoisier's laboratory work. In his publications in the *Mémoires de l'Académie Royale des Sciences* Lavoisier often made claims, of which the chronology does not seem completely honest, and the notebooks enable us now to put the record straight.

Another nineteenth-century scientist to draw attention to Lavoisier was the organic chemist, Adolphe Wurtz, Academician and Professor at the Paris Faculty of Medicine. It was in 1868 in the Introduction to his *Dictionnaire de chimie* that he made the bold claim: "La chimie est une science française. Elle fut constituée par Lavoisier, d'immortelle mémoire". This claim, in the years leading up to the Franco-Prussian war, created a storm. But Wurtz did not make any serious study of Lavoisier. His real interests lay elsewhere and in his most famous and influential book, *La théorie atomique* (1st edition, 1879) he adopted a completely different perspective. It is a book with a profoundly historical approach but the history here starts with John Dalton. The name of Lavoisier is only mentioned in passing but great emphasis is given to the work of Richter, Avogadro, Gerhardt, Laurent, *et al.*

It may seem that in the late nineteenth century Lavoisier had become more of a symbol than a historical figure, a symbol supported by establishment figures in the French scientific hierarchy; but if Dumas and Berthelot represented the establishment, Grimaux certainly did not. He was known from his schooldays as an independent character, even as a rebel. He was among the first to sign a protest in 1898 against the treatment of Captain Dreyfus and the price he paid for his activity was the loss of his chair and his laboratory at the Ecole Polytechnique, which came under the administration of the Ministry of War.

Although Dumas, as dean of the Paris Faculty of Sciences and later permanent secretary of the Academy of Sciences, appears as the embodiment of the French scientific establishment of his day, he did not concern himself with Lavoisier exclusively. He admitted, for example, the great debt he owed to Lavoisier's colleague, C.L. Berthollet. Dumas' history of chemistry was focused on the important part played by French chemists, and Lavoisier for him was merely the principal actor in a play with many characters. Marcellin Berthelot, on the other hand, spoke of "les méditations solitaires" of Lavoisier.

He insisted that "les conceptions qui ont fondé la chimie moderne sont dues à un seul homme, Lavoisier" and again: "les idées qui ont triomphé ne sont pas une oeuvre collective". Such a view, he thought, would discourage individual effort. Berthelot was therefore encouraging the cult of the hero, a role viewed with some reserve by modern historians.

THE TWENTIETH CENTURY

The publication in 1890 of Marcellin Berthelot's *La Révolution chimique. Lavoisier* was the last major contribution to Lavoisier studies in the nineteenth century, the last major contribution even in the period before the First World War. Science was changing rapidly at that time with the discovery of X-rays and radio-activity. There was little time for historical musing. After the war too there were major social and political changes. The Bohr theory of the atom made people think more about spectra and atoms than about the kind of questions Lavoisier had been concerned with. After thirty or forty years of great neglect, we find a new interest in Lavoisier in the 1930s. A.N. Meldrum's study of 1930 was followed in 1935 by the first of Douglas McKie's two books on Lavoisier. The outbreak of a second World War provided another interruption of Lavoisier studies but this time there was a difference. There is nothing like an anniversary to focus attention, and, if the anniversary is a centenary, it positively demands attention. Such a centenary came in 1943, the bicentenary of the birth of Lavoisier. Since this occurred during the Nazi occupation of France, it was difficult to mount more than a token exhibition, and the bicentenary was therefore celebrated only in a very restrained fashion.

But outside France there was another movement which was to have a more permanent effect, the arrival of history of science as a university subject in the United States and Britain. At University College London, one of the oldest university departments of the history of science, McKie took up again his old love, Lavoisier, producing in 1952 his book *Antoine Lavoisier, Scientist, Economist, Social Reformer*, thus placing his former studies in a more social context. Meanwhile in the United States Henry Guerlac at Cornell University was doing research for his book: *Lavoisier - The Crucial Year*, a title which refers to 1772, the year Lavoisier focused his attention on combustion and thus laid the foundations of the oxygen theory. The well-endowed American universities were encouraging research in history of science and (with rather less money) many British universities too adopted history of science as a study suitable for both science and arts undergraduates and for postgraduate research. Although there were useful books on different aspects of Lavoisier by Daumas (1955) and Scheler (1966), most serious Lavoisier research in the 1950s, 60s and 70s, depended on individual and independent American and British universities, which provided employment and research facilities. Meanwhile in France history of science suffered since it was not recognised by the Ministry of Education as a full subject to be studied on a national scale. Small groups in Paris and elsewhere have done their best with limited facilities but the major initiative affecting work in France since World War II on the Lavoisier front was the decision made with international support to publish the letters of Lavoisier. This was entrusted to a retired chemical engineer from Clermont Ferrand, René Fric, but the three fascicules produced before his death did not meet the standards of post-war scholarship and were the subject of serious criticism. We understand that, if the resources are available, a new start will be made in the 1980s to produce a definitive edition of Lavoisier's *Correspondance*. Had not the Italians included Galileo's letters in the monumental national edition of the *Opere* (20 volumes) published under the patronage of the President of Italy between 1890 and 1909? Had not the British with the help of the Royal Society concluded a seven-volume edition of the correspondence of Isaac Newton with full scholarly apparatus in 1977? It is certainly fitting that the memory of Lavoisier should be honoured by a parallel publication.

The position of history of science in France provides a contrast with the situation in philosophy, which is strongly represented both at school and university levels. Turning to another subject well established in the educational world, namely history, one might expect French historians to be interested in repairing historical gaps. But many historians in the 1980s, like sociologists, are less interested in the study of "great men" than in the study of social classes, like the peasantry or urban populations. As far as historians are concerned, this is an understandable reaction to excessive concentration in the past on kings and queens. After the monarchy, presidents, ministers and certain other famous individuals have been the subject of detailed study for several generations, it is understandable that modern scholars should wish to cast their net more widely for material to study. The pity is, however, that this movement in history has come *before* scholars have had an opportunity to study leading scientists properly. There is, therefore, a gap, and at the present moment the detailed study of individuals does not fit easily into the prevailing fashion. However, the conflict here is probably more apparent than real since we need not only to know more about Lavoisier's life and work but also that of the embryo scientific community to which he belonged.

THE EXAGGERATION OF LAVOISIER'S CLAIMS

But in claiming that Lavoisier has suffered remarkably from neglect, we must beware of repeating the over-enthusiasm of a Dumas. We must not exaggerate or distort the historical record and we must be fully aware of the exaggerations of the past. There are at least three ways in which the claims of Lavoisier have been distorted. In the first place it has been suggested that chemistry began with Lavoisier. This is implicit, for example, in Wurtz's claim that chemistry was founded by Lavoisier. We should not forget that a study called chemistry had been in existence long before the work of Lavoisier. In the middle of the eighteenth century there were even several theories of chemistry, showing that chemistry already had some claim to be a science. One theory was a theory of chemical affinity which in one form drew a parallel between the laws of attraction advanced by Newton between the bodies of the solar system and between particles of ordinary matter. Even the chemists, who found the theory of gravitational attraction impossible to apply to the micro-world, appreciated the value of affinity tables, in which different substances were listed in order of reactivity. A further theory (or series of theories) was centred on phlogiston, a supposed flame-stuff contained in inflammable bodies and given off in combustion. This was of course a theory which Lavoisier attacked, but in its own day the phlogiston theory did something to give chemistry coherence. Even before the eighteenth century there were theories of chemical composition and reaction based on 'elements' and atoms. Aristotle's theory of four 'elements' (air, earth, fire and water) had been prevalent for nearly 2000 years before it was attacked by Robert Boyle in his *Sceptical Chymist* (1661). Boyle had even been called the founder of modern chemistry but unfortunately, after his attack on the idea of the four 'elements' (or the three 'principles' of Paracelsus), he was not able to provide a useful alternative.

Enough has been said to remind the reader that chemistry had flourished long before Lavoisier. Yet the French chemists of the 1780s did something that helped to destroy the work of previous generations of chemists - they changed the language of chemistry. Although their stated aim was to provide a rational nomenclature in conformity with the new theory, their success was so great that the indirect result was that chemistry books and papers written before 1787 became largely unintelligible to subsequent generations. We can understand *oxide of zinc, sulphuric acid* and *oxygen* but not *pompholix, spirit of vitriol* and *dephlogisticated air*. I do not think it necessary to adopt a conspiratorial theory of history but, if Lavoisier and his collaborators had wanted to present themselves as the founders of chemistry, they certainly could not have hit on a more effective method, much better than the destruction of whole libraries of books written by their predecessors.

This brings us to a second way in which Lavoisier's achievement can be exaggerated and here Lavoisier himself is largely to blame. He took little trouble to acknowledge his debt to his predecessors. One of the names used for the new chemistry was "la chimie pneumatique" or "la doctrine des gaz" because it depended so much on the new understanding of the gaseous state of matter and on the gas oxygen in particular. Lavoisier clearly made use of the work of Joseph Priestley who, in 1774, isolated and examined the properties of the gas later called oxygen. When news came to Paris in 1783 that Henry Cavendish had obtained water by the burning of 'inflammable air' (hydrogen) in air, Lavoisier immediately repeated the experiment and used it to establish that water was not, as had previously been thought, a simple substance but a compound of hydrogen and oxygen. Lavoisier also owed a debt to Stephen Hales, who had examined a number of 'airs', and especially to Joseph Black, who had discovered 'fixed air' (carbon dioxide) by the careful use of a chemical balance. The French savant never adequately acknowledged his enormous debt to the British school of pneumatic chemists.

A further debt was to Lavoisier's French colleagues and here he was more generous:

> "Si quelquefois il a pu m'échapper d'adopter, sans les citer, les expériences ou les opinions de M. Berthollet, de M. de Fourcroy, de M. de Laplace, de M. Monge, et de ceux, en général, qui ont adopté les mêmes principes que moi, c'est que l'habitude de vivre ensemble, de nous communiquer nos idées, nos observations, notre manière de voir, a établi entre nous une sorte de communauté d'opinions, dans laquelle il nous est souvent difficile à nous-mêmes de distinguer ce qui nous appartient plus particulièrement." (*Traité*, 1789, Introduction.)

This then is the third way in which Lavoisier's achievements have been exaggerated. They have sometimes been presented as the work of an isolated individual. Lavoisier, living in Paris and surrounded by interested colleagues, was never isolated. Any new assessment of Lavoisier must consider the exchange of information and ideas in Lavoisier's laboratory at the Arsenal and with colleagues of the Academy of Sciences, of which Lavoisier was a devoted member. Lavoisier's phrases in the above quotation: "l'habitude de vivre ensemble" and "une ... communauté d'opinions" will undoubtedly be seized upon later by historians of science wishing to minimise the contribution of any single individual.

It seems to me that in the forthcoming major reassessment of the life and work of Lavoisier we must avoid the two extremes of either treating him as an isolated individual or of considering him as no more than the spokesman of a group. It was probably Lavoisier's association with Guyton de Morveau, Berthollet and Fourcroy in the *Méthode de nomenclature chimique* of 1787 that first gave rise to the idea of the new chemistry being the work of a collective. But one must distinguish between the new *nomenclature*, which was the work of a committee and the new *theory*, which was largely the work of one man. But the study of that one man will be richer if it includes a study of his intellectual milieu.

CONCLUSION

Of the four French chemists who collaborated in the reform of chemical nomenclature in 1787, the most studied in some ways over the past 20 years have been Fourcroy (two books, Smeaton and Kersaint) and Berthollet (one biography, Sadoun-Goupil and a study of the Berthollet-Laplace circle known as the 'Society of Arcueil', Crosland). We are still waiting for the book on the life and scientific work of Guyton de Morveau, and the only modern original and fully documented biography of Lavoisier is that written for the 16-volume *Dictionary of Scientific Biography* by Guerlac (1973) and republished with many illustrations as a small book. The scale of the biography was limited by the requirements of the encyclopaedia for which it was written. Certainly there have been new studies of aspects of Lavoisier's work. It is no longer possible to see him simply as the advocate of a new theory of combustion. Recent detailed research by historians of science has emphasised his new view of chemical composition, his caloric theory and his oxygen theory of acidity. According to the old historiography, if a scientist had done something which later scientists had discredited, the 'mistake' was quickly passed over. It is more honest not to hide such 'mistakes' but rather to *use them* to throw more light on the other work. Historians of science now try to do a more contextual history. Yet it is sad to report that the majority of these isolated studies of Lavoisier have been undertaken by British and American scholars and very little of this work of the past twenty years has been translated into French.

In 1983 there were celebrations in America and Britain to commemorate the 250th anniversary of the birth of Joseph Priestley (1733-1804), sometimes described misleadingly as "the discoverer of oxygen". 1987 will mark the 200th anniversary of the introduction of modern chemical nomenclature, which owes much to Lavoisier. Will this be remembered? 1989 will be celebrated in France as the bicentenary of the French revolution but it is also the anniversary of Lavoisier's textbook of chemistry, the *Traité élémentaire de chimie*, which could be considered the first textbook of modern chemistry. Finally I wonder whether France and the rest of the world will be ready in 1994 to commemorate in a suitable way the anniversary of Lavoisier's death. It may be necessary to train some new researchers. (It is a rather specialised research that is required.) Also few scientists can imagine the delay and frustration sometimes experienced by historians in trying to track down documents, often in private hands. Since there is so much to be done, scholars will need a full ten years to produce the fundamental documentation on which any authoritative assessment of Lavoisier must rest.

In some aspects of the Lavoisier story we still have to escape from myths which have grown up, for example, that he was a martyr of science. Because of historical research it is much more difficult today to believe this. The phrase "la république n'a pas besoin de savants" has been shown to be a fabrication - it was *not* uttered at Lavoisier's trial. But we still need much more research to distinguish fact from fiction. It is not enough that the *name* of Lavoisier is well known. It only deserves to be well-known if he made a major contribution to science, and what this was has still to be evaluated. The traditional simple picture of Lavoisier's work focused purely on the study of combustion is one which certainly requires revision. What we need is not

14

more general surveys and popular biographies but more serious and authoritative studies of different aspects of Lavoisier's work and the publication of relevant documents.

Ideally the publication of documents would come first. A start could be made with the fourteen large volumes of Lavoisier's laboratory note-books which are still known only by the brief summaries of the contents published by Berthelot in 1890. Then there is the publication of the manuscript *procès-verbaux* of the Academy of Sciences of the ancien régime. The volumes for the years 1768-1793 will have much to reveal about Lavoisier as well as about the whole question of science in eighteenth-century France. In some ways these projects are even more urgent than the *Correspondance*.

But there is a wider question than the study of Lavoisier or indeed of any one individual and that is the study of a subject area. The history of French literature and the history of French philosophy are well studied, not only in French universities but even in French schools. Racine and Descartes receive every attention but men of science are largely overlooked. Foreigners, who are favourably impressed to find in Paris and many other French cities such evocations of the past as a *Rue* Gay-Lussac, an *Avenue* Lamarck, and a *Boulevard* Arago, might hope that the commemoration of great achievements should not be limited to the names of streets and the occasional statue. The fact that the statue of Lavoisier, erected in 1900 in Paris by international subscription, has disappeared is less of a scandal than the neglect of his work. Perhaps some of the concern which has been directed in France to the use of the French language by scientists at international conferences should be directed to studying the very real contribution made by Frenchmen to the history of science. One trusts that common sense will prevail here over extremes of nationalism. But it is not chauvinistic to study aspects of French civilization which are now only known superficially.

Finally I would claim that the neglect of Lavoisier is only a symptom of a general neglect of history of science in France or rather the lack of official recognition of history of science as an autonomous discipline. If one of France's most famous scientists has not been sufficiently studied, what about dozens of others, less famous, but all playing their part in shaping modern science? At the end of the eighteenth century and beginning of the nineteenth century France led the world in most branches of science. Is the evidence for this to be left in the archives or can steps be taken in high places to put an end to this shameful neglect?

BACKGROUND READING

W.A. SMEATON
'New light on Lavoisier: The research of the last ten years', *History of Science, 2* (1963), 51-69.

Henry GUERLAC
'The Lavoisier Papers - A chequered History', *Archives Internationales d'Histoire des Sciences, 29* (1979), 95-100.

Maurice CROSLAND
'Chemistry and the chemical revolution', in G.S. Rousseau and Roy Porter (eds.), *The Ferment of Knowledge. Studies in the Historiography of Eighteenth-Century Science, Cambridge University Press, 1980, pp.389-416.*

Maurice DAUMAS
Lavoisier, théoricien et experimentateur, Paris, 1955.

(For an up-dated bibliography on Lavoisier see Preface endnotes.)

V

EXPLICIT QUALIFICATIONS AS A CRITERION
FOR MEMBERSHIP OF THE ROYAL SOCIETY:
A HISTORICAL REVIEW

T is well known that in the early years the Royal Society accepted as Fellows men of social standing, often with only the most superficial nowledge or interest in the natural world. The idea that candidates for election to the Royal Society should have a special claim to scientific knowledge as formalized in a statute of 1730, which introduced the idea of a certificate escribing the rank and qualifications of the candidate and signed by at least aree members (1). The immediate reason for this statute was the unease that ad been felt about the laxity of admission which had resulted in a large roportion of members having no serious title to membership of a supposedly ientific society. Indeed, approximately two-thirds of the members at that me had little knowledge of science or even any interest in it.

Although the financial problems of the early Royal Society have often been ressed, probably the greatest threat to its existence as one of the world's most nportant scientific societies in the first two hundred years of its existence lay the dilution of membership so that those with some commitment and nowledge of science remained in a minority. Hence a study of the formal quirement of the Society that members should have qualifications is of ntral importance to the history of the Society. In practice we shall see that any of these 'qualifications' were of a very general kind, but the fact that a ritten statement was required exercised a potentially powerful influence to ise standards, which finally resulted in a society in which the majority of embers were men of science. It was a gradual process which had to wait until e nineteenth century for significant advances but the seeds for this were sown far back as 1730.

After the death of Sir Isaac Newton, P.R.S., it was the new presidency of r Hans Sloane in March 1727 which gave rise to a discussion of a number of lministrative reforms. The Council Minutes reveal that the Society was preccupied at this time with the recovery of subscriptions from members but it so discussed the problem of limiting membership since it realized that andards were too lax. As an interim measure the Council insisted on approv-g all candidates for election before being balloted for at an ordinary Society

meeting (2). There were, however, some doubts as to the legality of th
Council taking upon itself these additional powers and it was decided to tak
legal opinion on this and other questions affecting the Society. One of th
questions submitted to the Attorney General was:

Q.8. Whether the Council cannot by virtue of their general power c
regulating the body limit the number of members thereof; or at least mak
such laws as may *check the too great increase of the body with new members unf
for answering the ends of the institution?* (3)

The legal opinion they received was that the Council has no special right i
the matter of election. The answer given to question 8 quoted above is wort
quoting in full:

Considering that the Charter hath left the body at large without limitin
the number of Fellows and considering also the nature of his foundation,
think the Council cannot make a statute to limit the Fellows to a certai
number; but they may make reasonable statutes or by-laws *to describe ar
ascertain proper qualifications* of persons to be elected Fellows in such manne
as may best answer and promote the ends of an institution so useful to th
learned world. (4)

In response to this advice the Society introduced a new statute presented to th
Council on 7 December 1730 and approved at an ordinary meeting c
members three days later:

Every person to be elected Fellow of the Royal Society shall b
propounded and recommended at a meeting of the Society by three c
more members, who shall then deliver to one of the secretaries a pape
signed by themselves with their own names, signifying the name, additio
[i.e. rank or title], profession, occupation and *chief qualifications* (5) of th
candidate for election, as also notifying the usual place of his habitation. (6

This was not to be, however, the end of all privilege. Whereas in ordinar
elections a minimum of three supporters was necessary and proper notice wa
to be given by displaying the certificate at ten ordinary meetings of th
Society, the 1730 Statute ended as follows:

... Saving and excepting that it shall be free for every one of h
Majesty's subjects, who is a Peer or the son of a Peer of Great Britain c
Ireland, and for every one of his Majesty's Privy Council of either of th
said kingdoms, and for every foreign prince or ambassador to b

propounded by any single person, and to be put to the ballot for election on the same day . . . (7)

Although one must remember that such privileged categories of membership continued until after the reforms of 1874 and 1902, this paper will be concerned with those who were not privileged, and including nearly all of those who made contributions to science.

The certificates drawn up for candidates for election to the Royal Society constitute a valuable documentary source not hitherto exploited (8). In the first place it is a matter of some interest to see how major figures in the history of science were seen by their contemporaries at a particular time in their careers. Thus Benjamin Franklin (elected 1756) is recommended as 'a gentleman who has very eminently distinguished himself by various discoveries in natural philosophy and who first suggested the experiments to prove the analogy between lightning and electricity', whereas the retiring Henry Cavendish (elected 1760) is simply described as 'having a great regard for natural knowledge and being very studious for its improvement'. Very few candidates were as 'studious' as the wealthy Cavendish. In a different social situation was Joseph Priestley (elected 1766), then writing his *History and Present State of Electricity*. Priestley was able to add the letters 'F.R.S.' after his name on the title page when it was published, thus improving his status as an author. Priestley already had several members among his friends and membership of the Society was valuable to him in a number of ways. At this stage in his career Priestley still had most of his books before him and he had not even begun his serious chemical studies. He is, therefore, presented as 'the author of a chart of biography, and several other valuable works, a gentleman of great merit and learning, and very well versed in mathematical and philosophical enquiries'.

The French savant Buffon was considered 'very eminent for his learning in mathematics' when he was elected in 1739, just the time when he was to embark on a second career in natural history. Similarly, when Joseph Banks was elected in 1766 at the age of 23, having succeeded to his father's estates, his great voyage with Captain Cook was still in the future but he had already singled out botany for special study. He is therefore described as a gentleman 'versed in natural history, especially botany, and other branches of literature'. Nine years later, but not yet P.R.S., he was to propose 'Captain James Cook of Mile End, a gentleman skilfull in astronomy and the successful conductor of two important voyages for the discovery of certain countries by which geography and natural history have been greatly advantaged and improved', a popular choice supported by the signature of 25 Fellows. It is interesting that

V

William Herschel, the discoverer of the planet Uranus, had not yet identified the object in the night sky as a planet at the time of his certificate (10 May 1781). He is therefore described as 'well versed in mathematics, mechanics and astronomy and having communicated several papers to the Royal Society, particularly those relating to the present comet, which he first observed'.

Terms used to describe scientific accomplishment

The principal value of the information to be gleaned from the certificates goes far beyond the 'great names' of the history of science to be found among the early Fellows. Since in its first two hundred years the Royal Society included so many men who were not distinguished for their contribution to science, the certificates provide a broader entry into social history than would be possible from the records of a more narrowly focussed society. The history of the Royal Society exemplifies a fascinating tension between social and intellectual values and the certificates reveal something of the way in which British society in the eighteenth and nineteenth centuries came to terms with the commitment and specialization which were increasingly necessary to obtain knowledge of the natural world which we call science.

It is all too easy for the modern scientist to read the present into the past, to assume standards of sophistication and specialization which are comparatively recent developments. A good corrective to this is provided by a study of the documents of an earlier period. The vocabulary forces the researcher to come to terms with the period and indeed the terms used can provide a valuable insight into the values of the time.

In many cases it is not possible to say who actually drew up the certificate. Obviously the strongest evidence is in the handwriting but many of the early certificates are all (including the 'signatures') in the same clear hand, probably that of a clerk. The form of words used in the certificate would often be that of the proposer and/or seconder and in some certificates the handwriting of the text corresponds clearly to one of the first two signatures. In some other cases it clearly does not; there is also a problem of certificates with a crowded group of signatures where it is not clear who actually proposed the candidate (9). It has, therefore, been decided in this paper not to emphasize sponsorship (10) but rather to study the language as a reflection of the social and intellectual values of the period and as providing information about candidates.

Among the terms of approbation found in the early eighteenth-century certificates one often finds the term 'curious'. Thus Rose Fuller M.D. of Sussex (elected 1732) is described as 'a gentleman well skilled in all [sic] parts of the mathematicks, natural and experimental philosophy and most branches of curious and useful learning', and in the same year the notable figure in the history of electricity, Stephen Gray, is simply described as being 'well known by his many curious experiments and observations laid before this society'. He was so well known to the Society that no title or address is mentioned. Charles Frederick (elected 1733) was described as a 'gentleman of great curiosity' and Richard Arundel (elected 1740) was recommended as 'a gentleman of great merit, knowledge and curiosity' a rather vague recommendation but apparently sufficient considering that he was 'Master of His Majesty's Mint'. The term 'curious' was applied more widely than to mean an interest in natural science, since the York antiquarian Francis Drake (elected 1736) was presented as the author of 'a very learned and curious work, the history and antiquities of that city'. The same term was applied to Voltaire (elected 1743), described as 'a gentleman well known by several curious and valuable works' and 'well skilled in philosophical learning', presumably on the basis of his earlier reputation for the popularization of Newton in France.

Yet while the Royal Society was willing to honour some figures of the French Enlightenment such as Voltaire and Montesquieu (elected 1744), it refused absolutely to admit Diderot, even though it was assured that 'il a donné les preuves de la superiorité de son esprit et de la plus grande capacité, tant par le grand ouvrage de l'Encyclopédie que par différens autres traités de mathématiques et de physique, de belles lettres et d'arts'. Not only was Diderot known as editor of the great *Encyclopédie* and of several works relating more particularly to science but his certificate of 1752 was supported by ten French signatures including Buffon and D'Alembert. Because of his reputation as enemy of Church and State it was not so easy to obtain British signatories and it is ironic that one of the four supporting signatures was that of the Catholic priest John Needham, whose own contributions to science included work on spontaneous generation. When the ballot was held on 8 February 1753 Diderot received 18 white and no less than 50 black balls and was, therefore, rejected.

Very few of the certificates relate to candidates who failed to be elected. When they were, it was usually because it was felt that they had neither sufficient social standing nor scientific eminence. Thus the surgeon John Wreden, described simply as 'a person well versed in natural knowledge', was rejected in 1734. Surgeons were often suspect as men with a minimum of formal education seeking to use the Royal Society for social advancement (11). This is illustrated by Archibald Cleland 'surgeon to General Wade's regiment of horse, now residing in London'. His certificate, dated 14 March 1738/9, stated that 'This gentleman is well versed in natural knowledge, particularly in anatomy, and hath invented several new instruments for the improvement of chirurgery, some of which have been shown to the Royal Society'. Despite seven signatures, including Sir Hans Sloane, P.R.S., Martin Folkes and Desaguliers, he was rejected. But the certificates show that the Society was prepared to recognize practical talents (12). In 1738 it elected the watchmaker, John Ellicott—'a person well acquainted with the principles of mechanics, astronomy and experimental philosophy, very ingenious in contriving and improving useful instruments for promoting natural knowledge'. It also occasionally elected apothecaries such as William Watson (elected 1741) and William Hudson (elected 1761). In the case of physicians occasional reference was made to their professional ability. Thus George Douglas, M.D. (elected 1732) and Peter Sainthill (elected 1734) were said to be skillful or eminent in their own profession, this being seen as a reason for election.

In 1742 both the Reverend Nathaniel Bliss and the Reverend Charles Mason were elected on the basis of certificates which mentioned no explicit qualifications at all but, as it was stated that the former was Savilian professor of geometry in the University of Oxford and the latter Fellow of Trinity College Cambridge and Woodwardian lecturer, this suggests that an academic position could sometimes imply a qualification. Considering the state of the two ancient universities in the eighteenth century, this might sometimes be an unwarranted assumption (13). However, when one George Mitchell, mathematical master at the Royal Academy at Portsmouth, was presented in 1767 without explicit qualifications, he was rejected. If Martin Clare, Master of the Academy in Soho Square, London was successful in 1735 it was presumably not only because he was supported by eight signatures (including Desaguliers) but also because it could be said of the candidate that he was 'a good mathematician, well skilled both in natural and experimental philosophy and a great promoter of the same'. 'Promoter' is obviously a term much used at this time and in the same year Alvaro Lopez Suasso, residing in London, was elected as 'a gentleman of universal correspondence who is well able and much disposed to promote the interests of the Society'. Here is also a reminder of the continuing importance of correspondence to scientific societies of the eighteenth century.

In eighteenth-century England birth was often considered more important than achievement in any evaluation of a person. It was an obvious advantage to a candidate if he could claim superior social status. It also helped if he was a close relative of a former Fellow of the Society. Thus in 1734 John Winthrop of New England was introduced as the grandson of the learned John Winthrop one of the first members of the Society. In 1772 we find a certificate for 'Martin Folkes, Esq. of Queen Square, Middlesex, nephew of Martin Folkes

Esq. late president of the Society, a young gentleman who passed [*sic*] a university education at Emmanuel College in Cambridge and spent a small time abroad, being studious and curious after natural knowledge'.

In an analysis of the certificates some distinction has to be drawn between conventional phrases and the use of language to provide information. The claim that a candidate was 'likely to become a useful [and/or valuable] member of the Society' or that he was 'deserving of that honour' was usually no more than an empty formula. The assurance that a candidate was 'conversant' or even 'well versed' in various branches of natural philosophy also meant very little. It was a compliment which was understood to mean that the candidate had some interest in that area. Such a phrase was used in the certificate of William Cruickshank (elected 1802), professor of chemistry at the Woolwich Academy but it was followed by the informative phrase 'particularly distinguished in his chemical discoveries'. It was the use of superlatives like 'highly distinguished' (Thomas Hope, elected 1810) and 'well skilled' (George Peacock, elected 1818) which indicated a candidate out of the ordinary. George Biddell Airy (elected 1836) was a 'gentleman *profoundly acquainted* with mathematics, astronomy and other branches of physical science'. On the other hand Charles Darwin (elected 1839) was no more than '*well acquainted* with geology, botany, zoology and many other branches of natural knowledge', admittedly a wide area. Another phrase used in the 1830s and suggesting the early stages of the professionalization of science was that a candidate was 'devoted to science'. This phrase is found in the certificates of Thomas Graham (elected 1836), professor of chemistry at the Andersonian Institution, Glasgow and William Sharpey (elected 1839), professor of anatomy at University College London. The other clue in the identification of scientific Fellows is to be found in the specification of a subject. For Peacock it was astronomy; for his friend Charles Babbage (elected 1816) it was mathematics; for Francis Beaufort, R.N. (elected 1814) it was hydrography.

The third clue to a real man of science was some mention of authorship. Understandably prominence was given to any paper that had been published in *Philosophical Transactions* or read before the Society. John Herschel, recently graduated from Cambridge and proposed for membership in 1813 at the unusually early age of 21, obviously had little chance of having had a paper published in *Philosophical Transactions*. His certificate reminded members that he was the 'author of a paper lately read at the Society on a remarkable application of Cotes' Theorem'. The criterion of authorship became increasingly important some time after 1800. Thus John Barrow (elected 1805), secretary to the Admiralty was described simply as 'advantageously known by his publica-

V

tions' and Thomas Thomson (elected 1811) was accepted on the basis of 'his chemical and philosophical knowledge and writing' as if the fact that they had published at all was remarkable. It was not thought necessary to mention that, for example, Thomson was the author of a highly successful text-book on chemistry. But William Whewell (elected 1820) was presented as 'the author of a treatise on mechanics'. By 1835 John Hammett M.D. was taking care to place in the library of the Royal Society his book on cholera and in 1840 authorship became explicitly one of the principal qualifications for membership.

SCIENTIFIC AND OTHER QUALIFICATIONS

In the early nineteenth century there are signs of the qualification criterion being taken more seriously, at least in those cases where qualifications existed. Thus John Brinkley, professor of astronomy in the University of Dublin, elected to the Society in 1803, is described, in one of the longest of such descriptions found in the certificates up to that time, as the author both of a practical method of determining longitude and as a mathematician who had contributed a new method of solving equations. Captain William Bligh (elected 1801) is described as a 'gentleman well versed in astronomy and other sciences connected with his profession and whose voyages to the Pacific Ocean have established his character as a navigator'. Humphry Davy (elected 1803) was not only 'a gentleman of very considerable scientific knowledge' but also the author of a paper published in the *Philosophical Transactions,* and James Ivory (elected 1815), who taught mathematics at the Royal Military College at Sandhurst, was the author of several papers which had been published in *Philosophical Transactions* prior to his election. A contrast is provided with other major scientific societies, notably the French Académie des Sciences which would only publish papers in their *Mémoires* if the author was already a member (14). By the 1830s the titles of various scientific journals began to feature more prominently in the qualifications of candidates. This reflects in the first place the increasing importance of journals as opposed to books in the publication of original research. But it also reflects a desire to provide more specific information and this feature deserves our attention.

Although the qualification criterion was still often met by the most vague and general statement—that the candidate was a 'gentleman well versed in various branches of natural knowledge' or a 'gentleman conversant in various branches of science', in appropriate cases the interest might be more specific. Thus John Pond (elected 1807), the translator of Laplace (15) was introduced as a 'gentleman conversant with the branches of natural philosophy and particularly attached to astronomy'. Incidentally all candidates, unless foreign, were normally described as 'gentlemen', including the potter Josiah Wedgwood (elected 1783), the manufacturer Mathew Boulton (elected 1785) and the engineers John Rennie (elected 1798), James Cockshutt (elected 1804), a former employee of John Smeaton, and Marc Isambard Brunel (elected 1814). One or two cases in which the candidate was described as a 'person' rather than a 'gentleman' were also those which were rejected. It is interesting that James Gambier Esq., described in 1734 simply as 'a person well versed in natural knowledge', was unsuccessful but three years later appearing as 'a gentleman very well skilled in natural philosophy' was elected.

At a time when trade and practical work were regarded as necessarily socially inferior activities, it is significant that the Royal Society tended to gloss over such social distinctions. However, if a theoretical interest complemented practical knowledge there was an obvious advantage. Thus Cockshutt was introduced as a 'gentleman well versed in many theoretical as well as practical branches of natural knowledge' and Brunel as a 'gentleman well versed in several branches of natural philosophy and particularly in theoretical and practical mechanics'.

Contemporary concern with qualifications serves to throw some light on intellectual interests and abilities in areas outside 'science' or 'natural knowledge'. Often science was linked with literature, as in the title of the Manchester Literary and Philosophical Society and other such societies founded in England in the late eighteenth and early nineteenth centuries. Thus Sir John Cox Hippisley was elected in 1800 as a 'gentleman versed in various branches of literature and science'. In the same year Charles Dickinson was described as a 'gentleman well versed in various branches of natural science and distinguished for his knowledge in polite literature'. In both these cases it should be noted that a slightly stronger claim is made for the candidate's knowledge of literature than of science. There were also cases of purely literary claims. Thus the Reverend William Douglas, Canon residentiary of Salisbury was elected in 1800 as a 'gentleman well versed in various branches of literature' without any mention of an interest in the natural world. Yet another country parson proposed in the same year, the Reverend James Hook, A.M., a 'gentleman conversant with various branches of literature', was one of the few candidates to be rejected. This was probably due less to his marginal qualifications than his weak personal support, having only four signatures not including any important Fellows or members of the Council.

In 1808 Alexander Hamilton, professor of oriental languages at the East India College was elected as a 'gentleman well skilled in several branches of literature', following a tradition which gladly accepted retired army officers and civil servants from the Indian sub-continent who had some knowledge of oriental literature, laws and customs (16). This was an area of specialization which in France at that time would have been assigned to a separate Academy (17). The Royal Society, however, continued with a much broader recruitment; there was, for example, some overlap of membership with the Society of Antiquaries, and even in 1840, when certification was made more rigorous, one possible category of membership was literary and another artistic, as we shall see in due course.

One factor influencing admission to the Royal Society was the policy of the President, especially in the case of a forceful personality like Sir Joseph Banks. Historians of the Society have often divided their chronicle into chapters describing changes in the Society under successive presidents. Soon after Banks was elected P.R.S. in 1778 he spoke about elections and announced to the Officers and members of the Council that he was prepared to challenge candidates whom he thought unworthy (18). He is also said to have given some guidance to the Society on suitable candidates: 'that any person who had successfully cultivated science, especially by original investigations, should be admitted, whatever might be his rank or fortune' (19), but he also wanted to admit 'men of wealth or status' as patrons. If one looks at the average number of Fellows elected during the four decades of Banks's presidency, it does not differ much from the numbers in the preceding decades (20) and, although it is true that Banks did pay some attention to qualifications, the proportion of non-scientific to scientific Fellows actually rose slightly during his period of office. The only marked change was a decline in the number of Foreign Fellows, which had reached a peak of 170 in 1766 but fell to 40 by the time of Banks's death in 1820.

FOREIGN MEMBERSHIP

There was a steady increase in membership of the Royal Society throughout the eighteenth century but a particularly heavy increase in Foreign Members. In the first four decades of the century the number of Foreign Fellows admitted was 24, 48, 63 and 83 respectively. This tended to distort the membership of the Society with the proportion of foreign members rising from 24 to 49 per cent and, given this clear trend, the Council was encouraged to place some restriction on their numbers. The standard administrative history of the Royal Society attributes the increase in the number of Foreign Fellows to the fact that 'some of them were being accepted without sufficient regard eing paid to the value of their scientific qualifications' (21), but the author oes not connect this with the introduction of certificates in 1730.

The traditional form of recommendation was by letter and many of the ertificates of the 1730s relating to Foreign Members referred to such letters nd often transcribed or translated part of their contents. News, however, of he new regulations of the Society gradually became known abroad and one of he earliest certificates to have been actually drawn up in another country was hat relating to the abbé de la Grive, dated 1 October 1733. Four French ellows of the Society including Fontenelle, secretary of the Paris Academy of ciences, signed a document stating that, as members of the Society, they ertified that de la Grive was a mathematician who had drawn a very exact nap of the city of Paris. They ended with the assurance that he was 'an cclesiastic of good morals'. This certificate was counter-signed by three 3ritish Fellows including the President, Sir Hans Sloane. Often a candidate's ellow countrymen were in the best position to assess a candidate's work and it s notable that when Count Algarotti of Venice, who was to trivialize Newton's theory of gravitation in his *Il Newtonianismo per le dame* (1737) was lected in 1736, he was sponsored by the newly-elected Philip, Earl Stanhope 22), who passed him off as 'a gentleman of great knowledge in all parts of 'hilosophical and mathematical learning' and 'exceedingly well qualified'. In 761 it was agreed that foreign candidates must have their certificates signed by t least 'three Foreign Fellows' as well as by at least 'three Fellows named in he Home List' (23). The Society accordingly did rather better in the late ighteenth century, for example with the election of Laplace, one of a batch of 'oreign Fellows to be elected in 1789, described as 'a gentleman well known to he learned world for his skill in analytical calculations and his useful applica-ion of them to the theory of astronomy'. Sponsors were now always asked to ecommend from personal knowledge. In this case they used the formula personal knowledge of him or his works'.

If one looks at the early nineteenth-century pattern of membership one inds that foreign candidates are almost invariably presented with more impres-ive qualifications than ordinary members. A striking example can be found in 822 when John Kidd, professor of Chemistry at the University of Oxford nanaged to be elected without his sponsors even hinting at any (further) ual ifications he might have. His certificate is remarkable for having the mallest ratio of supporting text to signatures (24). In the same year the andidacy was advanced of A. P. De Candolle, professor of natural history and lirector of the Botanical Garden at Geneva. The Swiss candidate was described s:

Author of the *Flore Francaise*; of the History of Succulent Plants; of an Elementary Theory of Botany: and of the *Systema Naturalis Regni Vegetabilis*—being not only distinguished for his superior knowledge in Botany, but likewise conversant with various branches of Natural Science. (25)

The contrast was not always quite so stark. Thus another Oxford professor Baden Powell (elected 1824) is mentioned as the author of a paper on the infrared spectrum 'lately read before the Royal Society', while several Foreign Members were introduced only in the vaguest terms. Fourier (elected 1823) for example, was 'the author of several important mathematical and physical investigations' which are not specified and Thenard (elected 1824) had 'greatly distinguished himself by his valuable and numerous contributions to chemical science'. Considering that Fourier had recently published his historic *Théorie analytique de la chaleur* (1822) and Thenard had discovered and investigated hydrogen peroxide in the period 1818–20, one might have expected some reference to these researches. However, both Fourier, as Secretary of the Paris Academy of Sciences, and Thenard, as Dean of the Paris Faculty of Science were sufficiently senior figures for it not to be thought necessary to go into the detail of their researches. This is probably the greatest difference in the nineteenth century in the standards for the admission of ordinary and foreign members. British members could be elected while still relatively junior, whereas Foreign Fellows had to have made their mark on the international stage. It is interesting that Oersted was proposed for Foreign Membership in November 1820 *after* he had received the Society's Copley medal for his discovery of a connexion between electricity and magnetism (26). One sometimes finds the phrase 'in the first rank' (in their particular field) used to describe Foreign Fellows, e.g. Blainville (elected 1832), and Elie de Beaumont (elected 1835), a compliment absent from the certificates of ordinary Fellows who at best were described (as in the cliché of the chairman introducing a speaker) as 'well known' (27).

Certificates giving details of status and qualifications had originally been introduced in the early eighteenth century partly to stem the flood of Foreign Fellows. They came to be used for all candidates for election who were not in the privileged class of Peers of the Realm, etc. But by 1765, because of renewed alarm about the great number of Foreign Members, it was proposed to make it more difficult for them than for those on the home list. It was accordingly resolved that 'no foreigner be proposed for election that is not known to the learned world, by some publication or invention which may entitle the Society to form a judgement of his merit' (28). Not more than two foreign member were to be elected in any year until the total number had been reduced to eighty.

In the late eighteenth century or early nineteenth century one might almost say that there was a double standard for entry (quite apart from the privileged membership). More definite qualifications were required in the case of foreign candidates than home candidates. However, by about 1840 there is evidence of a marked change. This was the time, as we shall see in the next section, when a new more detailed kind of certificate was introduced. But it was also a period when very few foreign members were being elected (29). Thus the more searching questionnaire was applied almost exclusively to home candidates. It even came to be considered as inappropriate and possibly discourteous for the sponsors of Foreign Members to search out all the details necessary to complete the new type certificate. Plain sheets of paper were substituted in which the candidate would often be described in one sentence, most of the page being left for the signatures of his supporters. Thus Claude Bernard was elected in 1864 simply on the basis of his eminence as a physiologist and Louis Pasteur (elected 1869) was described simply and briefly as 'eminent as a chemist'. The wheel had now turned full circle and we have the irony of a procedure introduced with Foreign Members in mind being used almost exclusively for the screening of home candidates, while overseas candidates, admittedly senior and distinguished, were spared the more rigorous enquiry.

THE 1839/40 REVISION AND THE EMPHASIS ON AUTHORSHIP

In the late 1820s and the 1830s a number of reforms were introduced. For example a committee was appointed in 1827 to 'consider the best means of limiting the members admitted to the Royal Society . . .'. In its report it emphasized the importance of 'including men of high philosophical eminence' even though an improvement of standards would lead to a drop in membership and thus a loss in income (30). In the election for President of 1830 a group of scientific Fellows put up their own candidate, John Herschel, in opposition to the Duke of Sussex. Although Herschel was unsuccessful, some reforms did take place under the new President. Future elections could only be held at quarterly intervals (31) and the certificate was to be signed by six instead of only three Fellows as supporters. A number of other reforms were put in hand by the Treasurer Lubbock. In 1838 specialist committees were established for different branches of science but it was the 1839 committee which is most relevant here. At the meeting of 9 May 1839 under the presidency of Lord Northampton it was resolved:

> That a committee be appointed to draw up a set of forms for certificates of persons desirous of being proposed as candidates for admission into the Society, and also to recommend such measures as may be desirable to be adopted on such occasions and that such committee consist of all the members of the council. (32)

Two months later it was reported that:

> A form of certificate, agreed upon by the Committee appointed for drawing up such a form, was presented and adopted by the Council. (33)

The foolscap printed form of certificate gave directions on how it was to be completed:

> It is recommended by the Council that the Members proposing a candidate for election should also state in the certificate his claim for admission under one of the following heads, applicable to the case . . .
>> As one who had made discoveries in some branch of science which should be specified.
>> As the author of a work or paper of merit, connected with science; the title of which should be stated.
>> As one who has invented or materially improved any astronomical, mathematical or philosophical instrument, or chemical process, which should be specified.
>> As one distinguished for his acquaintance with some branch of science which should be specified*.

Attached to this there was an asterisk with a footnote reading: 'This *single* testimonial will not be considered sufficient for any candidate with reference to his professional attainments alone'. At the end of the list was the most general 'qualification' of all:

> As one who is attached to science, and anxious to promote its progress,

a formula which, we shall see, was resorted to in the case of several candidates with no particular qualification. The longstanding tradition of the Society in including attainments outside science in the narrow sense is perpetuated by three further categories:

> As a person eminently distinguished in one of the learned professions.
> As a distinguished Engineer, Architect, Painter, Sculptor or Engraver.
> As one distinguished for his literary or archaeological attainments.

This permitted the nomination of Alfred Tennyson, Poet Laureate (elected 1864) as well as several professional men of lesser distinction.

On the form itself the relevant qualifications were reduced to five, namely:

The Discoverer of . . .
The Author of . . .
The Inventor or Improver of . . .
Distinguished by his acquaintance with the science of . . .
Eminent as . . .

The form also asked sponsors to sign in different columns according to whether they had personal knowledge of the candidate or only general knowledge. Very few candidates could claim to be the discoverer of a new phenomenon or a new substance so it was mainly the second category, that of authorship, to which attention was directed. The notes specified that it must be some *scientific* work, yet we find Joseph Phillimore elected in February 1840 on the basis of 'various works on civil and ecclesiastical law', although he did also claim to be a 'patron of science'. Another candidate, J. P. Gassiot, merchant, who claimed to be a 'lover of experimental knowledge' had not actually published anything but his paper on electricity had been accepted for the *Philosophical Transactions* and he was elected. Some like Thomas Henderson, professor of astronomy at Edinburgh, were represented vaguely as the author of 'various papers on astronomical subjects' and it is true that there was only one line on the form to describe publications. Nevertheless, most candidates had a more detailed description provided of their publications, even occasionally the title of an important one.

The details of the new form presented some candidates with a number of problems. The sponsors of James Cosmo Melvill, Secretary to the East India Company (elected 1841) could think of no memoir, invention or other claims to fame for their candidate. Hence all the categories of qualification had to be left blank but the Society was assured that he was 'attached to science and anxious to promote its progress'. This formula was applied in the case of many other weak candidates. The fact that the idea of qualifications might even be irrelevant is suggested by the certificate of Hart Davis, Deputy Chairman of the Board of Excise (elected 1841), where all five printed categories of qualification are actually deleted in pen and the phrase that he was 'attached to science and anxious to promote its progress' was inserted in inverted commas. For the entry 'Profession or Trade', Samuel Pearce Pratt, a Fellow of the Linnean and Geological Societies (elected 1842), gave 'Gentleman' (34), while Norman Lockyer's sponsors gave 'Gentleman' as his 'Title'. The entry 'Title or

V

Designation' gave candidates an opportunity to list their membership of major scientific societies (35) and from the 1850s, one observes a rash of degrees. Some like Joseph Dalton Hooker (elected 1847 at the age of 30), regretfully had to leave a blank for 'Profession or Trade' since he did not obtain an official appointment at the Royal Botanic Gardens Kew (where his father was Director) until 1855. The entry which caused the greatest heart-searching was probably the last: 'Eminent as . . .' which was originally intended for candidates distinguished in fields other than science. On many certificates it was deleted or ignored but the surgeon William Sharp, President of the Bradford Philosophical Society (elected 1840), proposed by William Scoreseby, apparently felt no embarrassment about its use. The certificate reads: 'Eminent as: Promoter of science in Yorkshire and founder of the Bradford Philosophical Society'. From 1846 the certificate assumed an even greater importance since the number of new Fellows was to be limited to fifteen a year, excluding still the privileged classes. This not only brought about a reduction in membership, it introduced a form of selection. For the first time in its history the Fellows would have to give serious consideration to the different merits of candidates who were now in some sense rivals.

Despite the instruction that for a candidate who was the author of a book or memoir 'the title . . . should be stated' we find William Robert Grove (elected 1840) given simple as 'author of various papers on the subject of electricity' although it was also mentioned that he was 'Improver of the Voltaic battery'. Similarly when A. W. Hofmann, professor at the Royal College of Chemistry, was elected in 1851 it was partly as the author of 'a series of memoirs on the volatile organic bases published in Liebig's Annalen, in the Transactions of the Chemical and Royal Societies and of various other chemical papers'. However, the certificate for Joseph Lister (elected 1860), which mentioned that he was the author of three memoirs published in the *Philosophical Transactions* for 1858, took the trouble to give their titles, providing a total entry under authorship running to seven cramped lines, although only one was offered on the printed form. It was circumstances like this which led to a change in the forms introduced in December 1863. No longer was one line left for each of the entries: Discoverer, Inventor, etc., but a space of approximately three inches was left, marked 'Qualifications'.

It is a matter of crucial importance to understanding the growth of professional science in the nineteenth century that in the majority of cases the space for 'qualifications' was used to describe authorship (36). Thus the astronomer William Huggins (elected 1865) has the following entry: 'Author of a paper on the spectrum of some of the chemical elements, on the spectra of nebulae and joint author of a paper on the spectra of some of the fixed stars, all of which are published in the *Philosophical Transactions*'. Also the certificate of the civil engineer, Fleeming Jenkin, elected at the same meeting, obviously mentioned practical qualifications but interestingly emphasized publication. William Henry Perkin, discoverer of Perkin's mauve, who had abandoned his academic career to do research in dyeing, was presented primarily as 'Author of numerous articles on organic chemistry' although, of course, his great discovery was also mentioned. Sometimes the qualifications given were simply the titles and dates of books and articles, as with the economist William Stanley Jevons (elected 1872), but the proposers of the astronomer Norman Lockyer (37) thought it important not only to give authorship but also to explain to the Society the significance of his research.

CONCLUSION

In the first place this paper has drawn attention to an important documentary source. It may be that because the certificates did not come into existence until the eighteenth century, while most social studies of the Royal Society have been confined to the seventeenth century, the potential of the certificates as a source of information has escaped the attention of many historians of science; they may now wish to go further than has been possible in this limited survey. As certificates provide information on residence and occupation as well as qualifications, there is scope for a sociologist of science supported by a suitable grant to undertake a detailed study going far beyond the confines of the present paper. Such a researcher might well wish to quantify his results, which could be valuable if the production of statistics were preceded by an intensive study of the historical background so that some confidence could be placed in the numbers produced. Because of the dangers of superficial quantification this paper has deliberately avoided this approach. As the first general study of the Society's certificates, of which many of the early ones have not survived, it is sufficient to present qualitative evidence.

The paper has drawn attention to the introduction of certificates in 1730/31 and their refinement in 1839/40. These dates may be regarded as minor landmarks in the history of the Royal Society, not because they changed the Society overnight but because they called attention to the desirability of sponsors stating the grounds for election. Given the financial problems of the Society, there was always the danger in the early years that a candidate's willingness to pay a subscription regularly would be regarded as more important than an intellectual qualification. Fortunately the financial criterion did not apply to Foreign Members after 1730 since their position was honorary. It is interesting that, before the 1730 statute was passed, newly-elected Fellows sometimes informed the Society of their scientific accomplishments *after* being elected, by way of gratitude, rather than before by way of relevant information (38).

Although the practice of admitting non-scientific Fellows continued, it was to the benefit of the Society that even with money and social position, membership should not be automatic. The stipulation of the certificate was a hurdle which might embarrass some with little true concern for science, while scientific Fellows could surmount it without difficulty. For them it was an encouragement not only to do scientific research but also to publish it, particularly in the *Philosophical Transactions.* Thus the Society's journal benefitted from the contributions of comparatively junior men of science who had original contributions to make.

In the eighteenth and early nineteenth centuries there were no standard qualifications which, even with the best will in the world, the Society could insist on, such as the holding of a university degree, since the curricula of the English universities were largely irrelevant to science (39). Membership of a profession was often a better guide and to be a reputable physician was probably one of the best recommendations in the earlier part of the period studied (40). It came to be replaced increasingly by the criterion of authorship of scientific articles or books.

By 1860–70, when this survey ends, the Royal Society for the first time in its history had a majority of scientific Fellows. It had finally reached the modern situation of being a high-level scientific society. In this transition the certificates played no small part. If the institution of certificates helped in a quiet way to change the course of history, now they may also serve to illuminate it.

NOTES

(1) Increased to six in 1830. Also signatories were supposed to be personally acquainted with the candidate.
(2) From 1728 to 1730 all candidates were subject to this double scrutiny.
(3) Council Minutes (Copy), Vol. 3. (1727–47), p. 83, 20 August 1730 (my italics). The original unsystematic capitalization in primary sources has not been retained.
(4) Ibid., pp. 87–88, 27 October 1730. (My italics).
(5) (My italics). At the meeting of the Society on 10 December after the words 'chief qualifications' were added the words: 'the inventions, discoveries, works, writings or other productions'. This provides an interesting elucidation of what were considered relevant qualifications at the time. These words were, however, omitted in later versions of the statute.

(6) Ibid., pp. 92–93, 7 December 1730. See also Journal Book (Copy), Vol. 13 (1726–31), p. 529, 10 December 1730.

(7) *The Record of the Royal Society of London,* 4th edn., (London, 1940), p. 91.

(8) C. R. Weld, (*History of the Royal Society,* London, 1848, 2 vols., vol. 1. p. 461) in over a thousand pages of text merely mentions the certificates in passing, saying that 'as they generally give an account of the scientific and literary (sic) attainments of the candidates, [they] are highly valuable records', whilst although Sir Henry Lyons (*The Royal Society, 1660–1940,* Cambridge, 1944, p. 153) describes them as 'a most valuable series of documents', he also makes no use of them.

(9) Much more research would be required, by someone with expertise in handwriting to make authoritative assertions about the authorship of certificates. There is occasionally useful evidence in the Journal Book in the form 'Mr. A. was proposed for a Fellow by Mr. B. and recommended by Mr. C. and the President' but unfortunately elections are not recorded systematically, especially after 1730 when the certificate was thought to provide all the information necessary for the records.

(10) Obviously there is further work to be done on sponsorship and patronage in the Royal Society. Not surprisingly the President often played the most prominent part in nominating candidates. For a discussion of nominations in the seventeenth century see Michael Hunter, 'Reconstructing Restoration Science: Problems and Pitfalls in Intellectual History', *Social Stud. Sci.,* **12,** 451–66 (454–5) (1982).

(11) Yet up to 1860 the medical profession formed the largest group of scientific Fellows, physicians and surgeons *together* numbering 63 in 1740 (63% of the scientific Fellows), 84 in 1800 (56%) and 80 in 1860 (24%)—Lyons, op. cit. (8), pp. 126, 342. The majority of these, however, were physicians, holding either the Fellowship of the Royal College of Physicians or an M.D. degree. They were normally university educated in contrast to the more practical training of the surgeons. It has been noted that no surgeon was knighted until 1778 and that, although the surgeons separated from the barbers' company in 1745, it was not until 1800 that they obtained a royal charter—Roy Porter, *English Social History in the Eighteenth Century* (London, 1982), p. 91.

(12) According to the first two charters the objects of the Royal Society were 'to further promoting by the authority of experiments the sciences of natural things *and of useful arts*' (my italics), *Record* (op. cit. (7)), pp. 226, 251.

(13) For the case of Richard Watson, appointed professor of chemistry at Cambridge in 1764 despite his complete ignorance of the subject, see L. J. M. Coleby, 'Richard Watson, Professor of Chemistry in the University of Cambridge, 1764–71', *Ann. Sci.,* **9,** 101–123 (1953).

(14) The subsidiary *Mémoires des Savants Etrangers* for selected papers of non-members never had the same prestige and in the nineteenth century was nearly always seriously in arrears.

(15) *The System of the World by P. S. Laplace,* translated from the French by J. Pond, F.R.S., 2 vols., (London, 1809).

(16) E.g., C. E. Carrington (elected 1800), a 'gentleman well versed in oriental

literature, particularly on laws and customs of the Mohammedans and Hindoos'. Warren Hastings, Governor-General of India was elected in 1801 as a 'gentleman of great and extensive knowledge in various branches of science'.

(17) From 1795–1815 the Academies were replaced by the concept of a single National Institute which, however, in practice was divided into separate Academies or 'Classes'. The First Class was concerned with science, the Second Class (temporarily suppressed by Bonaparte in 1803) dealt with moral and political sciences and the Third Class with literature and fine art.

(18) Lyons, op. cit. (8), pp. 198–9, 204. Weld, op. cit. (8), Vol. 2. pp. 152–3.

(19) Henry Brougham, *Lives of Men of Letters and Science who flourished in the time of George III*, (London, 1845, 46), Vol. 2. pp. 363–4.

(20) Some circumstantial evidence of Banks's interventions, at least in the early years of his presidency, is given by Weld (op. cit. (8), Vol. 2. p. 154n), who notes that from 1778–84 eleven candidates were rejected. As 127 Fellows were elected in this period this represents a rejection rate of 8%, which seems to be an unusually high proportion.

(21) H. Lyons, op. cit. (8), p. 126. Lyons divides the early eighteenth century not around 1730 but at 1741, when Sir Hans Sloane retired as P.R.S.

(22) Not to be confused with Charles, the third Earl Stanhope (elected 1772), a valuable member of the Society and the inventor of two calculating machines. George, the second Earl of Macclesfield, (P.R.S., 1752–64) was another eighteenth-century peer of the realm of some scientific distinction.

(23) *Record*, op. cit. (7), p. 95.

(24) The five lines of text are followed by a record 23 signatures, more than double the average.

(25) The description of De Candolle and his qualifications occupies fifteen lines of manuscript. Another very long description of publications was that for Cauchy (elected 1832).

(26) Just as Benjamin Franklin became a Fellow in 1756 at the age of 50, three years *after* he had been awarded the Society's Copley medal.

(27) E.g., the geologist Sir James Hall (elected 1806) and the chemist John Dalton (elected at the age of 56 in 1822).

(28) *Record*, op. cit. (7), p. 95.

(29) In 1776 the proportion of Foreign Members to Ordinary Fellows was one to three; by 1810 it had fallen to one in eleven and by 1823 it was one to seventeen. The maximum number of Foreign Members was then fixed at fifty. H. Lyons, op. cit. (8), p. 233.

(30) Ibid., p. 245.

(31) In 1846 this was reduced to one occasion each year.

(32) *Minutes of the Council of the Royal Society*, Vol. 1. (1832–46), p. 227. The names of members of the Committee are not given and there are no further relevant records in the archives of the Society.

(33) Ibid., p. 237.

(34) Because several candidates had no paid occupation a note was soon added to the certificates informing sponsors that if candidates did not have a profession or trade, they should insert 'None'.

(35) Already in 1830 Charles Babbage was ridiculing 'the custom to attach certain letters to the names of those who belong to different societies', *Reflections on the decline of science in England and some of its causes,* (London, 1830), p. 42.

(36) On the use of publications in the assessment of the emergent professional scientist of the nineteenth century see Maurice Crosland, 'Scientific Credentials: Record of Publications in the Assessment of Qualifications for Election to the French Académie des Sciences', *Minerva,* (1983) (in the press).

(37) See A. J. Meadows, *Science and controversy. A biography of Sir Norman Lockyer,* (London, 1972), p. 72.

(38) Thus at the meeting of 7 May 1730 a letter was read from Dr Johann Heucher— 'He herein makes his compliment of thanks for being elected a member of the Society. He mentions several tracts whereof he is the author that have been already published, the titles of which he subjoins to his letter'. Journal Book (Copy), Vol. XIII (1726–31), p. 473.

(39) An important exception to this generalization is provided by the mathematical teaching available as part of a liberal education at the University of Cambridge even in the eighteenth century. The mathematics taught was, however, based on the geometrical approach of Newton's *Principia.* For the 'analytic revolution' of the 1820s, deriving from the Continental developments of the previous century, see Harvey W. Becher, 'William Whewell and Cambridge Mathematics', *Hist. Stud. Phys. Sci.,* 11, Part 1 (1980), pp. 1–48.

(40) Yet Sir Joseph Banks (P.R.S., 1778–1820) vetoed the election of Dr Vaughan because he was no more than a *fashionable* physician, Weld, op. cit. (8), Vol. II, p. 153–4.

VI

The Development of a Professional Career in Science in France

A professional career entails full-time and remunerated employment entered into after a course of training.[1] Such careers and the prerequisite training became available in the educational structure of post-revolutionary France. They constituted a crucial phase in the establishment of science as a profession. The large number of scientific posts in official institutions in the early nineteenth century and the eminence of the men who filled them gave France unparalleled distinction in most branches of the physical and biological sciences. France was envied by men of scientific bent in other countries where there were neither educational facilities nor any significant number of posts which could employ scientific talents. The education and the mode of recruitment of scientists cannot be understood by confining attention to the study of any one institution or type of institution such as the university. Universities in eighteenth-century France were moribund and the national university established in 1808 with various faculties of science came only gradually to embrace the new professional scientists. Universities did not play a great part in laying the foundations for the profession of science in France.

One of the secondary characteristics of a profession is a claim to high standards of competence, standards imposed by the profession itself. If one remembers that the Paris Académie des sciences had begun to impose certain standards under the *ancien régime*,[2] one naturally thinks of its successor, the First Class of the Institut, as the body which would continue and extend this tradition into the nineteenth century. Undoubtedly the Institut did include among its functions the imposition of standards, particularly for work for which it recommended publication, but because educational institutions became so important after the Revolution, it was only at the highest level that the Institut played a major role. From the point of view of education and entry into the profession, it was the École polytechnique which played the most prominent role in the creation of standards—in the first place, the standard of mathematical competence required for entry,[3] but, more pertinently, for the degree of attainment in physical science and mathematics required for graduation. Both the staff

---

[1] For analyses of the profession of science, see, for example, Ben-David, J., " The Profession of Science and its Powers ", *Minerva* X, 3 (July 1972), pp. 362–383 ; and Shils, E., " The Profession of Science ", *Advancement of Science*, XXIV (June 1968), pp. 469–480.

[2] See Hahn, Roger, *The Anatomy of a Scientific Institution: The Paris Academy of Sciences, 1666–1803* (Berkeley: University of California Press, 1971).

[3] In 1794, it was merely stated that candidates should have a knowledge of arithmetic and the elements of algebra and geometry, but already in 1795 the syllabus of the entrance

and the governing body—the *conseil de perfectionnement*—were influential here. A major influence was that of the final examiners.[4] In a system which started with highly selected students, who were subjected to an intensive course of instruction and frequent tests, it was the external examiners who put their vital seal of approval on the final stage of graduation. They exercised influence not only over the student body but over the teaching.

Once the Paris faculty of science got into its stride, it tended to take over from the École polytechnique, particularly in the 1820s and 1830s. I would suggest that from the 1830s the doctorate became more and more the expected qualification for the aspiring academic scientist. The juries for the doctoral theses, drawn from the staff of the faculty and the *grandes écoles*, thus administered a diploma vital to the career of aspirants to the profession. The École normale, too, played an increasingly important role in the training of scientists who would make a career in higher education.

The Ancien Régime

France in the eighteenth century was one of the most powerful states in Western Europe. Its population of about 28 million was nearly twice that of Britain. It inherited from Louis XIV ideas of a centralised state very far from the tradition of local autonomy common in Britain. The British traditions of independence and self-help [5] may be contrasted with measures of governmental control in France which affected both industry and science. The ideas of the French enlightenment brought to a focus the general French attitude that action should be based on reason and practice on theory. This respect for the intellect, combined with the impossibility of a political career, helped to turn many French minds towards science. A church which had lost power over intellectual activity, and the idealisation of the possibilities of science, both contributed from different directions to produce a climate of opinion in which a new approach to nature was likely to flourish.

Much important science was done in France in the mid-eighteenth century, but a crucial change came with the Revolution when—if the event can be determined at all precisely—science became a profession. This happened in several different ways; one path was by differentiation and

---

examination included trigonometry, conic sections, and the solution of algebraic equations up to the fourth power. See Fourcy, A., *Histoire de l'École polytechnique* (Paris: 1828), pp. 30, 82.

4 In the period of the Directory the external examiners were Laplace and Bossut. The Danish astronomer, Thomas Bugge, attended some of these examinations in 1798 and said that Laplace was likely to fail several candidates that year, which meant that they could not proceed to the higher schools of specialised vocational training. Crosland, Maurice (ed.), *Science in France in the Revolutionary Era described by Thomas Bugge* (Cambridge, Mass.: Massachusetts Institute of Technology Press, 1969), p. 42. Laplace undoubtedly had a powerful influence on the content and standard of the courses.

5 Morrell, J. B., " Individualism and the Structure of British Science in 1830 ", *Historical Studies in the Physical Sciences* (Philadelphia: The University of Pennsylvania Press, 1971), vol. III, pp. 183–204.

specialisation. Under the *ancien régime* the man of science had hardly been distinguished from the man of letters. The terms *savant* or *philosophe* could be applied to both, and each, if fortunate, might receive the modest pension granted to *gens de lettres*. That there was no sharp demarcation between science and literature is illustrated by the work of Buffon for whom style was as important as content.[6]

Of course there were signs of a change even before the Revolution. As well as general books, men of science wrote technical monographs and memoirs more suitable for an academy than a *salon*. The difference between science and letters was recognised by the existence of the Académie des sciences as well as the literary Académie Française. Yet it was the Revolution and the war which forced a more absolute differentiation. Scientists provided quite different services for the revolutionary armies than did literary men. Playwrights could not advise on the casting of cannon; the imagination of poets was not much use in inventions of war.[7] Science, or a certain interpretation of it, was also in the ascendant in providing a comprehensive outlook on life.[8] Under the Napoleonic regime, science was specially favoured[9] at the expense of literature. This may be seen by a comparison of the treatment of Laplace, Berthollet, and Cuvier with that of the greatest literary figures of the time: Chateaubriand, Benjamin Constant, and Madame de Staël. One can detect signs of the beginning of a bifurcation of knowledge into " two cultures ", officially recognised in the Napoleonic Université de France with its faculty of science on a parallel with a faculty of letters. Science now existed in the educational system in its own right and not as a part of an arts course or of philosophy or medicine. Although the *Journal des savants*[10] continued to give news of science as well as literature, it was to the new specialised scientific journals, whether of mathematics,[11] physics and chemistry,[12] or in the biological

[6] Buffon, Count Leclerc de, "Discours prononcé à l'Académie Française" in Piveteau, J. (ed.), *Oeuvres philosophiques de Buffon* (Paris: Presses Universitaires de France, 1954) pp. 14, 500–504.

[7] A standard source for the part played by science and scientists in the war effort is Richard, C., *Le comité de salut public et les fabrications de guerre* (Paris: F. Rieder, 1922). See also Gillispie, C. C., "The Natural History of Industry", *Isis*, XLVIII, 154 (1957), pp. 398–407. For the role of artists in political propaganda, see Dowd, D. L., *Pageant Master of the Republic: Jacques Louis David and the French Revolution*, University of Nebraska Studies, New Series, no. 3 (Lincoln, Nebraska: University of Nebraska, June 1948).

[8] Condorcet strongly advocated the teaching of science as a means of eradicating superstition.

[9] For an account of the patronage of science in Napoleonic France, see Crosland, Maurice, *The Society of Arcueil: A View of French Science at the Time of Napoleon I* (London: Heinemann Educational Books, 1967), pp. 4–55.

[10] The *Journal des savants*, which published book reviews and general articles covering the whole field of learning, began publication in 1665, two months before the *Philosophical Transactions of the Royal Society*.

[11] For example, the *Annales de mathématiques pures et appliquées*, 21 vols., 1810–31 edited by Gergonne.

[12] For example, the *Annales de chimie*, 96 vols., 1789–1815. Its distinguished editorial board was originally headed by Lavoisier and included most leading French chemists. In 1816 the journal became the *Annales de chimie et de physique* under the joint editorship of Gay-Lussac and Arago, and for at least another generation it was the leading French journal in both chemistry and physics.

sciences,[13] that men would now turn for anything more than a superficial account of scientific developments. Specialisation was one of the features of science in early nineteenth-century France.

Even under the *ancien régime* work of a high standard was possible, as the achievements of Lavoisier clearly demonstrate.[14] But Lavoisier was not a professional scientist unless the word " professional " refers simply to work of a high standard. By other criteria, Lavoisier was an amateur, although this does not mean that his experiments were not most carefully planned, executed and reported. I prefer to use the term " professional " in an occupational rather than in an evaluative sense.

Salaries and Science

Lavoisier provides a convenient example of a stage in the growth of the profession of science. In the emergence of modern science, it was at first usual for men with scientific interests to be paid or employed to do something quite outside science. Lavoisier's occupation as a tax-farmer related to science only in so far as he used his earnings to meet the expenses of his laboratory. It was a significant advance towards the full professionalisation of science when men could be employed to do work related to their scientific interests. The teaching of elementary science provides a good example of this. In the *ancien régime* the mathematician Laplace improved his financial position by obtaining the post of examiner to the artillery. He explained that the great attraction of this post, apart from the salary, was that his duties would not take up more than three to four weeks of the year.[15] Examining, like teaching, could make all the difference between a bare livelihood or none; it also was closely connected with science but did not call for the performance of research as part of its stipulated duties.

The second stage in this simplified model of the establishment of science as a profession came when scientists were paid salaries to do scientific work which overlapped with the advancing frontiers of knowledge. They could be paid to do research, as at the Observatoire, or they might be given posts in advanced teaching.

The payment of an adequate salary was a significant step in the growth of the profession of science. In the first place it implied recognition of the value of the work done by a scientist. This was particularly important if the employer was not a wealthy private individual but a government. Condorcet was one of those who saw most clearly that with the Revolution the time had come for science to be a profession (" *une sorte d'état* ")

13 For example, the *Annales du Muséum d'histoire naturelle*, the " house journal " of the Muséum.
14 See Guerlac, Henry, " Lavoisier " in Gillispie, C. C. (ed.), *Dictionary of Scientific Biography* (New York: Scribners, 1970), vol. VIII, pp. 67–91.
15 Letter from Laplace to Lagrange, 11 February, 1784, in Serret, J. A. (ed.), *Oeuvres de Lagrange* (Paris: Gauthier-Villars, 1867–92), vol. XIV, p. 130.

and not the leisure-time occupation of those for whom money was a matter of indifference.[16] When the Institut was established in 1795, all members of the official body of science were for the first time paid salaries, and as a matter of principle they were bound to accept them.[17] There was no room in post-revolutionary France for the traditional wealthy amateur of the eighteenth century. Similarly, the abolition of the class of honorary members in the new Institut marked the end of a situation in which science could exist as an activity of a leisured nobility.[18]

The receipt of a salary implied a certain contractual obligation and a certain responsibility, whether to engage in the teaching of science, or research, or both. The most important consequence of a scientist receiving an adequate salary for his scientific work was that it enabled him to pursue science as a full-time occupation which was recognised as such by the government.[19] This stage was reached in the institutions established in France in the period of the Revolution.

Fourcroy, in one of his reports to the Convention, was quite explicit about the need to pay reasonable salaries to professors of science and medicine:

Your intention of reviving the useful sciences and of favouring their progress requires that professors and their assistants who have the responsibility of giving students lectures both theoretical and practical should be attached exclusively to those duties and that no other private occupation should distract them from them. It is therefore necessary that their salaries should be sufficient for their needs and that they should not be obliged to look for other posts as a means of completing their livelihood. Men who have spent twenty years of their lives in study in order to acquire profound knowledge and be able to transmit it to others, should be treated by the country which employs them in such a way that they are not tormented by domestic anxieties. By the exercise of their useful talents they should be able to draw on resources sufficient for their own maintenance and that of their families.[20]

Teaching

One of the changes which took place in teaching at the time of the Revolution was the raising of the standard. Whereas under the *ancien*

---

[16] Condorcet, Marie-Jean-Antoine-Nicolas Carifat de, *Oeuvres de Condorcet* (O'Connor, A. C. and Arago, D. F. J. [eds.]) (Paris: F. Didot frères, 1847), vol. VII, p. 423.

[17] "Rapport fait au Conseil des 500 par Villers au nom de la Commission des Dépenses, dans la Séance du 2 Prairial au 4" quoted in Aucoc, L., *L'Institut de France, Lois, statuts et règlements* (Paris: Imprimerie nationale, 1889), p. 38.

[18] The category of *Académicien libre* was also abolished. The reintroduction of this category by Louis XVIII in 1816 was merely a manifestation of his blindness to what had happened in France in the revolutionary and Napoleonic period.

[19] Such a situation is, of course, quite different from that of the wealthy amateur, who is free to spend as much time as he wishes on science and can suddenly decide to employ his leisure in some other way.

[20] Rapport et projet de décret sur l'établissement d'une École centrale de santé à Paris", *Réimpression de l'Ancien Moniteur* (Paris: 1794), vol. xxii, p. 665. Although the reference in this passage was to medical education (hence " *les sciences utiles* " above), Fourcroy was also concerned about the same time with the establishment of the École polytechnique.

régime an *abbé* would have been glad to present the elements of science to boys in the final year of college, and the mathematics teacher in the military academy would have helped cadets with their Bézout text book,[21] a change came abruptly—perhaps too abruptly—when at the École normale of 1795 the leading savants tried to present their own ideas on science.[22] It was impossible for the great mathematician Lagrange to come down to the level of his audience; nor would the original ideas of Berthollet on chemical affinity and the nature of acids have been appreciated by the majority of his students. The École, although short-lived in its first form,[23] provided an admirable precedent. It showed, among other things, that it was possible to establish educational institutions where the most learned scientists could present their own views of their respective subjects.

I distinguish here between teaching at an elementary level and teaching which overlaps with research, the former being today what one expects in a school and the latter what one hopes for in a university. The revolutionary situation in France enabled institutions to be established where teaching was at the level of research. Although Berthollet had little personal success as a teacher either at the École normale or at the École polytechnique, he was at least able to inspire his protégé Gay-Lussac [24] with a view of teaching as an extension of research. Even at the faculty of science where teaching was not at a particularly high level, Gay-Lussac and Biot divided the physics course into two parts according to their interests in research.[25]

These educational innovations had wider social implications both for teachers and students. We are all familiar with early nineteenth-century English novels where the governess, usually a gentlewoman by upbringing, is embarrassed by being treated as one of the servants. In eighteenth-century France there was some sort of parallel, although the vast bulk of education was in the hands of the various religious orders. The social standing of the Oratorian or Jesuit was much less as a teacher than as a cleric.

The social position of a teacher depended partly on the contemporary view of education and partly on the social status of students. As regards the first aspect, teaching obviously meant more in a society which prized achievement than in a society in which social position depended almost

---

[21] See, for example, Bézout, Etienne, *Cours de mathematiques à l'usage du corps royal de l'artillerie,* 4 vols. (Paris: Imprimerie nationale, 1770–72). The bibliography of Bézout's textbooks is complex and is complicated by the fact that, even after his death, further textbooks for military and naval cadets based on his work were published.

[22] The lectures were published as *Séances des Écoles Normales, recueillies par sténographes et révus par les professeurs,* 2nd ed., 10 vols. (Paris: Imprimerie du Cerce Social, 1800, 1801). A further three volumes of *Débats* was also published.

[23] It lasted only three months but was reconstituted by Napoleon in 1808.

[24] A full biography of Gay-Lussac by the present writer is in the course of preparation.

[25] Gay-Lussac and Biot asked in 1815 for permission to divide the course, the former giving lectures on gases, heat, etc. while the latter assumed responsibility for lectures on optics, sound and magnetism (Archives nationales, F17, 1933).

entirely on birth. Educational qualifications replaced birth and personal favour as criteria for selection for employment. Higher education imparted additional skills of value in government, administration, or various technical services. Education furthermore became a right of all men once it was thought that all men should have a voice in the choice of government. The provision of education thus became the duty of the state.

The other aspect which changed significantly was the relative social status of students and teachers. Before the Revolution, scientific subjects had been taught in the military academies [26] to officer cadets who were recruited almost exclusively from the nobility. Mathematical teachers like Monge and Lacroix were therefore instructing their social superiors. They themselves were neither officers nor did they belong to even the minor nobility. It was not until the Oath in the Tennis Court that the *Tiers État* asserted any political power. Lacroix,[27] teaching marine cadets under the *ancien régime,* contrasted the humiliation and insults he had formerly received from students and officers alike with the newly-found independence and authority he enjoyed at the École polytechnique where students came from all social classes. Lacroix ended his career as dean of the Paris faculty of science.

Some German visitors [28] to France just after the revolutionary period were amazed to find among professors of science, government ministers and councillors of state, most notably Chaptal and Fourcroy. In the new France there was no lack of dignity accorded to the professorial function. To succeed such lecturers must have been the ambition of many a young man in their numerous audiences. It was as a teacher that a scientist was most exposed to the public, and the scientist was probably thought of as much as a lecturer as a research worker.

Before the Revolution the teaching of science [29] was usually conducted in one of three contexts. It was either at an elementary level, as in the schools and colleges, or at a popular level, as in the famous courses of natural philosophy of the abbé Nollet. It often reached its highest level in a military or naval context. Artillery or military engineering cadets in the 1770s and 1780s received a very sound mathematical training, a training from which others might have benefited in a different system. The Revolution, in creating a much wider view of scientific education, was nevertheless able to draw on some of the traditions of the *ancien régime* and, most important, there was available a staff able to teach and examine; this staff included Monge, Lacroix and Laplace.

The Revolution marked both a quantitative and a qualitative change in

[26] See Taton, R. (ed.), *Enseignement et diffusion des sciences en France au XVIIIe siècle* (Paris: Hermann, 1964).

[27] Lacroix, S. F., *Essai sur l'enseignement en général* (Paris: chez Courcier, 1805), p. 128.

[28] Reichardt, J. F., *Un hiver à Paris sous le Consulat, 1802–1803* (Paris: Nourrit et Cie, 1896), p. 456.

[29] See Taton, R., *op. cit.*

the position of science. There was an increase in the scale of scientific activity, but there was also a change of pattern. Science not only had a significantly greater institutional support, it was now a nationally recognised activity inviting talented young men. The Revolution had replaced an aristocracy of birth by an aristocracy of achievement, or, as it has come to be called, a " meritocracy ". Young men who, a generation earlier, would have followed in their fathers' footsteps, now thought out afresh their place in society. The tradition of respect in which science was held by the *philosophes* turned many in that direction. Science could now provide not only an intellectual challenge but a career. A young man who acquired a scientific education by availing himself of the free courses of lectures given in Paris could aspire to fame and fortune in this newly opened field. The establishment of a system of grants enabled students to benefit from a higher education without the traditional financial barriers. The École polytechnique went further by providing not only grants but by inviting applications for places on a national basis, using the mechanism of a competitive examination to select the most talented. This institution therefore combined a staff of some of the best known men of science of the *ancien régime* with exceptionally able students. It was not long before some of the best of these students were themselves recruited to the staff. The École polytechnique thus played a dual role in the fostering of the profession of science. Not only did it provide the training by which Biot, Arago, Gay-Lussac, Poisson and others of their generation became scientists, but it provided posts for them afterwards.

Certification

In so far as there was any systematic training at a high level in the physical sciences in the early years of the nineteenth century, it was provided largely by the École polytechnique. As entrance to the Polytechnique depended on a competitive examination, it was in itself an achievement to have been accepted as a student. But to have undergone in addition an intensive course of study over two or three years was to have received one of the best scientific educations available at the time. The strong mathematical basis of the curriculum was supplemented by instruction in physics and chemistry, and there was a final examination. Unlike the military academies of the *ancien régime,* the science taught was not in a strictly vocational context. One of the aims of the École polytechnique was to provide a broad scientific education.[30] Purely vocational training took place later in the *écoles d'application,* where students really learned the craft of the military or civil engineer, or some other

---

[30] C. A. Prieur, one of the founders of the École polytechnique, had insisted that the school should produce not only engineers but also architects, industrialists, science teachers and simply " enlightened citizens ". " Mémoire sur l'École centrale des travaux publics ", Messidor an 3, Guillaume ed., *Procés verbaux du Comité d'Instruction Publique,* vol. vi, p. 302.

technical skill which was taught in the school specially devoted to it.[31] To be a *polytechnicien* was not necessarily to be a scientist. Many graduates went into the army, became engineers or administrators; but, at least in the early years, many of the best became scientists. When they did so it was by virtue of their training, their publications, and the posts to which they were appointed in recognised institutions. Their training as *polytechniciens* not only gave them a common social bond; it provided them with a recognised certification.

Certification was one of the primary functions of the university faculties established under Napoleon. The faculties of science, established in the years 1808–12 in Paris and various provincial centres,[32] suffered at first by being created later than the *grandes écoles,* having poorer facilities and lower prestige. The best faculty of science was that established in Paris and its posts were filled at a stroke of the pen by giving chairs as second appointments to established scientists who already held teaching positions in the *grandes écoles.* More posts had been created and hence, in principle, there were more opportunities for young scientists to seek a professional base, but it took a generation for these opportunities to be realised.

In medicine, the revolutionary period had witnessed a few years of anarchy with virtual freedom to practise without qualifications. A natural reaction came in the early years of the nineteenth century when the value of certified qualifications came to be doubly appreciated. This reaction affected science, drawing it into the higher educational system which was oriented towards examinations.

It was an initial weakness of the faculties that they had been principally examining bodies with the power to award the grades of *bachelier, licencié, docteur*—and later *agrégé* [33]—in science. Yet in a society increasingly conscious of the value of formal qualification, it became a positive advantage of the university system that it offered a series of grades which were useful and often necessary in gaining employment. Under the Restoration, the *baccalauréat* became a basic qualification in any profession [34]; the *licence*

---

[31] The principal *écoles d'application* (also called *écoles de services publics*) included the École des ponts et chaussées, École des mines, École des géographes, École du génie, École d'artillerie, and École des ingénieurs de vaisseaux.

[32] Faculties of science were established at Paris, Aix, Caen, Dijon, Lyon, Strasbourg, and Toulouse. By 1812 faculties were established at Besançon, Grenoble, Metz, and Montpellier. Further faculties of science were established in towns which soon reverted to Italy, Switzerland, and Belgium and these are therefore not listed. For another assessment of the faculties in the Restoration period, see Fox, R., " Scientific Enterprise and the Patronage of Research in France, 1800–70 ", *Minerva,* XI (October 1973), pp. 442–473.

[33] The *agrégation* was a very difficult competitive examination taken after the *licence* but before the *doctorat.* Although the origin of the *agrégation* (candidates were " *agrégé* " or attached to the university) goes back to the eighteenth century, it was only in 1821 that an examination was held. There were then three sections : grammar, letters, and science. In 1840 the *agrégation* in science was divided into two, with candidates choosing between mathematics and experimental sciences. By 1842, 58 per cent. of teachers and administrators in state secondary schools held the *agrégation.* Prost, Antoine, *Histoire de l'enseignement en France, 1800–1967* (Paris : Armand Colin, 1968), pp. 72–73.

[34] The *Ordonnance* of 13 September, 1820, stated that the *baccalauréat* was intended " ouvrir l'entrée à toutes les professions civiles et devenir pour la société une garantie essentielle de la capacité de ceux qu'elle admettrait à la servir ".

was a licence to teach required of all teachers except those in elementary schools. Crowning the university system was the doctorate by thesis, the official qualification of the faculty professor.[35] The first generation of French nineteenth-century scientists tended to be dominated by graduates of the École polytechnique—Biot, Gay-Lussac, Arago, Poisson—but this famous school was often by-passed by the next generation, that of J. B. Dumas (1800–84). It was with this generation that the doctorate really emerged as a professional qualification of the scientist.

In the 1830s doctoral candidates were not only more numerous than in the first 20 years of the faculties, but they were also more distinguished.[36] Among those submitting theses for the doctorate in the 10-year period 1832–41, were the mathematicians Bertrand and Liouville, and the physicists Regnault, Despretz and A. E. Becquerel. Among those who later distinguished themselves in the biological sciences and mineralogy were Milne-Edwards, Quatrefages and Delafosse, while the chemists included such names as Dumas, Persoz, Boussingault, Pelouze, Peligot, Malaguti, Henri Sainte-Claire Deville, Laurent and Gerhardt. All these did important scientific work and 13 out of the 17 became full members of the Académie des sciences. It was to the benefit of French science that, in a country when the emphasis in the educational system was on the transmission of a culture, there was a specific incentive in the examination system to do research. It was only a pity that, instead of harvesting young men's early creative talents, the research was delayed until the final stage in the system when the candidate would usually be in his thirties or forties.

Research

Appointment to a teaching post usually meant not only a salary but the incidental facilities to do some research in a laboratory. Even though research might not be regarded as part of the post, access to a laboratory, however simple, was a precious perquisite to many young teachers. It is right to emphasise the growth of science in the post-revolutionary era in the context of teaching, but there were other institutions which placed special emphasis on research. In addition to the Collège de France, the two institutions of particular importance were the Bureau des longitudes and the Muséum d'histoire naturelle, dealing respectively with the physical sciences and the biological sciences.

The Bureau des longitudes [37] was much broader in its interests than its British counterpart. Admittedly it was the utilitarian aspect which had

[35] Two theses were in fact required. Article 24 of the decree of 17 March, 1808, on the establishment of the faculties, states that for the doctorate " on soutiendra deux thèses, soit sur la mécanique et l'astronomie, soit sur la physique et la chimie, soit sur les trois parties de l'histoire naturelle, suivant celle de ces sciences à laquelle on declare se destiner."

[36] Maire, Albert, *Catalogue des thèses de sciences soutenus en France de 1810 à 1890* (Paris: H. Welter, 1892).

[37] A brief account of the establishment of the French Bureau des longitudes is given in Crosland, Maurice, *op. cit.*, 1967, pp. 209–213.

been emphasised to the Convention when a case was being made for its establishment [38]: the Bureau was to be a great help to shipping and navigation. But the influential role of Laplace in its early years, and the appointment of two mathematicians and four astronomers who constituted a majority of the Bureau,[39] provided a nucleus of men with strong interests in the physical sciences, who were able to pursue many of these interests at the Bureau. These salaried scientific civil servants had remarkable freedom for research within a broad framework. An inspection of the minutes of the Bureau reveals that they functioned rather like a scientific society holding regular meetings at which not only astronomical data and navigational work, but also other pieces of research were presented and discussed.[40] In engaging officially in further research on the metric system —both the prolongation of the meridian and the comparison of standards —the Bureau was continuing work which had been initiated by the Académie des sciences under the *ancien régime.*[41]

The Jardin du Roi had been a key centre for the study of natural history under the *ancien régime,* but when at the time of the Revolution it became the Muséum d'histoire naturelle and added significantly to its collections, it became something of a research centre, embracing chemistry, mineralogy, geology, botany, zoology and anatomy.[42] The staff of specialists not only had extensive collections on which to draw, but were provided with houses within the grounds of the institution. They thus became a scientific community in a very special sense of the term, and their sense of community was helped by several other factors, notably the democratic administration. Instead of being under the control of an administrator appointed by the government, they all took a share in administration and were indeed referred to as *professeurs-administrateurs.* The weekly meetings of the professorial staff were held under the chairmanship of one of their number elected for one year only. The professors also published a journal, the *Annales du Muséum d'histoire naturelle*[43]; its existence greatly strengthened the Muséum's orientation towards research, and there must have been informal pressure on the staff to contribute to it.

Stages of the Scientific Career

Anyone looking at the institutions organised in Paris or expanded after

38 *Le Moniteur,* XXV, 84–87 (11 Messidor an 3 = 29 June, 1795).

39 The staff of the original Bureau of 1795 was as follows: mathematicians: Lagrange, Laplace; astronomers: Lalande, Cassini, Méchain, Delambre; " *anciens navigateurs* ": Borda, Bougainville; geographer: Buache; technician: Carochez.

40 Examples of pure scientific research carried out under the auspices of the Bureau des longitudes include work in optics on the polarisation of light and interference, and on the velocity of sound.

41 Bigourdan, C., *Le système métrique des poids et mesures* (Paris: Gauthier-Villars, 1901).

42 Deleuze, M., *Histoire et description du Muséum d'histoire naturelle,* 2 vols. (Paris, 1823).

43 1st series, 20 vols., 1802–1813. Continued as *Mémoires du Muséum d'histoire naturelle,* 20 vols., 1815–1832.

the Revolution is immediately impressed by the quality of the staff. The men who taught in these institutions are famous among persons with scientific training, and their students are now being studied.[44] However, the persons in the middle of the hierarchy—the senior students and the junior staff, who have received less attention—also had a significant part in the establishment of science as a profession. Whether at the École polytechnique and the École normale, where their duties were largely teaching, or at the Muséum and the Bureau des longitudes, where they were mainly research assistants, their posts have a fundamental significance in any assessment of the formation of the idea and structure of a scientific career. These young men received a salary and had the benefit of association with the most eminent scientists of France. In an establishment of higher education, they could deputise for the professor in his absence, thus acquiring valuable experience. Many aspiring scientists were called upon in this way. These assistants were paid to do scientific work and they were obvious candidates for any vacancy which might arise in the scientific hierarchy. At a school like the École polytechnique, concerned with producing military and civil engineers and administrators as well as scientists, it was important for the future of science that some of the more able students should be retained at the school in a scientific atmosphere for further training, so that their basic scientific education should not be " lost " to more immediate and practical concerns. The fact that they received a small salary meant that they did not have to spend their time doing work unrelated to science in order to earn a livelihood.

Before the Revolution it was always theoretically possible for the bottle-washer to succeed his master, but there was an immense gulf between the two. Such a succession would entail a very long period of close personal relationship. Thenard,[45] who began his scientific career as bottle-washer to Fourcroy, was very much the exception in a generation which included scientists like Gay-Lussac, Biot and Arago, who all underwent formal courses of training. A major criticism of entry as a servant was that only occasionally did it develop into a system of apprenticeship. It was not an education but a substitute for one.[46] To speak of " day-release " at this time might seem a terrible anachronism, yet if it were possible for a youth to obtain a general or scientific education at the same time as helping in a laboratory, he would have the basic requirements to become a scientist.

[44] Bradley, Margaret, *The École Polytechnique: 1795–1830: Organisational Changes and Students* (M. Phil. dissertation, University of Leeds, 1974).

[45] Thenard, Paul, *Un grand francais: le chimiste Thenard, 1777–1857* avec introduction et notes de Georges Bouchard (Dijon: Jobard, 1950).
Although Thenard began under the old apprenticeship system, he was able to take advantage of the new organisation of science in as much as he was appointed *répetiteur* at the École polytechnique in December 1798, no doubt through the good offices of Fourcroy. Thenard thus represents a transitional case in the recruitment of scientists.

[46] For an example in the Napoleonic period of a laboratory assistant who was encouraged to take university examinations and become a professional scientist, see the case of J. E. Bérard (1789–1869) in Crosland, Maurice, *op. cit.*, 1967, p. 134.

Such an opportunity was provided by the École polytechnique, which appointed boys of 14 and upwards as laboratory assistants (*aides laboratoires*) to help the students with their practical work.[47] The École polytechnique was one of the first institutions to offer practical work in chemistry for students.[48] This called not only for teaching laboratories but also for staff. There were at least 10 laboratory assistants but they were not required to be on duty all the time. It was intended that these young men should " find in this service a means of instruction ",[49] and several hours each week were set aside for the study of mathematics. Although the boys were to be selected on grounds of intelligence, the position was in fact considered a valuable enough training by both deputies and members of the staff of the school for them to solicit such posts for young relatives or friends.[50] Unfortunately, in 1797 a short-sighted effort at economy abolished the 10 established posts of laboratory assistant.[51]

The creation of such posts was an enlightened step in the history of scientific education, but it could only provide basic instruction. Some boys managed to benefit from the École at a lower level than the entrance examination, but how did it provide opportunities for postgraduate education for the better students at the end of their course? When in 1798 the ambitious original programme of a three-year course was cut to two years as a measure of economy,[52] the possibility of remaining a third year was left to several categories of students including " those who having done two years' work wish to devote themselves to the study of a particular science of their choice...", and " students who, although not wishing to attach themselves to a public service,[53] wish to increase their knowledge of science and technology...". Because the extra year was a special privilege, students wishing to stay were required to take yet another competitive examination, and were also liable to be called upon to help in the education of the more junior students.

From the foundation of the École polytechnique there were to be section leaders or *chefs de brigade*, senior students who were rather like monitors since they were responsible for aiding the other students academically, keeping the register of attendance, and helping generally to maintain discipline. At first it was intended that these should be graduates of the school, but, as there were no graduates in 1794, some of the senior and more able students—Malus and Biot among them—were given a concentrated course of preliminary training to enable them to become *chefs de*

[47] Fourcy, A., *Histoire de l'École polytechnique* (Paris: chez l'Auteur à École polytechnique, 1928), pp. 56–57.

[48] Smeaton, W. A., " The Early History of Laboratory Instruction in Chemistry at the École Polytechnique, Paris and Elsewhere ", *Annals of Science*, X (1954), pp. 224–233.

[49] " Organisation de l'École polytechnique ", 30 ventose an 4, Titre IV, Art. XVIII, *Journal de l'Ecole polytechnique*, cahier 3, Prairial an 4.

[50] Fourcy, *op. cit.*, p. 92.

[51] *Ibid.*, p. 109.

[52] *Ibid.*, p. 137.

[53] *E.g.*, the École des ponts et chaussées.

brigade, their final appointment depending on votes by the senior students themselves.[54] The monitorial system did not work well, particularly with regard to discipline.[55] Nor was it satisfactory academically, since the section-leaders were spending so much of their time helping other students with their work instead of pursuing their own studies. It did provide, however, something of a stage in the scientific career, not only because of the duties of the post but because it provided an additional source of income.[56]

However, in 1798 the council of the École polytechnique took a step which, although intended partly to contribute to the instruction of students, had its most important effect on the careers of the more able students of the school. The step was the introduction of the post of *répétiteur*—sometimes translated inelegantly as " repeater "—to go over the work in mathematics with the students.[57] These were to be annual appointments given, as far as possible, to former section-leaders who wished to take up teaching as a career. One of the first of the two such appointments was given to Francoeur, one of the original students of 1794 who eventually became a mathematics professor at a Paris *lycée*. It was also decided to change the permanent post of demonstrator—*préparateur*—in chemistry to a similar annual appointment of two assistant demonstrators, preferably former students who were interested in practical chemistry. They recruited Desormes—a student of 1794—and also Thenard, who had not had the advantage of being a student there.

The post of *préparateur* for chemistry was soon re-named *répétiteur,* and a post was also created for physics.[58] Thus, mathematics, physics, and chemistry each had a junior post with an annual salary of 1,500 francs. Although the appointments were intended to be annual, the appointees were re-elected each year until 1804 when other vacancies arose.[59] The two *répétiteurs* for mathematics moved on to the more senior position of entrance examiners; Thenard was appointed to the chair of chemistry at the Collège de France, and Desormes resigned to set up in business as a chemical manufacturer. The vacancies so caused gave a first chance to other bright young men. We thus find Ampère, formerly holding a *lycée*

---

[54] The first *chefs de brigade* were taken from former students of the École des ponts et chaussées, the École des mines, or on the basis of their marks in the entrance examination. In 1794 the 50 senior students were called *aspirans-instructeurs*; it was the intention to divide them into two groups, the most mature becoming *chefs de brigade* immediately while the remainder might do so later. It was Monge who decided that the 50 students themselves should vote for the *chefs de brigade*: Fourcy, *op. cit.,* pp. 60–61.

[55] *Ibid.,* pp. 161, 241–242.

[56] Under the Directory, students usually received 360–500 francs per annum. For section-leaders this increased to 700 francs: Crosland, Maurice, *op. cit.,* 1967, p. 201.

[57] Fourcy, *op. cit.,* pp. 159–160. Under the *ancien régime* the use of advanced students to help in teaching had been introduced at the École du corps royal du génie at Mézières and at the École des ponts et chaussées. See Artz, F. B., *The Development of Technical Educations in France, 1500–1850* (Cambridge, Mass.: Massachusetts Institute of Technology Press, 1966), pp. 84, 100.

[58] *Ibid.,* p. 257.

[59] *Ibid.,* pp. 272–273.

post at Lyon, given the opportunity of a post in the capital. Gay-Lussac, who had already been ear-marked for a future vacancy with the title of *répétiteur-adjoint*, became a full *répétiteur* for chemistry. His duties were of two main kinds. As the title indicated, he was expected to help students both with difficulties in lectures and in practical work. But he was also the demonstrator for lectures, setting up apparatus and acting as store-keeper. He thus acquired good practical experience in an academic environment and when Fourcroy, the professor of chemistry, died in 1809, Gay-Lussac seemed the obvious successor to the post.

Other professors in the first half of the nineteenth century at the École polytechnique, who began as *répétiteurs*, include the mathematicians and physicists Poisson, Cauchy, Mathieu, Liouville, Savary and Regnault. Of course promotion was by no means automatic. Auguste Comte was able to obtain the post of *répétiteur* but not a higher one, despite his many solicitations. Le Verrier was promoted not within the school but to the Observatoire, of which he became, in time, a most distinguished director.

By the 1830s the *répétiteur* was generally recognised as a grade in the scientific career. An American visitor who made a study of the French educational system at this time remarked: " The utility of this latter class of teachers is well established in France, and they are found in every institution in which lecturing is practised...." [60] He went on to explain that this system had a double advantage. As it was clearly understood that the professors only had to deliver formal lectures—and not actually to teach students—eminent scientists were willing to accept chairs. The student was the richer both in consequence of hearing lectures by experts and also by being questioned on the substance of the lectures by the *répétiteur*. The *répétiteurs* also benefited. It was reported that " young men of talent seek the situation of *répétiteurs* as the best method of showing their particular qualifications, and the most certain road to a professorship ".

Of the many careers followed by graduates of the École polytechnique, one institution which offered only a few opportunities of entry but which was high in quality was the Bureau des longitudes. In the course of a discussion with Cassini in the opening years of the nineteenth century, Delambre made the following observations, which are of some importance in our study of the structure of careers in French science:

Is not the École Polytechnique a forcing ground [*une pépinière*], where one will always find students talented in mathematics and physics to fill the positions of *adjoint* [at the Bureau des Longitudes. M.C.]. But in spite of the confidence which must be inspired by the ability of these students, they are still required to justify their admissions. They enter the Observatoire as assistants. There they make observations and calculations, and when their talent is assured, they proceed to the rank of *adjoint*, which gives them a living. The *adjoints*

[60] Bache, A. D., *Report on Education in France* (Philadelphia: Lydia R. Bailey, 1839), pp. 568–569.

VI

have the chance of becoming full members [of the Bureau. M.C.] which finally determines their future, and provides them with double their salary, as well as the right of being chairmen at the meetings of the Bureau and of voting on all its decisions.[61]

We find that the post of astronomer at the Bureau des Longitudes—a senior post with the considerable salary of 6,000 francs—was consistently filled from the ranks of the assistant astronomers. Lefrançais Lalande, Bouvard, Burckhardt, Biot and Arago, who were the first five young scientists to hold posts as assistants, were all in turn promoted to the post of astronomer.

At the Muséum d'histoire naturelle, the post of assistant naturalist, with a curatorial function, was created very early in its new post-revolutionary organisation. In June 1794 the Committee of Public Safety agreed to establish, in addition to the professorial posts, two additional posts described as *naturalistes conservateurs*.[62] Their duties were related particularly to the arrangement and conservation of specimens. They were to be responsible to the professors, and they were to have a salary about half that of a professor. When this suggestion was approved, the professors asked for three posts at a slightly lower salary. The Committee of Public Safety[63] agreed, and the title of assistant—*aide-naturaliste*—began to be used for these three posts, soon be be increased to four.

It might have been possible to have increased the number of assistants gradually without any great deliberation about their precise function. However, certain events in 1803 were to lead not only to an increase in the number of such assistants but to an appraisal of their value, and in particular of their place in the structure of a career in the natural sciences in France. It began when the professors of chemistry, Fourcroy and A.-L. Brongniart, at a meeting on 1 June, 1803, made a case for an additional junior post for chemical analysis. This would be a new post of assistant chemist (*aide-chimiste*), comparable to the assistant naturalists, but he would be " attached to the professors of chemistry and responsible in particular for the analysis of different natural history specimens ".[64] It was agreed that special funds should be sought for this post. However, this brought the whole question of assistants under discussion, and at what

[61] Bigourdan, M. G., " Le Bureau des longitudes: Son histoire et ses travaux de l'origine (1795) à ce jour ", in *Annuaire du Bureau des longitudes*, 1928, A 39, quoted by Cawood, J. A., in *The Scientific Work of D. F. J. Arago (1786-1853)* (Ph.D. thesis, University of Leeds, 1974), pp. 8-9.

[62] Aulard, F. A. (ed.), *Recueil des Actes du Comité de Salut Public*, vol. XXIV (1901), p. 153 (17 Prairial an 2 = 5 June 1794). Under the *ancien régime*, at the Jardin des plantes the position of demonstrator had been recognised in addition to that of professor, but this was primarily—as in Renaissance faculties of medicine—a distinction of function between the practical and the theoretical.

[63] *Extrait des Registres du Comité de Salut Public* (26 Prairial an 2), discussed at a meeting of the Muséum professors on 28 Prairial (Archives nationales AJXV, 577).

[64] Archives nationales AJXV 590, Séance du 12 Prairial an XI, Lettre au ministre, 15 Prairial an XI.

must have been a lively meeting it was agreed to establish a committee [65] to report on the question of number, duties, and salaries of assistants. The number required might depend not only on the development of a particular science but the collections in that department and the needs of teaching.

The report of this committee is an important document in the development of the profession of science in France since it analyses in detail the professional role of assistants in the Muséum. It first pointed out that the title of *aide-naturaliste* had hitherto been given to young men of quite different educational background and function. There were those whom we would call scientists and a second group whom we would call laboratory assistants or technicians. In the words of the report: " The first group are true naturalists, whose work is almost entirely scientific; the others are technicians [*des artistes*] whose talent, although undoubtedly precious, does not however presuppose any formal education." The committee recommended that only the former group should be called assistant naturalists. If proposed the creation of two additional positions in this category.[66] The report continued:

Together with the three existing posts and with that which has just been created for chemistry, these two assistants would form a kind of intermediate body between the professors and the other employees, an honourable training ground [*une pépinière*] where excellent naturalists will be produced and where [future] professors can with advantage be recruited.

With the acceptance of this report by their colleagues, the designation " assistant naturalist " was no longer a miscellaneous description among the lesser employees, gardeners, and wardens of the Muséum; it had become a rank. From then on staff-lists of the Muséum gave first the professors, secondly the *aides-naturalistes*, and thirdly the miscellaneous staff required by the Muséum. The post of assistant naturalist was to have a tenure of five years, but in compensation for its temporary nature it now had a definite status, a position in the hierarchy calling on definite skills and providing valuable experience. In the early nineteenth century we find among their ranks Valenciennes, Laugier, Audouin,[67] and Isidore Geoffroy Saint-Hilaire,[68] all future professors at the Muséum where they had been trained. Promotions were by no means automatic but for the naturalists mentioned and others it was a definite grade in the career of

[65] At the meeting of 19 Prairial (8 June, 1803), it was agreed that the committee should consist of Lacépède, Lamarck, Geoffroy Saint-Hilaire, and Cuvier—all naturalists. However, the committee which reported also included the chemist Fourcroy. The committee was thus an impressively large one, consisting of most of the leading figures at the Muséum.

[66] With responsibilities for quadrupeds and birds, reptiles and worms, respectively.

[67] Jean-Victor Audouin (1797–1841) was one of the founders of the Société d'histoire naturelle de Paris in 1822. In 1830 he replaced Latreille as assistant naturalist at the Muséum and, on Latreille's death three years later, he succeeded him as professor of zoology.

[68] Isidore Geoffroy Saint-Hilaire (1805–61), the only son of his more famous father Etienne, was engaged as *aide-naturaliste* at the age of 19. He succeeded to his father's chair at the Muséum in 1841.

those who followed the new profession of science in nineteenth-century France.

A fourth institution which cannot be omitted in any discussion, however brief, of science in nineteenth-century France is the École normale, which in the Restoration became the leading institution to prepare for the competitive examination of *agrégation*; this was the qualification of the more highly paid teachers. As the qualifications for *lycée* and university posts were raised, what was needed in science now was not so much an inducement to keep a graduate from the École polytechnique or a graduate with a *licence* in a suitable atmosphere of scientific research, but to do this for someone with the higher degree of *agrégation*. This meant the creation of another research position, that of the *agrégé préparateur*, which might be translated into English as graduate demonstrator. At the École normale, Pasteur became the first *agrégé préparateur* (1846–48),[69] and the position of demonstrator was henceforth reserved for graduates of the school who stayed on to prepare their doctoral theses in association and consultation with the professors of the school.

Pasteur later wrote in enthusiastic terms to the Minister of Education:

I know by experience how much the leisure of these modest positions are worth for a young man who has been touched by the fire of science ... [He benefits from] an atmosphere of healthy study amongst well-endowed laboratories, under the benevolent direction of proven masters.[70]

The position of *agrégé préparateur* was really an extension of the post of *répétiteur* in the École polytechnique in the early nineteenth century. Both had responsibilities for helping more junior students with practical work and thus justified their small salaries in the eyes of the Ministry of Education. But they were also in immediate contact with the professors who could guide them with their research. The post of *agrégé préparateur* raised the demonstrator to a new level. It kept him in higher education and provided conditions which enabled him to obtain the qualification to become a university scientist for life.

In the 1830s, Dumas proposed that there should be a system for the attachment of *agrégés* to the faculty of science.[71] It was about time, he felt, that the faculties took responsibility for guiding research for the doctorate. Positive steps in the establishment of a sequence of stages of a career in French science were taken in the 1840s with the creation of the post of *agrégé préparateur* at the École normale; in the 1850s with encouragement for senior students at the École normale to stay on and do

[69] Dulou, R. and Kirrmann, A., " Le laboratoire de chimie de l'École normale supérieure ", *Bulletin de la Société des amis l'Ecole normale supérieure*, No. hors série (September 1973), p. 8.

[70] Pasteur to the Minister of Public Instruction, 24 October, 1858, Vallery-Radot, P., *Correspondence de Pasteur, 1840–1895* (Paris: Flammarion, 1940–51), vol. II, p. 37.

[71] " Extrait des proces-verbaux des deliberations de la Faculté des Sciences ", Seance du 15 Novembre 1837, Greard, O., *Education et Instruction, Enseignement Superieur*, 2nd ed., (Paris: Hachette et Cie, 1889), pp. 242, 248.

56

their doctoral research [72]; and finally, in the 1860s, with the establishment of the École pratique des hautes études.[73] The École pratique is sometimes seen as atypical of the French scientific scene—a belated move in 1868 after half a century of neglect. Professor Ben-David calls it " the first [French] experiment in post-graduate training." [74] I believe that a closer study will show it merely as a logical conclusion and consolidation of a system which had been implicit in the French higher educational system since the very beginning of the nineteenth century. Unfortunately it was not sufficiently general; it depended on private as well as state laboratories, and until the Second Empire it was not part of any national plan. The main weakness in the structure of the career in natural sciences was that whereas the *grandes écoles* were able to attract the best students, they were not connected with the network of faculties of science with their many teaching posts and the obvious possibility of a career within them.[75] Only the École normale cut across and joined these two systems. The École normale thus became increasingly important in the opportunities for a scientific career in France; it enabled French scientists, at the same time, to benefit from the prestige and conditions of a *grande école* and to obtain the qualification necessary for employment in the faculties.

Britain and France

The professionalisation of science in Britain came later than in France. In a sentence which has often been quoted, Charles Babbage in 1851 could write that " science in England is not a profession; its cultivators are scarcely recognised even as a class ".[76] I have tried to show here how, half a century earlier, in France science had already become a profession. It was a profession with its own specialised journals, so much so that a French writer in the 1820s could complain that in his country " the literary journals seldom concern themselves with the sciences ".[77] This situation contrasts with Britain, where periodicals such as the *Edinburgh Review* and the *Quarterly Review* gave serious treatment to science in a society where science was still a part of the culture of an educated middle class, rather than a specialised activity offering full-time employment and remuneration. Much reputable science in early nineteenth-century Britain was done in the setting of literary and philosophical societies,[78] where

[72] Law of 22 August, 1854, Ponteil, F., *Les institutions de la France de 1814 à 1870* (Paris: Presses universitaires de France, 1966), p. 452.

[73] Duruy, V., *L'Administration de l'instruction publique de 1863 à 1869* (Paris: 1878?), pp. 644–658.

[74] Ben-David, Joseph, *The Scientist's Role in Society: A Comparative Study* (Englewood Cliffs, N.J.: Prentice-Hall, 1971), p. 103.

[75] Gilpin, R., *France in the Age of the Scientific State* (Princeton: Princeton University Press, 1968), p. 112.

[76] Babbage, Charles, *The Exposition of 1851* (London: 1851), p. 189.

[77] Baron de Ferussac, *Bulletin général et universel des annonces et des nouvelles scientifiques*, vol. I (1822), Prospectus, p. 1.

[78] One of the most prominent of these societies was the Manchester Literary and Philo-

science was still formally associated with literature as polite culture rather than a specialised intellectual activity. It was said of the early nineteenth-century English geologist William Smith that "geology had kept him poor all his life by consuming his professional gains;" [79] for other British men of science, instead of "geology", one could write "chemistry", or "astronomy", or just "the pursuit of science". After John Dalton had published his atomic theory, he continued in nineteenth-century Britain to give the sort of tuition in elementary science and mathematics which reminds one of the situation in France under the *ancien régime*.

This almost menial employment of men of science contrasts with the situation in France after 1800. David Brewster gave the following account of the employment of British men of science:

> Some of them squeeze out a miserable sustenance as teachers of elementary mathematics in our military academies, where they submit to mortification not easily borne by an enlightened mind. More waste their hours in drudgery of private lecturing, while not a few are torn from the fascination of original research and compelled to waste their strength in the composition of treatises for periodical works and popular compilations. [80]

Thus, in some ways Britain in the 1830s invites comparison with France in, perhaps, 1780.

Of course Babbage and Brewster tended to exaggerate the lack of recognition of science in Britain. If there was a cultural lag, there were also important cultural differences, and the British tradition of individualism, local initiative, and independence from governmental control conferred benefits in other ways. When eventually science became a profession in Britain, it had a French model not necessarily to copy but to adapt to the contemporary state of science, British national traditions, and the different social conditions of Victorian Britain.

sophical Society, where John Dalton found encouragement and in whose *Memoirs* he published much important work.

[79] Walker, W., *Memoirs of Distinguished Men of Science of Great Britain, Living in 1807–1808* (London: 1862), p. 169, quoted by Cardwell, D. S. L., *The Organization of Science in England*, 2nd ed. (London: Heinemann Educational Books, 1972), p. 17.

[80] *Quarterly Review*, XLIII (1830), p. 327.

VII

Scientific Credentials:
Record of Publications in the Assessment
of Qualifications for Election to the
French Académie des Sciences

THE modern scientist working in a university is accustomed to supporting an application for an academic appointment with a *curriculum vitae* and a list of his publications. The list of publications is of major importance in any consideration for promotion. Whether appointive committees are more impressed by the quantity than the quality of publications is a question still disputed but no one challenges the idea that a list of publications constitutes the principal evidence of the qualifications of a scientist. This practice is now nearly universal in matters of academic appointment. It did not, however, begin in universities. Its first appearance was in fact in the Paris Académie des sciences, which was founded in 1666. Although membership before the French Revolution sometimes conferred a pension and, after 1795, always carried a modest honorarium, election to the Académie was not so much an admission to a salaried post as a mark of recognition of scientific eminence. This was particularly the case after the Revolution, when the members of a young scientific community competed for a strictly limited number of places. The choice of new members was not made by a minister of the government or a civil servant but by members of the Académie. Thus we are considering the assessment of scientific publications by scientists themselves; the criterion they applied was the quality of scientific achievement.

The practice of accepting publications of original research as the credentials of a scientist first arose in France shortly after the French Revolution. It soon became an accepted part of candidacy for election to the Académie and from there the practice spread to candidacies for chairs in institutions of higher education. It replaced personal recommendations and royal and private patronage by a procedure with some claim to objectivity. By offering reward for industry and merit it offered encouragement to scientists not only to undertake research but to publish it. It has been suggested that the pattern of scientific work and of the academic profession of science originated in the German universities in the nineteenth century.[1] Nevertheless, the idea of estimating the merit of a scientist by examining the whole range of his publications and the practice of linking rewards with publication of original research had an independent origin outside German universities.[2]

[1] *E.g.* Ben-David, Joseph, "German Scientific Hegemony and the Emergence of Organised Science", *The Scientist's Role in Society* (Englewood Cliffs, N.J.: Prentice Hall, 1971), pp. 108–138. See also *ibid.*, pp. 88–89, for a reference to the French claim.
[2] R. Stephen Turner discusses the relevance of publications (among other criteria) to appointments in

606

The lists of credentials prepared by candidates for election to the French Académie des sciences since the beginning of the nineteenth century have been largely overlooked. Nearly all the lists were printed for a very limited circulation within the Académie.[3] They were certainly never obtainable at the appropriate time from booksellers,[4] although more recently an occasional *Notice* submitted to the Académie is to be found listed in the catalogues of antiquarian booksellers.

The Académie des sciences and the Institut national

Under the *ancien régime* there had been a small number of journals, of which the most eminent was the *Mémoires* of the Académie royale des sciences, a compilation reserved for members of that body. Much important scientific work, however, was still published in book form. Some scientists might do original scientific work without ever publishing. As for reviewing a person's publications, such activities might be left to someone composing an obituary rather than be the concern of the scientist himself. A decision to seek publication was a purely personal matter rather than an accepted part of a man's professional life. Indeed there was scarcely anything like a "professional scientist" before the Revolution. The nearest approximation to a body of men accepting common standards and practising science was in the Académie.[5] Entry, however, did not depend on prior completion of a course of study and institutionally organised instruction in scientific subjects at advanced levels did not exist. Indeed, one of the differences between entry to the Académie in the eighteenth century and entry in the nineteenth century is that in the latter candidates for election were expected to have

German universities in the eighteenth century but points out that this was publication aimed at establishing a literary reputation in the public eye and not the publication of original contributions to knowledge. It was not till about 1840 that publication by a professor had come to mean the results of original research addressed to specialists. See *e.g.* "The Growth of Professional Research in Prussia, 1818 to 1848, Causes and Context", *Historical Studies in the Physical Sciences*, III (1971), pp. 137–182, esp. p. 170; also Turner, R. Stephen, "University Reformers and Professional Scholarship in Germany, 1760–1806", in Stone, Lawrence (ed.), *The University in Society* (Princeton, N.J.: Princeton University Press, 1975), vol. 2, pp. 495–531, esp. p. 522.

[3] When very occasionally mentioned, the origins of the lists have often been misunderstood. Thus it is sometimes said that they were published by the Académie. An obvious starting point in any systematic search for the *Notices* are the individual dossiers of *académiciens* in the archives of the Academy of Sciences. Unfortunately there are no *Notices* for unsuccessful candidates, but for the successful it is occasionally possible for the patient investigator to find successive versions of a particular scientist's *Notice*. The single most valuable source used in this study was a collection of documents relating to the Académie built up by the veterinary scientist, Jean-Baptiste Huzard (1755–1838) and preserved in the library of the Institut. There are also some *Notices*, catalogued under the name of the author, in the Bibliothèque Nationale. In England there is an interesting collection of some two dozen *Notices* individually bound in the library of the Science Museum. These are all catalogued under bibliography in the classification 012. In addition there are two small collections in the British Library catalogued as "Tracts" with the press marks 733.g.18 and 1502. 79. Some show that they were originally addressed to the *académicien* Alexandre Brongniart.

[4] A modern bookseller who has reprinted one of these *Notices* can therefore claim that it was essentially an unpublished document. See *Note de l'éditeur* in Painlevé, Paul, *Analyse des travaux scientifiques jusque'en 1900*. (Paris: A. Blanchard, 1967).

[5] See Hahn, Roger, *The Anatomy of a Scientific Institution: The Paris Academy of Sciences, 1666–1803*, (Berkeley: University of California Press, 1971). Professor Hahn has modified his claim for "professionalism" in the old Academy in "Scientific Careers in Eighteenth-century France" in Crosland, Maurice (ed.), *The Emergence of Science in Western Europe* (London: Macmillan, 1975), pp. 127–138 (a different version was published as "Scientific Research as an Occupation in Eighteenth-Century Paris" in *Minerva*, XIII (Winter 1975), pp. 501–513).

qualifications conferred by formal training, whereas in the eighteenth century a person could be elected to the most junior grade—originally called "pupil" (*élève*) and later "assistant" (*adjoint*)—and acquire "training" by associating with members of the Académie. In due course the *adjoint* might be promoted to *associé* and finally to the senior position of *pensionnaire*. It was only for these senior grades that the statutes of the Académie mentioned the qualification of publications:

No one can be proposed to His Majesty for the position of pensioner or associate if he is not known for some considerable printed work,[6] for some course given with brilliance, for some machine of his invention or for some special discovery.[7]

It is interesting that publication is mentioned as one criterion for promotion among several. Brilliant popularisations of science or inventions were regarded as alternative qualifications. In practice, in considering promotion to higher grades, some embarrassment was felt in assessing the rival merits of men who were in the position of colleagues. On the whole the Académie preferred promotion through seniority.[8]

The Académie was dissolved simultaneously with the abolition of other learned institutions during the French Revolution. A new period began in 1794–95 when many important institutions such as the École polytechnique and the École normale supérieure were founded. In place of the various Académies a new Institut national was created in December 1795. The union of the interests represented by the former Académies into one large body was an application of the principle of the *Encyclopédie* which had stood for the unity of knowledge. The Institut national was divided into three sub-groups or "*classes*". The Première Classe of "mathematical and physical sciences" (which included the biological sciences) corresponded to the former Académie royale des sciences. It was the largest section of the Institut, being composed of 60 members compared with 36 in the Seconde Classe (moral and political sciences) and 48 in the Troisième Classe (literature and fine art). There were further divisions within each *classe*, the Première Classe, for example, being composed of ten sections of six members representing such specialities as mathematics, astronomy, chemistry and botany. It was recognised that in much of their work the various *classes* would need to meet separately but it was decreed that elections should be by the Institut as a whole. A defect of this electoral system was that more than half the electors were likely to have no knowledge of the fields of the candidates. A poet might find himself deciding between the merits of two chemists or a physician might be asked to decide between two painters.

---

[6] "*par quelque Ouvrage considérable imprimé*".

[7] Article XIII. Règlement of 1699. A rare example under the *ancien régime* of a fully documented dossier on a candidate for election has recently been discovered in the archives of the Academy. It was the work of Lavoisier in support of the candidacy of Baron de Dietrich as *associé libre* in 1786. It included reports by other *académiciens* on three papers by Dietrich and a letter (in the third person) from Dietrich to Lavoisier, describing his scientific qualifications. Perrin, C. E., "A Lost Identity: Philippe Frederic, Baron de Dietrich (1748–1793)", *Isis*, LXXIII (1982), p. 549. The combination of a powerful patron and a good case won Dietrich the place.

[8] Rappaport, Rhoda, "The Liberties of the Paris Academy of Sciences, 1716–1785" in Woolf, Harry (ed.), *The Analytical Spirit* (Ithaca, N.Y.: Cornell University Press, 1981), p. 246.

However, the initial choice was made by the section, the group best fitted to assess the relative merits of the contributions of the candidates within a specialised field. A list of candidates in order of merit was presented to the Première Classe which would then vote. The three candidates receiving the highest votes would then be presented to the next monthly meeting of the whole Institut for their final choice. This situation encouraged the candidates to produce printed statements of their qualifications.

These lists of qualifications, intended for the eyes of the entire membership of the Institut, would obviously include some by non-scientists. One of the earliest, dating from 1799, was by the lawyer Moreau St-Méry, who produced an eight-page octavo pamphlet dated *"10 brumaire an 8"* (*i.e.* 1 November, 1799).[9] The greater part of the pamphlet was devoted to the enumeration of legal works, but remembering his audience, he also had a section headed: *Sciences, Arts, Littérature* in which he mentioned his writings on his travels and his corresponding membership of the Société d'agriculture de Paris. A printed document signed by the man of letters Thiébault and dated according to the republican calendar: *"1 messidor an 9"* (20 June, 1801) was in the form of a letter. The circulation of such documents was obviously far from an established custom and Thiébault was suitably apologetic:

In allowing myself, Citizen, to remind you of the decision (of nomination by the Grammar section of the *classe* of Literature) so honourable of me, my intention is not to approach you with a kind of solicitation of which you would not approve. Also I shall not talk to you either of the career which I have had, nor of the principles which I have always professed and followed, nor even of my works, of which incidentally a note has been left at the secretariat of the Institut . . .[10]

Thiébault then allowed himself to mention one of his books: a treatise on style. When Gail, professor of Greek literature at the Collège de France, stood for a vacancy in the class of moral and political sciences and found himself one of the three final candidates recommended to the whole Institut, he drew up a one-page summary of his qualifications, which referred to ten published works.[11]

In the scientific section of the Institut, the first mention in the minutes of qualifications of candidates came as early as 1796 but there was no mention yet of the submission by candidates of their lists of publications. The relevant minutes read as follows:

On the motion of one member, the assembly decrees that its secretary should draw up a list of all the scientists [*savans*] who present themselves for a place as member or associate [*i.e.* correspondent] of the Class of Physical and Mathematical Sciences and that this list will include a statement of the qualifications [*titres*] which they claim.[12]

The first list of scientific qualifications seems to have been drawn up in

---

[9] *Note des travaux*, Institut, HR16**, No. 10. Nearly all the *Notices* published were of approximately quarto format. Where they were smaller or larger this is noted.

[10] Institut, HR5*, Vol. VI, No. 35.

[11] *Exposé des titres du Citoyen Gail* . . . n.d. Institut HR5*, Vol. III, no. 30.

[12] *Procès-verbaux des séances de l'Académie des sciences*, (10 vols., Hendaye, 1910–22), Vol. I, p. 25, 26 *Germinal an 4* (15 April, 1796).

1798 when there was a vacancy in the chemistry section, following the death of Bayen (1725–98). Altogether there were eight candidates but when the Première Classe voted, the three with the greatest number of votes were the chemical manufacturer and disciple of Lavoisier, Chaptal (1756–1832), a mineralogist of the old school, Sage (1740–1824) and a pharmaceutical chemist still loyal to the phlogiston theory, Baumé (1728–1804). Chaptal proudly recalled in his memoirs that he was elected without having anything printed on his behalf.[13] But Sage felt less secure and had a two-page octavo pamphlet printed. In this, he mentioned his various publications but he paid greatest attention to his career, for example, as the founder of the École des mines in 1783. He was also particularly bitter about his imprisonment during the Revolution and his lack of recognition[14] in the world of science after the Revolution.

The special circumstances in which specialists in one field were asked to decide on the relative merits of specialists in quite different fields would itself have been a sufficient reason to engender a whole new kind of literature: the *Notice des travaux* ... This electoral situation came to an end in 1802 when the Institut was reorganised, giving each *classe* full control over its own elections.[15] The *Notice* might have been no more than an ephemeral ripple on the surface of the internal organisation of the Institut, if it had not been for a second feature of the elections. The Première Classe found itself at the time of each election inundated with *"mémoires"* from prospective candidates. The candidates realised that the best way of enhancing their chance of election was to present a memoir to the *classe* to which they sought election. In this way their names would come to the attention of the *académiciens* and the Académie was expected to nominate a commission to appraise the memoir. There could be no better recommendation for a candidate than a favourable report by members of the Première Classe presented immediately before an election. Such a procedure, if employed by a large number of candidates, could, however, disrupt the smooth functioning of the *classe* and in May 1799 it was agreed that no candidate for a vacancy should be allowed to read a memoir before the election. A second resolution read:

In order that the candidates should be able to avoid the visit which they think they have to make to members of the Institut, the *Classe* decrees that the senior clerk of the secretariat of the Institut will keep a list of candidates who call there and he will receive a note stating the works which they claim to have written.[16] A list of the names will be read to the *Classe* as well as the accompanying notes (of publications).[17]

[13] Chaptal, Jean-Antoine, *Mes Souvenirs sur Napoléon*, (Paris: Librairie Plon, Imprimerie de l'Observatoire d'Attadia 1893), p. 53.
[14] Sage was finally elected to the mineralogical section in 1801.
[15] Decree of *3 pluviose an 11* (23 January, 1803). The Deuxième classe of the Institut was abolished as potentially subversive of the Napoleonic state and its members were divided between the other *classes*. On the Deuxième classe see Staum, Martin, "The Class of Moral and Political Sciences, 1795–1803", *French Historical Studies*, XI (1980), pp. 371–379.
[16] "*la note des ouvrages qu'ils déclarent avoir composés*".
[17] *Arrêté concernant les candidats aux places vacantes dans la classe, 18 floréal an 7* (7 May, 1799). Aucoc,

610

Both of these decisions soon fell into abeyance. Candidates continued to read *mémoires* and by October 1800 an election in the botany section was preceded by the feverish reading of *mémoires* by candidates as if there had never been a regulation proscribing it.[18] Nor did candidates often draw up lists of their publications; if they did, even for a short time, there is no trace of these manuscript lists surviving in the archives of the Académie.

Such a resolution soon became unmanageable. The best candidates had published so much that members of the Première Classe preferred to see the list of their publications rather than hear the list read out. Indeed one member of the Première Classe had previously proposed that a list of publications—described as *titres littéraires*—should be posted in their meeting place so that members could read it.[19] This proposal was not accepted and the only way that candidates could have their publications seriously considered was to have copies made and to present a copy to each member of the Institut. (Printing costs were quite low throughout the nineteenth century.) The regulation about presenting a list was, like so many other early decisions, soon forgotten. The early years of the Institut were indeed years of administrative experimentation. The important thing as far as this paper is concerned is the persistence of a literature continuing throughout the Napoleonic period, even if it was only in the 1820s that the practice was revived and extended so that it became by custom (though not by regulation) the rule rather than the exception. In so far as the *Notice* was compiled for a specific occasion, well understood by all the recipients, it was often not dated.[20] The type of publication we are examining was an ephemeral one but it presented the achievements of a scientist as seen by himself in a form intended to appeal to a select panel of judges. After the first few years this panel consisted exclusively of scientists.

The Early Notices

There was a change in the *Notices* supporting candidates which marked a change in what were considered relevant qualifications. In the very earliest *Notices* seniority was often stressed. Thus Sage in 1798 says he had been teaching chemistry[21] and mineralogy for 40 years. Also he had first been elected to the Académie royale of the *ancien régime* in 1771 "two years before Citizen Baumé", his rival. He also claimed that Chaptal, the other candidate, had been a former student of his. Jeaurat, campaigning for a position in the Bureau des longitudes, described himself as the *doyen* of

---

L., *L'Institut de France. Lois, status et règlements concernant les anciennes Académies et l'Institut de 1635 à 1889* (Paris: Imprimerie nationale, 1889), pp. 168–169.

[18] The election on 23 October was preceded by the reading of memoirs by the three leading candidates: La Billardière, Beauvois and De Candolle, together with the presentation of reports by commissions of the Première classe on other memoirs they had written. *Procès-verbaux*, Vol. I, pp. 241–62.

[19] *Ibid.*, Vol. I, p. 395 (26 *floréal an 6*) (15 April, 1798).

[20] Sometimes the date can be deduced from the title, indicating that it related to a vacancy caused by the death of X. Occasionally the approximate date has to be inferred from the date of the last publication listed.

[21] His spelling of this science as "*la Chymie*" is a reminder that he adhered to an older tradition, in opposition to the new chemistry of Lavoisier.

French astronomers who had been publishing for 50 years.[22] The former naval engineering officer, the Marquis de Chabert (1724–1805), aspiring to a place in the Première Classe, pointed out that he had read a paper to the Académie royale as long ago as 1748. He wrote:

General Chabert, octogenarian and almost the *doyen* of the former Académie des sciences, attaches a very high regard to the honour of returning to its bosom, since the Première Classe of the Institut represents it. He dares to hope for the benevolence of the members of this *Classe*, of the majority of whom he has so long been proud to be a colleague. It would be the most flattering recompense of his work and the sweetest satisfaction at the end of his career.[23]

The Première Classe, however, did not see itself as having any necessary obligation to reinstate former members of the Académie royale in their retirement. Even in 1823, by which time one might have expected the norms of the new professional science to have been fairly well accepted, Armand Séguin (1767–1835), standing as a candidate in the chemistry section of the Académie, began his solicitation: "I have the honour to send you a summary of my scientific work over a period of 36 years. The first memoirs I read to the Académie date from 1786 . . ."[24] He did not neglect to point out that he was a former collaborator of Lavoisier, but the members of the Académie were really more interested in what he had achieved in the nineteenth century. They decided he was not a suitable person to replace the eminent chemist Berthollet. Gastellier, an elderly provincial physician applying in 1821 for a vacancy in the medicine section, submitted a *Notice chronologique de mes ouvrages*,[25] which listed publications in a very prominent chronological order, making it clear that he had been publishing material since 1771. Once again the length of the candidate's career was not, in itself, considered grounds for election.

The *académiciens* laid greatest weight on commitment to science. Under the *ancien régime* dilettantish interest in science had been quite common; in the Institut the amount of time and trouble devoted to scientific work began to be regarded as more important. Thus Palisot de Beauvois, aspirant to the vacancy in the section on botany caused by the death of L'Heritier in August 1800, spoke of the 18 months he had spent in the worst possible climate of Africa on a scientific expedition. "His correspondence with the Académie through Citizen Jussieu as intermediary, his sending seeds to the Jardin des Plantes and his collection of insects and plants are proofs of the useful employment of his time."[26] Palisot mentioned that he had read six memoirs to the Institut since his return. His *Notice* referred to the presentation of *mémoires* both to the former Académie royale and the Première Classe, but did not list a single publication. While Palisot had braved the tropics in the service of science, Patrin had suffered the other extreme of climate. In 1777

[22] *Notice succincte des ouvrages de sciences*, n.d. (probably 1802) folio, 2 pp. Institut HR5 (no asterisk), Vol. I, No. 10.
[23] *Notice sur les travaux de Joseph Bernard Chabert*, n.d. Institut HR6*, Vol. XLI, No. 33, (16 pp.), p. 9.
[24] Printed letter from Séguin to Huzard, 21 January, 1823, Institut HR5*, Vol. LV, No. 22.
[25] Institut HR5*, Vol. LIII, No. 27, n.d. (32 pp.).
[26] *Aux membres de l'Institut National*, n.d. [1800] B.M. 733.g. 18 (51) 2 pp.).

612

he had gone voluntarily to Siberia and stayed there for six years. As a result of his privations, he had been able to bring back to France a collection of new minerals which he had described in the *Journal de physique*. He concluded: "I have sacrificed the major part of my patrimony for science; the Revolution has devoured the rest of my money, I rely on the justice of the members of the Institut."[27]

The adduction of important social connections had not been uncommon in the *Notices* of candidates in the early part of our period. For example, the *Notice* of Jean Pierre Joseph D'Arcet (1777–1844) not only mentioned his employment as *inspecteur général* at the Hôtel des monnaies de Paris, his membership of the Légion d'honneur and of the editorial board of the *Annales de Chimie*, but also pointed out that he was "Grandson of the celebrated Rouelle, great-nephew of Hilaire Marin Rouelle and son of Jean D'Arcet, former member of the Académie des sciences who died in 1801, being a member of the Institut and of the Senat."[28] He must have been aware that social status and connections would not be sufficient since he mentioned that he had obtained his first appointment at the Monnaie in open competition. His candidacy failed and when he applied again in 1823, the family and social connections were drastically cut and prominence was given to publication in the main scientific journals.[29]

The new criterion of scientific qualifications did not replace the old criteria immediately. The candidates' list of scientific publications did not at once become the sole or even preponderant criterion. One of the oldest scientific dynasties in France was that of the Cassinis, Jean-Dominique Cassini (1625–1712) having been brought from Italy to Paris in 1666 as one of the number of distinguished foreigners who were to grace the early Académie royale des sciences. There were four generations of Cassinis in French astronomy, the fourth Cassini being the director of the Observatoire up to the Revolution. When Henri Cassini, the fifth generation, stood as a candidate in 1827, his father wrote in a confidential letter to a colleague: "... I therefore make bold to ask you to examine in all justice if the unique fact in the history of letters of a devotion to science supported by five successive generations over more than 170 years, should not add some weight in the balance when you come to weigh the scientific qualifications which my son presents, and whether he would not deserve some preference on your part over his competitors."[30] Henri Cassini did not, however, neglect to support himself with a list of publications.[31]

The older criteria were deeply rooted and many candidates would not bow

[27] *Notice des travaux de Patrin*, n.d. Institut HR5*, Vol. VII, No. 25.

[28] *Notice des travaux de J. P. J. D'Arcet* (14 lines of titles, etc.), n.d. [1816?] Institut HR1 (no asterisk), Vol. XXI, No. 30.

[29] *Notice des travaux de J. P. J. D'Arcet* (11 lines of titles, etc.), dated January 1823. Institut HR5*, Vol. LV, No. 23.

[30] Letter from Cassini to Huzard, n.d. but marked *"reçu le 4 mai 1827"*. Institut HR5*, Vol. LXVIII, No. 10a.

[31] The list to support his candidacy for the botanical section in 1820 has been found: Institut HR5*, Vol. LI, No. 16.

before the new one. Many of them were extremely reluctant to prepare and circulate such bibliographical lists. Some thought that their qualifications were self-evident and that it would be derogatory to their status to submit a list of their publications. This was particularly true of the more illustrious candidates, and Chaptal later boasted that he was elected in 1798 despite the fact that he had refused to have anything printed on his behalf similar to what had been prepared by other candidates for the vacancy.[32] The veteran astronomer, Jeaurat, who was elected to the Institut in 1796 and who spent the period from then until his death in 1803 vainly trying to obtain a post in the Bureau des longitudes, was prepared to draw up a list of his qualifications for such a position,[33] but reminded his friend Huzard in January 1802 that his demands to be admitted to the Institut had only been made orally and not in writing.[34] The astronomer Alexis Bouvard (1767–1843) prepared a two-page *Notice* of his work[35] to support his candidacy for election to the Institut in December 1801 but concluded with the following apology:

Such is the work that has occupied me in the last nine years or so that I have been attached to the National Observatory. It was with the greatest regret that I was forced to render this notice public. But my reputation is compromised by harmful gossip about me which is circulating at the present moment. I owe it to my own honour, to learned men, and to the support of the class of physical and mathematical sciences of the Institut national which has done me the honour of placing me first on its list, to prove that my conduct during the Revolution has in no way rendered me unworthy of the favour which it has bestowed on me.

To confirm the propriety of his behaviour during the Revolution, Bouvard included in his *Notice* the text of a letter written a few days earlier by the former director of the Observatoire, Dominique Cassini (1748–1845), which said that Bouvard was in no way to blame for Cassini's exclusion from the Observatoire during the Revolution. In other words, although Bouvard's *Notice* is concerned largely with the career and contribution to astronomy of its author, he only felt it necessary to provide this documentation because recriminations were being made about the earlier treatment of the royalist astronomer Cassini, who obviously had friends within the Première Classe. In fact, Bouvard was not successful in this election but was more fortunate 16 months later.

Even in the 1820s, Blainville explained that it was his friends who had persuaded him to have a list of his qualifications printed to support his candidacy.[36] Finally there is the case of Joseph Pelletier (1788–1842), a persistent candidate for election. His *Notice* submitted in 1832, began:

In publishing a *Notice* about my work at a time when I present myself to the Académie des sciences as a candidate for a vacancy in its midst, I feel a keen embarrassment. It is true that I am only following a usage which has been introduced

---

[32] Chaptal, Jean-Antoine, *op. cit.*, p. 53.
[33] *Notice succincte des ouvrages de science* . . . n.d. folio. Institut HR5 (no asterisk), Vol. I, No. 10 (2 pp.).
[34] Letter of *11 pluviose an 9*, Institut HR5*, Vol. VII, No. 27.
[35] *Notice des Travaux Astronomiques* . . .·n.d., 4to, Institut HR5*, Vol. VII, No. 18.
[36] *Notice analytique sur les travaux physiologiques et zoologiques de M.H.M. Ducrotay de Blainville*, n.d. [1825?] Institut HR5*, Vol. LXIII, No. 40.

among us for some time and which will finish by being altogether one of our customs, but it is always painful to have to speak of oneself. On the one hand one fears not to make the most of oneself in all respects in so important a cause. On the other hand one is afraid of being accused of vanity or at least of egotism because one tries to obtain appreciation for one's work.[37]

Quite apart from inhibitions of modesty and some degree of shock in response to criteria which stressed scientific achievement exclusively, some prospective members of the Institut did not want to bother their prospective colleagues with an exhaustive list of their publications and sometimes they claimed they were presenting only a summary. Even if the list were reasonably complete, it was a gesture of deference to the members of the Première Classe in the early days to confine any list to one sheet of paper. Most would be printed on both sides and some would resort to the use of small type to compress as much information as possible on to the conventional single sheet. This provides an interesting contrast with the practice established by the 1820s of preparing a booklet. Margins became wider and by the second half of the nineteenth century a table of contents was quite common.

The standard account of elections to the Académie in the nineteenth century stresses the *visite*, in which candidates humbly went around to the private residence of the electors to solicit their votes in the forthcoming election. Certainly this practice was widespread but it was not necessarily indifferent to criteria of scientific merit. Often candidates would call on *académiciens* ostensibly to present a list of their publications. At the same time they might of course try to ingratiate themselves but the bibliography was not insignificant. In some cases the absence of *académiciens* was anticipated and so instead of an oral interview, which would be lost to the historian, there is documentary evidence of the visit and the message conveyed by the candidate.

Thus Sepmanville (1762–1817), a naval officer and candidate for the position of *correspondent* in 1801, left a one-page printed letter with members of the *classe* which began: "Citizen Sepmanville called for the honour of seeing you and asking you for your support for the place of non-resident associate in the astronomical section. . . . He asks you to glance at the note which follows."[38] The note on the same page provided a brief summary of his naval and astronomical careers.

A later document was provided by Puissant, who called on Huzard on 1 December, 1825 in support of his candidacy for the geography section and left the following note: "Sir, I have just had the honour of calling on you to ask you to grant me your support. Although my academic titles are known to you (since Puissant had previously been a candidate), I shall take the liberty of putting them before you again."[39] This letter was accompanied by a

[37] *Analyse succincte des divers Mémoires* . . . n.d. (32 pp.). Académie des sciences, dossier Pelletier.
[38] Printed letter, dated Paris, *4 floréal an 9* (24 April, 1801), Institut HR5*, Vol. VI, No. 27.
[39] Institut HR5*, Vol. LXIII, No. 60.

modest two-and-a-half-page *Liste des ouvrages de M. Puissant.*[40] The Première Classe had vainly hoped in 1799 that the list of publications might replace the visit. It could do this, at least in the winter months. A letter from Auguste Saint-Hilaire written on 25 January, 1830 to Huzard mentioned his candidacy and reviewed his career. It concluded: ". . . Such, Sir, are my qualifications. I would consider myself fortunate if so many works were able to deserve your support. I would have hastened to express to you in person the value I attach to your support but I dare to hope that the rigour of the season will provide an excuse . . ."[41]

Style and Presentation

In the early days the use of printed notices was not universal. In 1801 the astronomer Michel Le François de Lalande (1766–1839) gained admission to the Première Classe of the Institut after circulating a handwritten summary of his career and publications.[42] More than a trace of the *ancien régime* can be seen in the documentation submitted in 1822 by the American diplomat David Bailie Warden (1778–1845), seeking the place of corresponding member of the Académie in the geographical section. His solicitation was in the form of a letter, written by hand rather than printed. The qualifications listed were a brief *curriculum vitae* followed by a list of memberships of different scientific societies.[43] Not being successful, Warden tried again for a vacancy as a corresponding member four years later.[44] Again his solicitation was in the form of a handwritten letter but this time Warden wisely mentioned that he had written on the geography of the United States, that he had translated several scientific memoirs from French into English or vice versa. He even listed some publications. Warden was complying with the norms of the Académie and this time he was successful. The form of a letter was still used as late as 1823 by Armand Séguin, although by then most candidates understood that what was required from them was a less personal document.[45] This was most often called a *Notice* but in the earliest years it occasionally took the form of a handbill: "To the members of the Institut national",[46] or it might have no heading and simply begin with the candidate's name: "Louis Ramond, associate of the Institut national has obtained the most votes for the class of physical and mathematical sciences for the vacancy in the section of natural history and mineralogy. The qualifications which he submits to the judgement of the Institut are the following: . . ."[47] Another with no title or date (but prior to 1804[48]) simply began:

[40] *Ibid.*, No. 59.
[41] Institut HR5*, Vol. LXXV, No. 10.
[42] Institut HR5*, Vol. VII, No. 13.
[43] D. B. Warden to Huzard, 30 October, 1822, Institut HR5*, Vol. LV, No. 3.
[44] D. B. Warden to Huzard, 10 March, 1826, Institut HR5*, Vol. LXIV, No. 3.
[45] The latest *Notice* in the form of a letter found was one submitted by a mechanical engineer in 1847. The letter was signed E. Grimpé and applied for the vacancy caused by the death of Gambey; Science Museum.
[46] Palisot de Beauvois [1802], Institut HR5*, Vol. V, No. 49.
[47] Institut HR5*, Vol. VI, No. 24 [1802].
[48] Since it is dated according to the republican calendar: *pluviose*.

616

Eugène-Melchior-Louis Patrin, encouraged by the favour which the Institut has already shown him by admitting him as one of its associate members, makes bold to-day to present himself as a candidate for a place vacant in the section of mineralogy and natural history. If the Institut decides to do him the honour of choosing him, he will not hesitate to take up residence in Paris.[49]

Whereas one astronomer, Burckhardt,[50] presented his name in large capitals at the head of his leaflet, some, more modest, began quite informally, for example: "In 1806 I read a memoir . . ." the author's name only being given at the end: Jaume Saint-Hilaire.[51]

One of the earliest lists, that composed by Balthazar Sage (1798) was entitled: *Exposé des titres*,[52] but this formula never became general. In the 1820s one finds the titles: *Liste des mémoires*[53] and *Liste des ouvrages*[54] as well as *Ouvrages publiés par . . .*[55]. All these titles emphasized the fact of publication, but someone who had presented numerous memoirs to the major scientific societies of Paris without necessarily publishing them pre-ferred to promote his candidacy under the heading: *Mémoires lus à l'Institut, à la Société Philomatique . . .*[56]. Some candidates felt that they should not impose on members of the Académie more than a summary of their work and so we find the formulae: *Exposé succincte des travaux*[57] and *Analyse succincte des divers mémoires publiées. . .*[58]. Although early lists were often in the form of letters, the form of a *Notice* allowed the writer to over-come the embarrassment of speaking of himself by the use of the third person.

Even more modest than a *Notice* was a *Note*. Thus, Cloquet, who did not ask to be elected but merely to be added to the list of candidates considered to fill a vacancy in the medical section in 1829, called his submission a *Note*.[59] Cagniard de la Tour was modest enough to entitle his submission a *Note*.[60] But sometimes a *Note* disguised a cunning strategy. The young physiologist François Magendie compiled a *Note des travaux*[61] which was divided into a list of the principal memoirs which he had presented to the Première Classe of the Institut. Many of these had been recommended for publication in the Académie collection, the *Mémoires des savants étrangers* but, through no fault of the author, their publication had been delayed. Having related his

[49] B. M. "Tracts", 1820–38, 733 g. 18 (50).
[50] *Ibid.*, (52).
[51] 29 February, 1820. Institut HR5*, Vol. LI, No. 13.
[52] *Exposé des titres de B. G. Sage*, n.d. [1798], (2 pp.), Institut HR5*, Vol. III, No. 30.
[53] *Liste des mémoires de botanique publiés par M. Henri Cassini*, n.d. [1820].
[54] *Liste des ouvrages de M. Puissant*, n.d. (2½ pp.), Institut HR5*, Vol. LXIII, No. 59.
[55] *Ouvrages publiés par M. Chaussier*, n.d. [1821], Institut HR5*, Vol. LIII, No. 27.
[56] Laugier, 1829. (19 pp.), Institut HR5*, Vol. LXXIV, No. 101.
[57] *Exposé succinct des travaux de M. F. Cuvier en histoire naturelle*, n.d. [1825], Institut HR5*, Vol. LXIII, No. 33a.
[58] *Analyse succincte des divers mémoires publiés par M. Robiquet*, n.d. (31 pp.), Institut HR5*, Vol. LXXIV, No. 107.
[59] *Note des titres de M. Jules Cloquet à l'appui de sa demande pour être porté sur la liste des candidats à la place vacante à l'Académie royale des sciences*, n.d. (2½ pp.), Institut HR5*, Vol. LXXIV, No. 58 bis.
[60] *Note de M. Cagniard La Tour*, 1827 (10 pp.), Institut HR5 (no asterisk), Vol. XII, No. 26.
[61] *Note des travaux de F. Magendie*, n.d. (2 pp.), Institut HR5*, Vol. LIII, No. 29.

work to the Académie, Magendie in the second part of his document listed five actual publications.

Quite a number of *Notices* from practical men described their achievements and inventions rather than any scientific publications. Armand Séguin, who had developed a method of tanning leather during the Revolution, was happy to mention his work both practical and chemical but without listing a single publication.[62] A more serious contender for a place in the Académie, the physicist Cagniard de la Tour (1777–1859) circulated a ten-page brochure in support of his candidacy in 1827 which did not list a single publication.[63] Instead Cagniard described his various inventions, machines and his work on acoustics which he had made a point of presenting in succession to the Académie. He also quoted the favourable reports which had been made by commissions of the Académie appointed to examine his submissions, including one on the vapourisation of liquids under the combined effect of heat and pressure, which constituted the discovery of the concepts of critical temperature and critical pressure. Although Cagniard had published this work in the volumes of the *Annales de chimie et de physique* for 1823, he did not mention this fact. Similar apparent carelessness about the value of his record of publications is shown in the case of L. B. Francoeur (1773–1849), whose interests spanned the fields of mathematics, astronomy, applied science and education.[64] Although for his unsuccessful candidacy of 1828 he was prepared to conform to the Académie's procedures by submitting a *Notice*, he was hardly striving to impress his colleagues by his record of publications. His achievements are listed under 11 heads, several publications being grouped together, *e.g.*, "No. 8. Various small works and several memoirs inserted in scientific journals such as those of the Société philomatique, the *Revue encyclopédique*, the *Bulletin des sciences*, etc."[65]

Another example of a practical *Notice* is provided by the engineer Clapeyron (1799–1864), who supported his candidacy with a *Notice sur les travaux*.[66] There was deliberate ambiguity in the use of the term *travaux*. The *Notice* consists of a biography in almost continuous prose in which Clapeyron described his career as an engineer in Russia and in France, drawing attention to a memoir on the theory of the vault which had received a favourable report which he quoted from a commission of the Académie. He mentioned several publications in the *Annales de chimie et de physique*, the *Annales des mines* and the *Journal de l'École polytechnique*. Clapeyron not only built bridges but he did the appropriate calculations and he included a prominent triple integral in the text as an indication of his mathematical

[62] Letter from Séguin to Huzard, 21 January, 1823, Institut HR5*, Vol. LV, No. 22.

[63] *Note de M. Cagniard La Tour*, 5 November, 1827 (10 pp.), Institut HR5 (no asterisk), Vol. XII, No. 26.

[64] For the special problems of Francoeur as a non-specialist see Crosland, Maurice, "The French Academy of Sciences in the Nineteenth Century", *Minerva*, XVI (Spring 1978), pp. 73–102 esp. p. 81.

[65] *Notice sur les travaux publiés par M. Francoeur*, 22 October, 1828 (1 p.), Science Museum.

[66] *Notice sur les travaux de M. Emile Clapeyron, Ingénieur en chef des Mines* n.d. (11 pp.), Académie des sciences dossier: Clapeyron. After several unsuccessful attempts to enter the Académie, Clapeyron was finally elected in 1848.

ability. He concluded: "Here I end this rapid account of my career which has been both practical and scientific." Finally he mentioned that he taught a course at the École des ponts et chaussées, a reminder that he had some pretentions to a career in higher education. Indeed the practical men often stressed their participation in higher education. Thus the unsuccessful candidate Belanger (c. 1839) described himself not only as *ingénieur-en-chef* but also as *professeur de mécanique appliquée* at the École des ponts et chaussées and the École centrale des arts et manufactures. He explained that some of his lecture material had been circulated in the form of lithographed sheets and that he was hoping to have a book published later.[67]

The forms of presentation of publications show wide variation, at one extreme of which was a simple list of titles of publications. The fact that the candidates had published was allowed to speak for itself. However, this seemed to many candidates to be taking too much for granted and the standard practice in the nineteenth century came to be a list of publications in which each title was accompanied by some explanation of its scope and significance. A further variation which was soon adopted was the inclusion wherever possible of quotations from reports on the memoir by a commission of the Académie. In a few cases the report of the Académie was felt to be so important that it was reprinted *in extenso* for the benefit of the members of the Académie.[68] It was even possible to include excerpts from one's own publications or offprints of publications, as was done by Hachette in 1823.[69] Thus the twentieth-century practice of circulating offprints of scientific papers was foreshadowed in the selective procedures of the Académie. One candidate presented the members of the Académie with a collection of offprints. This was what the former Oratorian priest and meteorologist, Louis Cotte (1740–1815) did in support of his candidacy in 1803 for the position of corresponding member attached to the physics section.[70] Cotte brought together more than a dozen papers, originally published over the period 1790–1802 on meteorology and the metric system, forming a booklet of more than a hundred pages. The copy in the archives was headed in Cotte's own handwriting on the cover: *Recueil de quelques Mémoires sur la Météorologie que j'ai publié dans le Journal de physique.* This must be one of the first cases in academic history where the sheer weight of a candidate's publications probably helped to secure for him the desired appointment.[71] Another case has been found in the 1820s where a scientist

[67] *Note des travaux scientifiques de M. Belanger*, n.d. (5 pp.), Science Museum.

[68] E.g. *Rapport fait à l'Académie des Sciences par MM. Thenard et Chevreul sur un mémoire de M. Serrulas ayant pour titre: De l'action de l'acide sulfurique sur l'alcool et les produits qui en résultent.* Imprimé par Ordre de l'Académie (Séance du 19 Janvier 1829), (15 pp.), Institut HR5*, Vol. LXXIII, No. 18. This is a rare case where a friend of Serrulas had persuaded the Académie to print the report. It was more usual for the author to bear the cost.

[69] Hachette's candidacy in 1823 and *Notice des travaux* was accompanied by three offprints of previous publications, one relating to a course he had given at the École polytechnique, a memoir on alloys of steel reprinted from the *Bulletin de la Société d'encouragement* (1820) and some hydraulic experiments published in the *Bulletin de la Société philomatique* earlier in 1823. Institut HR5*, Vol. LV, Nos. 68, 68a, 68b, 68c.

[70] Institut HR5*, Vol. XXXI, No. 52.

[71] This is not to say that the work of Cotte, a pioneer meteorologist like Lamarck, was not of some value.

in search of patronage addressed copies of a monograph to at least a large number if not all of the members of the Académie.[72] Although not successful then, Marc Séguin, nephew of Joseph de Montgolfier, was eventually elected as a corresponding member in 1845.

The Final Establishment of the Notices

The *Notices* were introduced during the Revolutionary and Napoleonic periods but they were far from universal. Chaptal, placed first on the list of recommended candidates in the elections of 1798, scorned to set out his qualifications. Other well-known and well-recommended candidates of the Napoleonic period such as Biot, Poisson, Gay-Lussac and Thenard did not submit such lists. Those who had already established their positions in the centre of French science disdained to point to their achievements. In this small group who were already known and appreciated by those scientists who counted, it was not necessary for them to support their claims by detailed bibliographies.[73] The increase in size of the French scientific community and the consequent pressure of the increased number of candidates necessitated the innovation. As a result, candidates—even strong candidates—had to swallow their pride and present a review of their own work as a basis for the deliberations of the members of the Académie.

In the early years qualifications were sometimes discussed informally. Thus at an election in the botanical section in 1806, in which the candidates included A. P. De Candolle and Alexander von Humboldt's associate Bonpland, the minutes of the meeting stated: "The section provides information about the qualifications and the works of these six candidates. They are discussed by the *Classe*."[74] It is very doubtful whether more than two or three of the six candidates had gone to the trouble of providing a printed list of their publications, but the botanical section would obviously have had some knowledge of the specialised publications of most of the candidates. Neither of the better known botanists, De Candolle nor Bonpland, seem to have bothered to produce a *Notice*, whereas the successful candidate, Palisot de Beauvois, did submit one.

In the Napoleonic period and the early years of the Restoration, the compilation of a *Notice* was more the exception than the rule. In 1816 on the occasion of a vacancy in the chemical section, it is recorded that "The *Classe* hears a very detailed report on the qualifications and the publications of the different candidates."[75] This suggests severe competition and serious con-

---

[72] In the library of Gay-Lussac there are several memoirs sent to him by their respective authors including: *Des ponts en fil der fer* by Séguin Ainé d'Annonay (Paris, 1824). The flyleaf bears the inscription: *"offert par l'auteur à Monsieur* [sic], *Membre de l'Institut"*, the name of the recipient being omitted. From this we infer that Séguin had not specially favoured Gay-Lussac with this present but that he had prepared a large number of copies to present to members of the Institut who would have the opportunity of supporting his candidacy.

[73] For a few case studies of how members of the Arcueil circle managed to present themselves most favourably to the Première Classe of the Institut, see Crosland, Maurice. *The Society of Arcueil: A View of French Science at the Time of Napoleon I* (London: Heinemann, 1967), pp. 161–168.

[74] 10 November, 1806. *Procès-verbaux*, Vol. III, p. 446.

[75] *Ibid.*, Vol. VI, p. 19 (5 February, 1816).

sideration of the respective merits of the candidates, although there is no evidence of documents supporting candidacy. In 1818 Dupin did provide such documentation and was rewarded by being elected.[76] Yet in 1820, on the occasion of an election in the botanical section, the sponsor of Du Petit-Thouars could say that he would not bother to discuss the work of his candidate since it was so well known. He simply—and successfully—presented the botanist as a man of genius.[77] There were other cases in the 1820s when candidates might solicit the vote of members of the Académie without presenting their scientific credentials. Thus the physician Desgenettes hoped in 1822 and again in 1829 to obtain support without the necessity of compiling a *Notice*.[78]

Yet the trickle of *Notices* in the early years of the century was, by the 1820s, to become a stream in full flood. The *Notice* had now become the rule. For the election in November 1825 four out of the five serious candidates produced a *Notice* and the candidate who apparently did not bother to display his credentials, Anselme Desmarest, (1784–1838) a mineralogist and son of an *académicien*, received only one vote. The other candidates for this vacancy in the zoological section were Blainville, Serres, Férussac and Georges Cuvier's brother, Fréderic. Although Fréderic Cuvier could call on the powerful support of his elder brother, who was one of the permanent secretaries of the Académie and a leading scientific politician, the issue was decided not by nepotism but by the careful consideration of scientific credentials and seniority between the two main candidates, Blainville and Fréderic Cuvier. A comparison of Blainville's *Notice* with that of Cuvier shows not only that the former had published far more, but that he had been a candidate ever since 1814. Blainville reproduced the order of merit as proposed by the zoological section in the elections of 1814, 1816 and 1821 and the way he had steadily moved up the list in order of preference implied that it was now his turn to be elected. In fact he was elected at the third ballot with 30 votes to Cuvier's 24. This, therefore, is evidence of a strongly contested election in which the documentation supplied by the contenders played a significant part.

Another election strongly affected by *Notices* was that of the mathematical section in October and November 1828. Of the six candidates, five prepared special *Notices*,[79] and one seems not to have done so.[80] The election was won by Puissant (1769–1843), the candidate with the longest of the modest *Notices* produced by a rather weak field of candidates.

Evidence of the change in the importance attributed to the enumeration

[76] *Travaux théoriques et practiques de Charles Dupin*, n.d. 8vo. Science Museum.

[77] *Mémoires de l'Académie des Sciences de l'Institut*, Vol. XX (1849), p. xxii.

[78] Letter, Desgenettes to Huzard, 28 February, 1822. Institut HR5*, Vol. LIV. No. 13. Letter, Desgenettes to Huzard, 19 November, 1829, *ibid*,, Vol. LXXIV. No. 80.

[79] *Notice sur la travaux publiés par M. Francoeur*, 1828 (1 p.). Institut HR5*, Vol. LXXII, No. 57. *Note adressé à MM. les Membres de l'Académie royale des sciences par M. J. Binet*, 1828 (2 pp.), *ibid*., No. 58. *Liste des principaux ouvrages de M. Puissant*, 1828 (3 pp.), *ibid*., No. 59. *Note des ouvrages et mémoires présentées à l'Académie par M. de Corancez* (2 pp.), *ibid*., No. 60. *Note à l'appui de la demande adressée par J. L. Boucherlat* (1 p.), *ibid*., No. 61.

of scientific qualifications may be seen in the·case of the vice-admiral Roussin (1781–1854), a candidate for a vacancy in the navigation section, who began his solicitation: "Convinced of the importance of the qualifications [*titres*] with which it is necessary to support oneself . . .".[81] By the 1830s the *Notice* was standard, so that when Libri, the Italian refugee mathematician, asked for support in the forthcoming election in the mathematical section of the Académie, Lacroix told him to draw up a list of his publications. Although asserting that his qualifications were slight, Libri managed to spread them over three folio pages of manuscript and he asked Lacroix to pardon him for boring him with such a long list.[82]

The Pattern of the Notice *as a Document addressed to the Académie*

Every candidate had to appear as a man of science. A candidate whose reputation was built purely on practical skills, like a surgeon, had to present himself as a scientist.[83] But although the general support of the Académie as a whole was needed, the support of the section in which the vacancy had occurred was crucial. This affected the presentation of candidates whose interests crossed the boundaries of the sections or, whose fields, *e.g.*, physiology, were not explicitly recognised in the name of the sections. Most of the great physiologists of nineteenth-century France, like Magendie, Claude Bernard and Paul Bert, had taken a medical degree and entered the Académie in the medical section. However, when a vacancy arose in the zoological section in 1852, it was a temptation for Claude Bernard to present himself.[84] Bernard, aged 39 and therefore considered to be a young man, began uncompromisingly by telling the section what their subject was about: "Zoology considered as a whole should be conceived as embracing all the knowledge that can be acquired of animals. Therefore this vast science is founded first on natural history . . . secondly on anatomy . . . thirdly on physiology . . .".[85] Zoology was therefore a vast and all-embracing activity in which his own work was described as "animal biology". The zoologists in the Académie were far from convinced and when the zoological section drew up a list of five candidates Bernard's name was omitted, to be added only at the second stage of the election by the Académie. In the final vote he came second. Fortunately for Claude Bernard there was a vacancy in the medical section two years later and he was elected.

The need to show one's qualifications not just as a scientist in general but

---

[80] No *Notice* has been found for the sixth candidate, Parseval, who was on the list but did not receive any votes. The election was unusual in that Binet (1786–1856) withdrew at the last moment. He was finally elected on the death of Puissant in 1843.

[81] Letter from Roussin to Huzard, n.d. Institut HR5*, Vol. LXXIV, No. 90.

[82] Bibliothèque nationale, NAF 3271 f. 243.

[83] See *e.g. Note sur les travaux scientifiques publiés par G. Breschet, chirurgien ordinaire à l'Hotel Dieu.* Publications are divided into parts headed: *Anatomie humaine et comparée, Physiologie etc.*, and only toward the end: *Chirurgie et Médecine opératoire.* Institut HR5*, Vol. LXXIV, No. 67.

[84] The full title of the section was *Anatomie et zoologie*, which makes the position of a physiologist less anomalous.

[85] *Notice sur les travaux de M. Claude Bernard, candidat à une place vacante à l'Académie des sciences dans la section de zoologie*, n.d. (39 pp.), Académie des Sciences, dossier: Claude Bernard.

as a specialist in one particular science is illustrated by such *Notices* as that of Bouvard who obviously was a candidate for the astronomical section, when he presented his *Notice des travaux astronomiques du citoyen Bouvard, astronome de l'Observatoire national, adjoint au Bureau des longitudes*.[86] Even more explicit is a printed letter sent by Bory de St. Vincent, candidate for a vacancy in the botanical section in 1830.[87] He had heard that some members of the Académie were saying that he was not really a botanist but a zoologist or, alternatively, that he did not specialise in any one field. He insisted that he was a botanist and that his specialisation within that science was microscopical studies. This insistence that the candidate was a specialist is perhaps most strident in the case of candidates who were obviously not specialists. Thus the Baron de Férussac, a writer whose interests covered a wide area of natural history and who had been unsuccessful in his candidacy for a place in the zoological section in 1821, prepared a special *Notice* for a further candidacy in 1825.[88] This was divided into three parts which covered the work on molluscs, the application of his results to the geology and geography of animals, and brief indication of his unspecialised works. The first two parts were marked in large capitals: "Specialités". This word was repeated prominently later so that not even the most cursory glance at the *Notice* could fail to reveal that Férussac claimed to be a specialist. Unfortunately for him there was no lack of genuine specialists in zoology among his competitors. When Férussac was eventually elected, it was not in any of the scientific sections but as an *associé libre*, a category which had been introduced in 1816. Interest and contributions in natural history were never considered a sufficient qualification for election to the biological section of the Académie in the nineteenth century. This was evidently not understood by Jaume Saint-Hilaire when he stood for a vacancy in the botanical section in 1820. It did not help his case to point out that his general book with several hundred plates on French plants, which had been criticised by some specialists, had been welcomed by the public, who had bought every copy of the edition.[89]

Candidates liked to think that they already had some support among the members of the Académie and they wanted other members of the Académie to believe this. Thus, when quoting the favourable reports that had been made on his work by several commissions of the Académie, the pharmacist Serrulas, who presented himself as "professor of chemistry", displayed prominently in capitals the names of members of the Académie (*e.g.*, Thenard, Gay-Lussac, Chevreul, Dulong) who had served on these commissions, almost as if they were offering their support for his candidature.[90]

---

[86] n.d. [1802], (2 pp.) Institut HR5*, Vol. VI, No. 18.

[87] Letter from Bory de St. Vincent to Huzard, 28 February, 1830. Institut HR5*, Vol. LXXV, No. 26.

[88] *Notice analytique sur les travaux de M. de Férussac* . . . , n.d. [1825] (16 pp.) Institut HR5*, Vol. LXIII, No. 47.

[89] [*Note sur les travaux botaniques de Jaume Saint-Hilaire*, 1820] Institut HR5*, Vol. LI, No. 13. Jaume Saint-Hilaire was probably more distinguished for his plant illustrations than for his botany.

[90] *Analyse succincte des travaux de M. Serrulas, Professeur de chimie à l'Hôpital militaire d'instruction de Paris*, 1829, (20 pp.). Institut HR5*, Vol. LXXIV, No. 92.

There was also the practice of invoking the great names of the Académie. Thus Blainville remarked that he had succeeded to one of the teaching positions of Cuvier. Later he remarked: "I should perhaps also count among my qualifications for the Académie general co-operation in a part of the great work of M. Cuvier which is concerned with the mechanics of locomotion."[91] Carrying even more authority than Cuvier in the opening years of the nineteenth century was Laplace. Bouvard, a candidate for the astronomical section in 1803, said that in some of his research he had made use of Laplace's formula and that some of this work would be published in the third volume of Laplace's *Mécanique céleste*. Férussac appealed to the authority of both Cuvier and Laplace. Férussac said that his research on crustaceans fell into the second of the four general divisions of the animal kingdom established by M. Cuvier.[92] Of his geological ideas he said that they fitted in with the physical laws of the universe and continued: "The ideas of M. de Férussac in this respect have the advantage of being in harmony with the theories of MM. de Laplace and Fourier."[93]

The difficulty of establishing an exhaustive bibliography of the literature published by candidates for election to the Académie is that they sometimes had printed a supplementary note to answer some of the criticisms that had been made of their work in the preliminary discussion of scientific qualifications in the secret session of the Académie. Such confidential discussions came to be reported outside the Académie, something which is all the more understandable when one of the candidates was a relative of an *académicien*. Thus Adolphe Brongniart had an eight-page pamphlet printed: *Réponse de M. Adolphe Brongniart aux observations faites sur les travaux de physiologie végétale dans la séance de l'Académie des sciences du 1er Mars 1830.*[94] The younger Brongniart complained that he had been accused of plagiarising the work of the Scottish botanist Robert Brown. He refuted this by quoting a letter from Brown praising his work. He also gave four pages of parallel text showing the differences between Brown's work and his own. On the publication of the pamphlet, the *rapporteur* of the botanical section, Mirbel, felt that he had to circulate an open letter addressed to Alexandre Brongniart, his fellow-*académicien* and the candidate's father.[95] One should not conclude, however, that elections to the Académie were normally reduced to the level of a pamphlet war. Sometimes a *Notice* itself would contain some personal discussion, particularly if it were a second or third attempt. Thus Clément in 1823 answered the criticism that much of his work was carried out jointly with Desormes.[96] In 1825 when Férussac was standing as a candidate for the zoological section he produced a supplementary note in which he tried to strengthen his candidacy.[97] He explained, for example, why the publication

[91] *Note analytiques sur les travaux*, n.d. (18 pp.) Institut HR5*, Vol. LIII, No. 25.
[92] *Notice analytique sur les travaux de M. de Férussac*, n.d. (16 pp.) Institut HR5*, Vol. LXXIII, No. 47.
[93] *Note supplémentaire à la Notice des travaux de M. de Férussac*, n.d. p. 3. *ibid.*, No. 48.
[94] Institut HR5*, Vol. LXXV, No. 28.
[95] *Lettre de M. Mirbel à M. Alexandre Brongniart*, 15 March, 1830, Institut HR5*, Vol. LXXV, No. 39.
[96] *Notice sur les travaux de M. Clément*, n.d. [January 1823] (7 pp.), p. 6. Institut HR5*, Vol. LV, No. 24.
[97] *Note supplémentaire à la notice des travaux de M. de Férussac*, [1825] Institut HR5*, Vol. LXIII, No. 48.

of his work on molluscs had been temporarily discontinued. The attention paid to the *Notices* by candidates and electors show how prominent the criterion of scientific achievement had become since the turn of the century.

The *Notice* as a necessary condition for entry to the Académie was not without its critics. Jules Marcou, a prominent critic of the Académie, was strong in his condemnation of the *Notice*:

I can think of nothing more nauseating than reading a series of twelve or fifteen brochures, setting out the publications of candidates for the Académie des sciences. There is no other academy or scientific society in the whole world which imposes or accepts such an unworthy condition. What! You make your own criticism of yourself, you act as your own herald, let us say it, you blow your own trumpet. You have to show what you have done that is original in science, you have to hide your mistakes, you have to indicate and portion out your good points. But this is public announcement, bill-posting, advertising carried over into science. This is excusable for the laudatory historical orations of members of the Académie. At least it is after their death that they are deified. But those volumes made up of notices about the publication of candidates for the Académie, are the most glaring proclamation of the debasement of character . . . .[98]

Yet Marcou admitted that the *académiciens* did not have the time to read the original publications of candidates before each election. Even if the time had been available, the diversity of sciences represented in the Académie was so great that specialists in one area would have been quite unable to understand the significance and originality of work done. Any work which was highly mathematical might be understood by less than half the members. Marcou accepted that something like a *Notice* was necessary in order to form a basis for comparison between rival candidates. What he really objected to was the prominence given to particular pieces of research and the half-hearted attempts by candidates to appraise their own work. He therefore proposed a simple chronological list of titles of memoirs and books. He was, in fact, proposing a return to the form of the earliest *Notices*.

Notices *Presented by the Successful and the Unsuccessful*

The study of the literature of the *Notices* reinforces what we know about the difficulty of entering the Académie. One should not assume that it was enough for a candidate to be scientifically talented to be immediately successful. A number of other issues were involved, sometimes political, sometimes more personal. There was also the recurring problem of the vacancy arising in the wrong section. In such cases a candidate might not only change the direction of his current research but would be tempted to present his past research in a way which showed its relevance to the section in which the vacancy had occurred. There is more to be learned from candidates who were obliged to compose a succession of *Notices* to gain admission than from the fortunate few who were successful at their first candidacy.

One candidacy of special interest is that of Hachette, a republican friend

---

[98] Marcou, Jules, *De la science en France*, 2ème fascicule, (Paris: C. Reinwald, 1869), p. 161.

of Monge who had been expelled from the Académie at the Restoration of the Bourbons. Hachette had made his first attempt to become an *académicien* in May 1815 during the "Hundred Days" when Napoleon had temporarily returned to power. He produced a *Notice* consisting both of publications and a description of his services to education.[99] Standing for a further vacancy in the mechanics section in 1818 Hachette compiled a notice addressed rather unusually to the members of that section only.[100] He began, perhaps unwisely in those politically sensitive years, with a brief autobiography, before listing various text books he had written. He then rather significantly listed his original research and concluded with a personal note to the members of the Académie explaining what further research he would undertake if he had the encouragement of their support. In a further candidacy in 1823 Hachette produced a similar *Notice* and he was elected.[101] The king, however, refused to confirm his election because of his republican sympathies and it was not until after the revolution of 1830 that Hachette's politics no longer constituted a bar to entry to the Académie. When a vacancy arose in 1831 in the section on mechanics Hachette produced a further *Notice*. This omitted his *curriculum vitae* and concentrated on his publications which were listed with a brief explanation of each. One might say that the emphasis on scientific publications as opposed to career was not only appropriate for someone labelled as politically dangerous but also fitted in well with more modern ideas of appropriate qualifications. Yet as late as 1839 the mathematician Gabriel Lamé (1795–1870) presented his credentials as a *Notice autobiographique*.[102]

As a further example of a scientist producing several *Notices* in his repeated attempts to enter the Académie we may take the case of Louis Pasteur (1822–1895). As in many other cases, it illustrates that patience was needed as well as ability to gain admission to the society of the scientific élite. Pasteur first tried to enter the Académie in 1856 at the age of 34.[103] He listed his publications, all of them either from the *Comptes rendus* of the Académie or the *Annales de chimie et de physique*. Each was followed by a brief description, although some memoirs which he obviously considered less relevant to a candidacy in the mineralogical section were merely listed at the end. As would be expected, Pasteur gave due prominence to his work showing the relation between crystalline form and the effects of a solution on polarised light, but he passed quickly over work which might have been labelled unambiguously as chemistry. This is a reminder of the rhetorical aspect of the *Notice*.

Pasteur had to wait several years for a second attempt. In 1861, his patron

[99] "Etat des services de M. Hachette dans l'instruction publique", *Notices sur les travaux de M. Hachette*, May 1815, B. M. 733.g.18 (44).

[100] *Notice présenté à MM. les membres de la section de mécanique de l'Académie royale des sciences par M. Hachette* (3 pp.), 20 September, 1818, Science Museum.

[101] The *Notices* for 1823 and 1831 are also in the Science Museum library.

[102] Science Museum.

[103] The three *Notices des travaux de M. L. Pasteur* 1856 (16 pp.), 1861 (31 + 3 pp.), 1862 (35 pp.) are all to be found in the Bibliothèque nationale.

626

Biot advised him to stand for a vacancy in the botanical section on the strength of his work on spontaneous generation. In his *Notice* of 1861 the title of each important memoir was followed by an extract from the Académie report usually bestowing moderate praise. During the discussion of his candidacy, many *académiciens* must have queried his credentials as a botanist. As an example of his ignorance of developments in the biological sciences, it was even said that he did not know of Schwann's work in Germany and Pasteur accordingly published an appendix denying this.[104] Being unsuccessful as a botanist, Pasteur was fortunate that another vacancy arose in the mineralogical section in the following year. In the *Notice* of 1862 he gave appropriate prominence to his publications on crystallography although he did not neglect to list in a second section his various contributions to the study of fermentation and spontaneous generation. Thus, although the compilation of a *Notice* provides an early example in the history of science of the importance of how scientific credentials are presented as well as the intrinsic merit of the credentials themselves, it was always a matter of emphasis and never one of falsification of the record.

There were also a number of *Notices* from men on the margins of science. Of these probably the most interesting are those who saw themselves as contributing to science but whose "science" was rejected by the official body. Among these was the phrenologist, Franz Joseph Gall (1758–1828), who, having toured Europe, decided in 1807 to take up residence in France. In 1808 his doctrine of the special function of different parts of the brain had been rejected by the Institut and his ideas were associated with materialism. In 1821 he boldly tried to gain admission to the Académie with the support of Geoffroy Saint-Hilaire. He supported his candidacy by compiling a *Notice* in the form of a letter addressed to the members of the Académie des sciences:

My qualifications are of three kinds: discoveries which are numerous and, I would claim, important in the natural sciences; fairly extensive literary works; finally a record of twenty years teaching in the great capital cities of Europe and especially in Paris where I have had audiences totalling 15 to 20,000, among whom I have counted some most able scientists and a very large number of doctors and naturalists.[105]

Gall wisely refrained from an exposition of his theories, preferring to remind the Académie rather of the difficulties he had had to overcome to make his contribution to science and medicine:

I shall not enter into a more detailed exposition of my discoveries partly because they are well known to several of you and also because they are described in my works and there now exist several societies which have been founded with the object of propagating and extending these ideas. I ask you only to appreciate the difficulties of every kind which I have had to overcome in order to follow them; they required considerable expense and I was without money and without the support of any government or of any learned society; what is more I had to combat prejudices of all kinds from adversaries from all classes of society. So if you consider that my work

[104] *Addition à la Notice des travaux de M. Pasteur* (3 pp.).
[105] *A Messieurs les membres de l'académie royale des sciences* (3 pp.), 15 October, 1821, B.M. 733.g.18 (7).

leaves much to be desired, it is in large part to these obstacles that their imperfection must be attributed.

Gall received no more than one vote in the election—that of Geoffroy Saint-Hilaire. The fate of his candidacy provides us with a useful perspective on the position of the "outsider" in French science in the nineteenth century.

Publications as Criteria for Positions in Higher Educational Institutions

The requirement of a scientist's record of publications as a condition for entry to the Académie des sciences spread beyond its confines. It was followed in other types of scientific institutions, notably in higher education. One of the first lists of publications produced which was not written to support an election to the Académie was that drawn up by the astronomer, Jeaurat, in 1800. Whereas it was common for the Première Classe to receive one copy of a book or memoir from its author, the minutes for 1 April, 1800, record that Jeaurat distributed to all the members present "a notice of the articles which he had published in the *Connaissance des temps*".[106] This was a by-product of Jeaurat's effort to gain an appointment to the newly established Bureau de longitudes.[107] Jeaurat, who boasted that he had entered the Première Classe in 1796 without the support of any printed circular, nevertheless found this form of announcement attractive. A further *Notice* which had no connection with the Académie was compiled by Jeaurat in 1802 for an appointment at the Bureau des longitudes.[108] The *Notice* is confined to two sides of one page but a page of folio size and in small type. It is as much a *curriculum vitae* as a list of publications and it introduced the text of a letter to substantiate the accusation that his exclusion from the Bureau was attributable to the enmity of Lalande. Despite these idiosyncrasies, the document was claimed to be a *Notice succincte des ouvrages*, and was clearly modelled on the Académie's *Notices*.

There are few such examples of printed lists of qualifications and publications supporting applications for posts in institutions of higher education until the Restoration of the Bourbons and particularly the firm establishment of the *Notice* among the practices of the Académie.[109] The earliest used for this purpose was that of Alexandre Brongniart (1770–1847), candidate for the chair of geology at the Muséum d'histoire naturelle in 1819.[110]

---

[106] *Procès-verbaux*, Vol. II, p. 129. This *Notice* is referred to again in a letter by Jeaurat to Huzard on 31 January, 1802, Institut HR5*, Vol. VII, No. 27.

[107] The Bureau des longitudes was a research institute attached to the Observatoire which was principally concerned with astronomy applied to navigation. Jeaurat would have been well qualified to fill one of its well-paid positions. For an account of Jeaurat's position in the Observatoire in this period see Crosland, Maurice (ed.), *Science in France in the Revolutionary Era described by Thomas Bugge* (Cambridge, Mass.: M.I.T. Press, 1969), pp. 104–105.

[108] Institut HR5 (no asterisk), Vol. I, No. 10.

[109] Sometimes qualifications were simply presented orally. Thus when the Première Classe nominated Laugier to succeed Fourcroy to the chair of chemistry at the Muséum d'histoire naturelle in 1810, the minutes report that the chemical section "expose les titres de ce savant", *Procès-verbaux*, Vol. IV, pp. 297–298 (8 January, 1810).

[110] *Exposé des titres de M. Brongniart, membre de l'Académie des sciences pour être présenté comme candidat à la chaire de géologie au Jardin du Roi en 1819*, Institut HR5*, Vol. L, No. 26.

628

Although Brongniart used his title of *académicien*, his document of three pages was clearly based on what he had done rather than who he was. It was divided into three parts, the first being his publications in mineralogy and geology, the second his publications in zoology, and the third his teaching and field-work in geology. Another such *Notice*, not dated but probably of the 1830s, was prepared by the pharmacist, J. Virey, in connection with his candidacy for the chair of "medical natural history" at the École de pharmacie in Paris.[111]

At the Collège de France the practice of presenting a list of publications had become fairly well established by the mid-nineteenth century. The death of the physiologist François Magendie in 1855 created a vacancy for a successor as professor of medicine at the Collège de France. It was customary in such cases for candidates for the chair to write to the administrators announcing their candidacy. There were three candidates: Langet (who wrote only a brief letter), Brown-Séquard (1817–1894) and Claude Bernard (1813–1878). Brown-Séquard enclosed with his letter a note of his publications, 35 in all, arranged in approximate chronological order over three pages of manuscript. Claude Bernard on the other hand wrote:

The qualifications which may recommend me in the eyes of the professors of the Collège de France as a scientist consist of a series of works which I have published on medicine and physiology over a period of fifteen years. These works have several times obtained the highest recognition by the Académie des sciences and latterly they won me the signal honour of becoming a member of that illustrious company. I do not think it necessary to make a detailed enumeration of them here because they are known to those professors of the Collège de France who are members of the Académie des sciences. . . .[112]

Thus even at the very distinguished and long-established Collège de France, where professors were able to lecture freely on their own research interests, it was permissible to refer to one's career in the Académie as a basis for appointment. Claude Bernard excused himself from providing a list of his publications on the ground that he had already drawn up such a list for the Académie for his election the previous year.[113] Only scientists, such as those who were members of the Académie, would appreciate fully their significance. Claude Bernard's application was considered sufficient and he was appointed to succeed Magendie.

But if Claude Bernard, fresh from his entry to the Académie, thought he was sufficiently eminent not to have to review his whole career again for the benefit of his prospective colleagues at the Collège de France, Quatrefages, another member of the Académie, was a little more modest when he wrote

[111] Virey stated that he had previously been nominated to the same chair in 1825 but "for purely political reasons the minister refused to confirm the appointment". *Notice des travaux et des principaux mémoire de J. J. Virey*, n.d. [early 1830s?], (3 pp.), MS annotations, Science Museum 012.

[112] Letter of 18 November, 1855, Archives, Collège de France, V.IV.c. 32. Assemblée des professeurs, 25 November, 1855.

[113] Bernard was elected to the section of medicine on 26 June, 1854 after submitting his *Notice sur les travaux de M. Claude Bernard* (June 1854) (45 pp.). He had been unsuccessful in an attempt two years earlier to enter the zoological section. See *Notice sur les travaux de M. Claude Bernard, candidat à une place vacante à l'Académie des sciences dans la section de zoologie*, n.d. (39 pp.).

to the administrator of the Collège de France on 31 March, 1855 as a candidate for a chair of biology.[114] Quatrefages had been elected to the zoological section of the Académie in 1852 and enclosed with his letter of candidacy a copy of the *Notice* drawn up previously for the Académie. But whereas the Académie had been interested in his claims as a specialist, the Collège de France might be interested in other publications, so he mentioned in his letter various articles published in the journal *Revue des Deux Mondes* and contributions to popular works such as the *Dictionnaire universel d'histoire naturelle* and the *Règne animal illustré*. Quatrefages also mentioned research he had done since his election to the Académie.

So although the *Notices* for the Académie constituted an innovation, this brief sample has been sufficient to show they loomed so large in the minds of the Parisian scientists that everything else was referred to them. The *Notice* submitted to the Académie served not only as a precedent but the very same copy could be used in seeking an appointment in higher education. The extension of the practices of the Académie into higher education in Paris is all the more understandable in view of the fact that the power of the Académie in nominations extended to certain chairs, a power which started in Napoleonic times and extended to a few major institutions, notably the Collège de France and the Muséum d'histoire naturelle, and to all chairs in pharmacy throughout France.[115] As the Académie became accustomed to the submission of lists of publications by aspiring members, candidates for chairs in the faculties began to submit similar lists. Thus someone aspiring in 1826 to be the Académie's nominee for a chair of pharmacy at the University of Montpellier published a long *Notice* in the form accepted for candidates to enter the Académie itself.[116] A little later another aspirant to a chair in pharmacy, this time in Paris, submitted a similar list of publications to the Académie.[117] By the 1830s Auguste Comte could complain that the Académie was so used to basing elections to membership on lists of publications that it was using the same criterion in nominations to chairs, for instance at the École polytechnique. He complained that he gave up so much time to philosophy that he did not have the time to write mathematical or scientific papers such as would be presented by his competitors as qualifications for the vacant chair of *mathématiques transcendantes* at the École polytechnique.[118] Obviously Comte was writing as an amateur in a world of increasing professionalism.

[114] *La chaire des corps organisés.* Archives, Collège de France, G.IV, c.28B.
[115] Crosland, Maurice, "The French Academy of Sciences . . .", pp. 73–102. Limoges, Camille, "The Development of the Muséum d'histoire naturelle of Paris, 1800–1914", in Fox, Robert and Weisz, George (eds.), *The Organisation of Science and Technology in France, 1808–1914* (Cambridge: Cambridge University Press, 1980), p. 217.
[116] *E.g., Notice sur les travaux chimico-pharmaceutiques de P. Bories* (1826), (24 pp.), 8°, Science Museum.
[117] *Notice des travaux et des principaux mémoires de J. J. Virey*, n.d. (3 pp.). It is highly probable that a thorough search among the nineteenth-century archives of the many institutions of higher education in Paris would reveal a number of lists of credentials submitted by candidates for chairs.
[118] Auguste Comte, *Correspondance générale et confessions.* Textes établiés et présentés par Paulo E. de Berredo Carneiro et Pierre Arnaud (Paris: Mouton, 1973), vol. 1, p. 267, letter to General Bernard, 20 September, 1836.

Conclusion

The great progress of scientific knowledge in the nineteenth century owes a great deal to the decision, as scientific work became institutionalised, to give pre-eminence to the criteria of scientific achievement and publication as the basis for appointment and promotion. The ascent of these criteria understandably occurred first in France in the early part of the century when France was still the leading scientific country of the world.[119] The change took place first in the Académie des sciences of Paris. The Académie forced leading scientists to record and evaluate their own work in the categories of specialisation corresponding to its sections.

By the late 1820s, the interest in the publications of those entering the Académie was beginning to affect the evaluation of *académiciens* on their departure. The two secretaries had the duty of reading eulogies of académiciens after their death and it is significant that, although Cuvier's *éloge* of his fellow zoologist Lacépède in 1826 concentrated on his books, by the time the *éloge* was published in 1829 Cuvier had added a very long footnote listing the memoirs he had published in different journals.[120] By the early nineteenth century for most scientists the papers published in scientific journals were at least as important a part of the record of their professional achievements as their books. When it fell to Cuvier in 1829 to present an *éloge* of the *académicien* and agricultural scientist Louis Bosc, he concluded with a six-page supplement setting out in chronological order Bosc's many publications in the journal *Annales de l'agriculture française*.[121] It was therefore fitting that when Cuvier himself died in 1832, the *éloge* by the new secretary, Flourens, should conclude with a long section setting out Cuvier's publications, almost in the style of a *Notice*, according to subject.[122] The analogy with the candidate's credentials for entry to the Académie was complete. The inclusion of a detailed list of published works in the *éloge* of a deceased *académicien* became standard in the mid-nineteenth century, despite the large amount of labour required to compile such a list.

Many of the origins of modern science as a profession are to be found in France. Interest in specific lists of scientific publications gave further encouragement to scientists to write up their research. In Britain even in the eighteenth century some very eminent men of science, like Henry Cavendish (1731–1812), had not bothered to publish some of their major work and Joseph Black (1728–99), professor of chemistry at Edinburgh, published nothing after the dissertation for his medical degree. Publication of research as one of the duties of the professional scientist owes something to the encouragement given by the French Académie des sciences, both publicly in

[119] For a contrast with the state of affairs in Great Britain in the Royal Society in London, where qualifications of candidates for election were only enquired into superficially before the nineteenth-century reform, see Crosland, Maurice, "Explicit Qualifications as a Criterion for Membership of the Royal Society: A Historical Review", *Notes and Records of the Royal Society* (1983), (in press).

[120] *Mémoires de l'Académie royale des sciences de l'Institut de France*, VIII (1829), pp. ccxxxviii–ccxl.

[121] *Ibid.*, X (1831), pp. cxci–ccxviii.

[122] *Ibid.*, XIV (1838), pp. i–lxviii, *éloge* read on 29 December, 1834.

the well-known *Mémoires* and *Comptes rendus* and privately in its more personal *Notices*.

Special attention has been focused on the early development of the practice of submitting detailed bibliographies of scientific works as the basis for the application of criteria of achievement. This has become so much a part of modern science that it is now a matter of course and yet it is absent from the first major scientific societies. Unlike so many other practices of the Académie des sciences, it was introduced not by governmental decree but by the initiative of scientists themselves, mainly of the *académiciens*. From about 1820 the Académie was responding to a new situation in which a considerable number of well-qualified candidates from an increasingly large scientific community presented themselves for every vacancy in the most highly respected scientific society in the country. Although we cannot rule out political intrigue and nepotism entirely, most *académiciens* felt that acceptance should be dependent more on objective assessment than personal connections. In a few cases a man could be assessed on the basis of a well-known discovery of a new phenomenon, but in most cases the electors needed to look over an entire scientific career to decide how scientifically meritorious the candidate was. As the average age of election rose throughout the nineteenth century, by 1900 many candidates were submitting evidence of their achievements through practically the whole of their creative lives. Some of the late nineteenth-century *Notices* are particularly compendious. The chemist Schutzenberger (1829–1897) was elected in 1888, supported by a booklet of 110 pages, setting out his many publications. Such a *Notice* with its large format, paper covers, division into chapters, and table of contents may be considered an invaluable summary of what had been accomplished over almost an entire scientific career.

This introduction of and reliance on criteria of intellectual achievement were part of a larger movement in French society. At a lower level, examinations had been introduced in the Napoleonic Université de France. They served as a filter to separate out the most able. France made good use of the examination system long before Victorian England introduced examinations as regulators of admission to the civil service and the professions. But if the coveted doctorate in science represented the highest rung of the examination system, it was still below the lofty peaks of the Académie. In a society which prided itself on the use of the *concours* to decide on important appointments, election to the Académie represented a *concours* of a very special kind. What might be thought to have been a system belonging primarily to the world of universities turns out to have originated in quite a different context. The procedure established in the early nineteenth-century Académie des sciences was later quietly adopted in the university system, first in France and later in other countries and is still predominant today. It has become an indispensable part of the modern scientific community.

VIII

Assessment by Peers in Nineteenth-century France: The Manuscript Reports on Candidates for Election to the Académie des Sciences

BY THE LATE twentieth century we are well accustomed to the adoption of certain criteria for appointment and promotion of scientists, usually related to their research. One of the main objects of this paper is to study the emergence in the nineteenth century of certain criteria of assessment in the careers of scientists, including seniority and publication record. Candidates to the French Académie des sciences were usually among the most able in their field, excluding only the members of the section they were hoping to join. They were being assessed partly on their potential achievements but mainly on their past achievements, since one had to be fairly well advanced in a career in science before presuming to risk a candidacy to become a member of the Académie. In so far as many of these scientists had to stand several times for election before being successful, it is sometimes possible to follow the development of their careers and how they rose in the estimation of their peers.

The Académie des sciences must be unique in its documentation on the evaluation of the quality of scientific research of candidates and their qualifications for election to the Académie. In some scientific societies of international standing such as the Royal Society of London, social criteria were often more relevant than intellectual merit as qualifications for membership. At best a document might be submitted, stating that such and such a candidate for election to the Royal Society was the author of a book.[1] Any discussion of merit, if it took place, was never recorded. With no restriction on the number of members, election was usually little more than perfunctory. The Académie des sciences on the other hand had severely restricted numbers and election to membership was very competitive. Although there was a great tendency to document everything, oral discussions of the merits of candidates were never recorded. The minutes of meetings were always tactful in personal matters and in any case the discussions took place in secret sessions, for which a separate confidential register was kept. Even here it was not the details of discussion but decisions which were normally recorded, especially the order of merit in which candidates were placed by the section in which the vacancy occurred.

[1] Crosland, Maurice, "Explicit Qualifications as a Criterion for Membership of the Royal Society: A Historical Review", *Notes and Records of the Royal Society*, XXXVII (1983), pp. 167–187.

414

It seems, however, that there was increasing formalisation of procedures in the first half of the nineteenth century. As election to membership became more competitive, discussion of the rival merits of candidates might be less spontaneous. The senior members of the section would write notes on what they proposed to say. In some cases, a *rapporteur* would write out his statement in full; several such speeches found their way into the archives and thus provide the basis for the reconstruction of what probably happened in the elections.

"Notices" and "Rapports"

The manuscript *rapports* on candidates were responses to printed lists of publications submitted by candidates. These *notices des travaux* had become increasingly common from the 1820s onwards. Although such information was highly relevant in considering what a candidate had contributed to a particular branch of science, the *notices* were not left to speak for themselves. They provided the raw material for an evaluation, but did not in themselves constitute that evaluation. However, once candidates began to take seriously the presentation of their credentials in the form of *notices*, it became more incumbent on the members of the section to consider in detail the respective merits of candidates.

The *notices des travaux*, mainly printed, were compiled by candidates in support of their candidatures for election to the Académie.[2] These documents, however, hardly had an official status, being the individual responsibility of outsiders hoping for election. The *notices* were expressions of the point of view of the candidates. They tell what the candidates thought they needed to present about themselves and their work so as to appear before the Académie in the most favourable light. The members of the Académie could not possibly know personally all candidates, nor did they know much about the candidate's area of special expertise if it differed from their own.

It is, however, necessary to investigate candidacies not only from the point of view of the candidate but also from the point of view of the Académie, and more particularly from the point of view of the section of specialists in the branch of science in which the vacancy occurred. It was the responsibility of the section to present the list of candidates in order of merit. We are concerned principally with the difficult questions of how this order of merit was decided, since it became increasingly incumbent on the members of a section to argue the case to their colleagues in other sections to justify their preference. Whereas candidates themselves in their *notice* could list and explain their publications, assessment was necessarily left to the members of the Académie.

---

[2] Crosland, Maurice, "Scientific Credentials: Record of Publications in the Assessment of Qualifications for Election to the French Académie des Sciences", *Minerva*, XIX (Winter 1981), pp. 605–631.

Another feature of the *notice* is that the writer, embarrassed to claim too much for the quality of his own work, would tend to emphasise quantity. He would often provide an exhaustive list of his publications to demonstrate how much he had accomplished. The *rapports* were more selective. Thus Cauchy, reviewing candidates for a vacancy in the mechanics section in 1843, said that he would save time by concentrating on the most important works which could be considered as the more relevant qualifications ("*des titres sérieux*").[3] For the author of a *notice*, the fact that research had been accepted for publication usually implied that it had reached a certain standard, hence publication in itself became an assertion of merit. The *rapports sur les travaux* attempted to evaluate quality. The author was always a senior *académicien* within the general field of the candidate's expertise. He was therefore eminently qualified to pass judgement. A few examples will illustrate the position.

When the chemist Grimaux was a candidate in 1877, the *rapporteur* on his candidacy, Cahours, began by mentioning that over a period of 12 years he had published 27 papers.[4] What he emphasised, however, was not the quantity but the quality—all the *mémoires* were "of incontestable originality". Method was another aspect he mentioned: "M. Grimaux has always been directed in his researches by theoretical views. His discoveries are not the result of lucky chance observations made by a skilled experimenter. All were predicted by analogies, as the views which he has expounded demonstrate." On the other hand, the physicist Jamin was recommended in 1868 by Fizeau not for his theory but because, although he had done original work, he had "kept within the limits of the most careful procedures and without indulging in exaggerated conclusions which do not follow clearly and logically from observations and which are presented too often to the public in the most ambitious and exaggerated form".[5]

However, it is too much to expect complete objectivity. Vulpian, a practising physician, at the end of his report on the physiologist Paul Bert as a candidate in the medical section, thought it pertinent to point out that the candidate's research was hardly relevant, either directly to clinical medicine or to surgery.[6] But although he was no physician or surgeon, Paul Bert as a physiologist drew on a range of sciences from physics to zoology, as was pointed out by Marey in his report four years later,[7] and it was as a scientist and a follower of Claude Bernard that he was elected in 1882.

The *rapports sur les travaux* which have survived are mainly from the second half of the nineteenth century. The earliest *rapports* still available

---

[3] *Rapport de la section de mécanique sur les candidats à la place vacante par le décès de M. Coriolis*, Académie des sciences archives, Bertrand papers, Cauchy file.

[4] *Rapport sur les titres scientifiques de M. E. Grimaux, par M. Cahours, 12 février 1877*, Académie des sciences archives, Grimaux dossier.

[5] *Rapport sur les travaux de M. Jamin, ibid.*, Jamin dossier.

[6] *Rapport de M. Vulpian sur les titres de M. Bert*, Comité secret du 25 novembre 1878, *ibid.*, Paul Bert dossier.

[7] *Rapport de M. Marey sur les titres de M. P. Bert*, séance du 27 mai 1882, *ibid.*

date from the 1840s and 1850s and they become common only in the 1860s. In 1876 Dumas pointed out that the reports which had been drawn up for elections usually disappeared afterwards. On behalf of the administrative committee, he asked that henceforth all reports on the relative merits of candidates for election to the Académie should be deposited with the secretariat.[8] He appreciated, however, that the authors of some reports might feel it embarrassing—"*inconvénient*"—presumably if a report had been particularly critical of one or more of the candidates; in such cases the author would not be obliged to deposit his report. Thus at best we have only a semi-official and incomplete record, but one of immense value. Future investigators must be grateful to Dumas for foreseeing the possibility that "these reports would in future constitute the most complete history of the Académie". Such reports are less polished stylistically than the obituary or *éloge* which is traditionally one of the standard sources for the historian. Yet the *rapport sur les travaux* has the inestimable advantage over the obituary that it gives not a retrospective but a contemporary view of a scientist's life and work. It shows how frankly and without rhetorical flourishes his work was evaluated by leading scientists in mid-career.

In many cases the reports were never filed; it is even possible that in a few cases someone had a special interest in removing them from the files. On rare instances the information in a report on a candidate was later published, as occurred in the obituary of Henri Poincaré. Darboux as secretary was able to turn to the files and find the report on his candidacy for election to the Académie that Camille Jordan had composed in the 1880s. The *rapporteur* had concluded with the words:

> Such then in its essentials is the work accomplished by M. Poincaré. It is beyond all ordinary eulogy and forcibly reminds us of what Jacobi wrote about Abel—that he had resolved problems which no one before him had dared to imagine. Indeed we must recognise that we are witnessing a revolution in mathematics, comparable from every point of view to that which revealed itself half a century ago by the advent of elliptical functions.[9]

In 1913 the Académie was proud to remind the world of science of the prophecy that it had made a generation earlier.

The Historical Development of the "Rapports"

In the early years of the nineteenth century, discussion of the respective merits of candidates was largely informal. The development of the practice by which candidates submitted a *notice* of their publications helped to focus discussion and, by the 1820s, an oral report would be given by a senior member of the section in which the vacancy existed. The minutes of the Académie very occasionally give an indication of the thoroughness of the review and the criteria considered important. Thus in 1823 Prony, as the

---

[8] *Comité secret procès-verbaux*, 1870–81 (20 March, 1876), p. 272, *ibid.*
[9] *Mémoires de l'Académie des sciences de l'Institut*, LV (1914), p. cii.

senior member of the mechanics section, is described as making "an enumeration of the different researches undertaken by the candidates, the works that they have published, and the memoirs they have presented to the Académie and on which reports have been written".[10]

Again, in 1825, it was recorded that in an election in the astronomical section, after Arago had made a detailed report on the different qualifications of the candidates, there was a discussion about which criteria were the most relevant in guiding the choice of the Académie.[11] Discussion of the scientific merits and qualifications of rival candidates had become such an established practice in the Académie des sciences by the early 1830s that, when the Académie des sciences morales et politiques was founded in 1832 under the more liberal government of Louis Philippe, an article was inserted in the constitution (article 11) stating explicitly that prior to any election the qualifications of candidates were to be discussed in a secret session.

As the practice developed of commenting in detail on the rival candidates, brief notes were written for the presentation, although this still allowed for comments *ad lib*. The next stage in this reconstruction was the practice, by the mid-century, of making longer notes or even writing out an entire speech, particularly about a favoured candidate. By the 1860s it had become the practice for the senior member of the section to write a report on all candidates. The existence of a written record encouraged tact in assessment. Candidates were never derided, even if they were placed quite low in the order of preference which the section was asked to submit. Instead there were certain hints, for example that a candidate was still a young man, that he had only been working in the field for ten or 15 years, or the report might conclude that he was worthy of being considered; this was damning with faint praise. Once these reports became the rule rather than the exception and there was nothing harmful about the contents, it was thought that they should not be discarded after the election. It was under these circumstances that in 1876 Dumas, as secretary, suggested that the reports should be retained in the archives. Even then, however, some *rapporteurs* were more willing than others to hand over their manuscript notes. By the end of the century, a few reports were even cyclostyled, showing that the early concern for confidentiality had evaporated. Such semi-public reports tended to be eulogies of the living and are of less value for research than manuscript notes intended for confidential—and therefore more frank—assessment. Yet they are still useful, if not for adverse criticism, at least for the positive grounds offered for giving strong support to favoured candidates, on the basis of seniority, originality or sheer bulk of published work.

The first full reports on candidates found in the archives date from 1842 and 1843. It seems largely by chance that Anselme Payen (1795–1871) had

---

[10] *Procès-verbaux des séances de l'Académie des sciences*, 10 vols (Hendaye: Imprimerie de l'Observatorie d'Abbadia, 1910–22), Vol. VII, p. 582 (3 November, 1823).

[11] *Ibid.*, Vol. VIII, pp. 248–249 (25 July, 1825).

the honour of a full report on his candidacy for the agricultural section in 1842.[12] The report is anonymous, but one sheet is headed "*Motifs de préférence soumis à M de S*"—this presumably refers to Baron de Silvestre, the senior member of the section. The document gives six reasons for preferring Payen to his chief competitor, Vilmorin, and the information given is illuminating. It ran as follows:

1st Presented for the longest time (*plus anciennement*) by the section.

2nd Nominated earlier as a member of the Société royale et centrale d'agriculture (for the last ten years, his competitor [Vilmorin] for the last two years).

3rd All his time has been exclusively devoted to agricultural science. It is easy for him to be in regular attendance of meetings of societies concerned with it.

4th His extensive land under cultivation is nearer, under his eyes the whole year round and also it is at the disposal of his colleagues for trials and observations on a large scale. The same is true of his laboratories.

5th His discoveries, his *mémoires* approved by the Institut and his lectures at the École centrale and at the Conservatoire are more useful to science, to agriculture and to engineering than the results presented by his competitors.

6th He has won a larger number of prizes on important questions in the competitions of the Académie des sciences, of the Société centrale d'agriculture and of the Société de pharmacie de Paris.

As if this were not enough grounds for supporting Payen, it was said that since the last election in which he was a candidate he had persevered in both theoretical and practical work in agriculture. Payen had submitted a *mémoire* to the Académie on which Mirbel had reported favourably. Two paragraphs from the report were quoted. All this was preceded by a list of the candidate's principal qualifications and publications. Indeed this part of the presentation closely resembled the *notice* prepared by the candidate. The presentation was unusual in its explicit comparison between rival candidates. Such comparisons are not often found and suggest that procedures were still being worked out.

Another very early report on candidacies found in the archives dates from 1843 and is in the hand of Cauchy.[13] We owe the existence of these precious manuscripts, previously lost among the hundreds of thousands of documents in the archives of the Académie, in the first place to the conscientiousness of Cauchy, who wished to justify to Dumas the actions of the section on mechanics, of which he was the most senior member.

The previous member, Coriolis, had died in September 1843 and Cauchy began his presentation at the subsequent election with a tribute to his deceased colleague. These 15 lines of the report find no parallel in later reports and are evidence that the procedure had not yet become entirely standardised. A considerable number of candidates had presented themselves as possible successors to Coriolis within the section on mechanics. Cauchy began by eliminating those who were employed in "industrial mechanics", including two civil engineers and a mining engineer.

---

[12] Académie des sciences archives, Payen dossier.

[13] Bertrand papers, Cauchy file, *op. cit.*

They might have made important practical contributions, but their work fell outside the concept of "mechanics" held by most members of the section. On the other hand, the entire section had agreed that among the candidates were two outstanding applied mathematicians, Morin and Saint Venant, and their names were placed together in alphabetical order at the head of the list. Several other names were given, sharing second and third places, but Cauchy concentrated on the merits of the first two. His ten-page report on the work of Saint Venant was more than twice as long as that on Morin. It was clear which of the two Cauchy himself favoured.[14] But Cauchy was not a popular figure and, when the Académie voted, Morin was elected. Indeed, to have appeared as the protégé of the Catholic royalist, Cauchy, could be a serious disadvantage and Saint Venant, having failed, had to wait another quarter of a century to be elected.

There are also a few surviving reports for the 1850s. The first ordinary manuscript *rapport sur les travaux* for an election found in the archives is about the mathematician Chasles and his election in 1851. It was the misfortune of Chasles to be a brilliant geometer in an age when geometry was out of fashion. He was also unlucky in that, after the election of 1843 in which he had shared the first position with Binet—the successful candidate—he had to wait another eight years for a vacancy in the mathematical section.

The report by Sturm in 1851 was written in pencil and covered eight pages. It began:

The mathematical works of M. Chasles, which are considerable both in number and importance, are well known to the Académie and their reputation has already been established. He has long been a correspondent of the Académie and at the last election the mathematical section placed him joint-first with the candidate who was elected. We shall remind you first, but very rapidly, of the work on analysis and geometry by M. Chasles before that time . . .[15]

Chasles, now aged 58, was elected. He was the senior candidate and in a sense, considering his previous placing, the Académie might have thought it had an obligation to consider him favourably, all the more so as he had done important work since the previous election.

The fact that this report only survives in a pencil draft suggests that reports still depended very much on the *rapporteur*, and indeed only one other manuscript report for the 1850s has been discovered thus far. It is about Bravais (1811–63), who created special problems as a candidate, not because he lacked qualifications, rather the reverse. Bravais had made wide-ranging contributions to science, but when he entered the Académie in 1854, it was in the geographical section.[16]

The case of Bravais illustrates the problems of the scientist whose work

[14] It was against the accusation of partiality that Cauchy felt obliged to defend himself in his correspondence with Dumas. His letter to Dumas is dated "16 December, 1840" (=1843). Morin was elected on 28 December, 1843.

[15] Académie des sciences archives, Chasles dossier.

[16] Archives nationales, F[17] 3579.

was not highly specialised. At the time of his final candidacy to the Académie, he had worked in no less than 13 different fields of scientific activity, of which the twelfth was crystallography, the field in which he is best remembered. Bravais explained frankly some of his difficulties in a private letter of 1847:

I have often been told that I was wrong not to devote myself to a specialisation of my choice. People seemed to think that, after studying botany, I should have returned to my early studies of mathematics; but to pass on quickly from there to meteorology and then astronomy was quite foreign to the ordinary way of doing things. Persons highly placed in science have, to my knowledge, remarked that I have wasted my time and my future. By accepting the post of professor of physics at the École polytechnique, I undertook the moral obligation, to those scientists who have honoured me with their support, to devote myself exclusively to the study of physics. It is because of this that I am really obliged to decline all consideration of candidacy for the astronomical section of the Institut . . .[17]

When Bravais was finally elected to the Académie, it was in the section on geography and navigation, relating to his experience as a former lieutenant in the Navy. In the *notice* he compiled for this occasion, he naturally placed first his publications on hydrography and navigation. It was also appropriate to stress his different voyages and his contributions to meteorology and terrestrial magnetism. He admitted to the publication of works on physics, astronomy, mathematics, crystallography and the natural sciences. He was at pains, however, to stress that most of his work had had a practical motive. He indignantly denied the accusations of some of his critics that he was really a mathematician![18] Such were the tribulations of the scientist of broad interests and very diverse achievements. In order that *académiciens* should take seriously his credentials in one field, he himself had to suppress reference to achievements in other fields.

A third example of a report of the 1850s has survived because a copy was made and, even more important, because it was about Louis Pasteur. Although the original manuscript is no longer extant, the text has been published in a twentieth-century edition of Pasteur's complete works.[19] The report by the mineralogist Senarmont is by no means typical. Whereas the average report might be content to summarise a candidate's work, this one, made in 1857, went further in setting a historical context for Pasteur's work on crystallography, relating it to that of Haüy, the founder of the section to which Pasteur was to be elected. Moreover, Senarmont in his report included a discussion of scientific method to which Pasteur conformed. Pasteur was repeatedly praised for his originality—"*des horizons nouveaux*", "*un sujet d'études absolument neuf*", "*un fait absolument*

---

[17] Bibliothèque nationale, FR.3267, f.169, Bravais to Libri, 6 February, 1847.

[18] ". . . quelques personnes . . . m'ont accusé d'être un géomètre . . .". "Remarques générales sur les travaux de M. Bravais", *Notice des travaux scientifiques* (Paris, 1854), pp. 19–20.

[19] Pasteur, Louis, *Oeuvres*, ed. Pasteur Valléry-Radot, 7 vols (Paris: Masson, 1922–39), Vol. VII, pp. 435–440. Since the report makes no mention of the vacancy, the text as printed is probably incomplete.

inattendu", *"les découvertes si originales"*, etc., and also for his hard work—*"un nombre immense d'expériences"*, and *"de longues et minutieuses recherches crystallographiques"*. Since Pasteur was, at 35, comparatively young to enter the Académie, the *rapporteur* emphasised the reliability of the work which, he said, had been repeated and verified. Pasteur's work on the optical rotation and crystalline form of tartrates was described as *"un ensemble qui lui appartient tout entier"*. Indeed, Pasteur's research was presented as a model. It was usually the Académie which provided a model for aspirants, not vice-versa!

Senarmont's glowing report on Pasteur's work reveals the *rapporteur* as an advocate rather than a judge. There are a few other cases where the usual detachment is lacking. A generation later, the applied mathematician, Saint Venant, became the patron and the outspoken advocate of the younger Boussinesq (1842–1929). As the younger man failed in his early candidacies, his patron became increasingly emphatic in his praise. Of course this was not an arbitrary or baseless assessment. Boussinesq had a steady flow of publications at a high level and thus strengthened his qualifications for membership.

"Rapports" and Specialisation within the Académie

An important feature of the Académie since its early years and one which distinguished it from its opposite number in England, the Royal Society, was its division into a small number of recognised disciplines. In 1699 six subjects had been recognised, each represented by six members. By 1803 the number of sections had risen to 11; the normal complement for each section was six members.[20] The subjects recognised were: mathematics, mechanics, astronomy, geography and navigation, physics, chemistry, mineralogy, botany, agriculture, anatomy and zoology, and medicine and surgery. Some of the sections represented more than one field of science. Others, such as "mineralogy", corresponded to a late eighteenth-century view before geology had become recognised as a science. Disregarding nomenclature, however, most *académiciens* recognised that if geologists were to be elected to the Académie this was the most appropriate section. Candidates were rarely elected for their general scientific ability. To be successful they had to claim pre-eminence in one particular field.

The reports on candidates show how the sections interpreted the disciplinary specialisations they represented. They show the difficulties that physiologists encountered when they tried to enter the Académie, since there was no section explicitly concerned with physiology and the most appropriate section was medicine. Thus Etienne-Jules Marey (1830–1904),

[20] The exception was geography and navigation, which only had three members. This was because the subject originally belonged to another class of the Institute abolished by Bonaparte in 1803, but with the more scientific members being transferred to the "First Class of the Institute", as the Académie des sciences was called during the Revolutionary and Napoleonic periods.

who is well known in the history of experimental physiology for his development of techniques of graphical recording and cinematography—although the latter came only after his entry to the Académie—was a candidate in 1876 for a vacancy in the section of medicine and surgery. The report on him was the responsibility of the surgeon Gosselin who devoted only four pages to Marey in a long report on the various candidates. He complained that Marey was concerned principally with inventing instruments to record physiological processes; Gosselin said that Marey was a scientist too far removed from studies of the ill patient to be considered as one of the favourite candidates. When Marey was again a candidate for a vacancy in the same section two years later he was listed only third in order of preference. This time the *rapporteur* Vulpian was rather more diplomatic. While recognising Marey's scientific reputation, he again hinted that the section preferred candidates with a reputation for clinical work. He admitted, however, that Marey had won several of the Académie prizes, a standard means by which a younger scientist could draw the favourable attention of the Académie to himself; when the matter was put to the vote of the entire membership, Marey was elected.

The conflict between clinical medicine and science was a continuing problem in the medical section. When the physiologist Arsonval, after several unsuccessful attempts, finally entered the Académie in 1894, the *rapporteur* Bouchard could say that only some of the members of the section supported him. Although the manuscript report of 34 small pages was a long one and ranged in the traditional manner over Arsonval's research publications, an additional argument used was that he was the favourite pupil of Brown-Séquard, whose death had caused the vacancy. He was therefore the legitimate heir to the place of his teacher.

Officially a *rapport* was the work of the entire section but it usually devolved on one man to write it. In contrast to some reports, which might be delegated to a relatively junior member, this report was considered to be the duty of a senior member of the section, often the most senior. Where comparisons were to be made, it was useful to have reports by the same author, but when the number of candidates made the task particularly onerous, or there were achievements in quite distinct branches of science to be assessed, the duty might be shared with two senior *académiciens* acting as *rapporteurs*. In that case, each *rapporteur* would be close to the specialisation of the candidate in sections such as anatomy and zoology, where both branches of science were explicitly represented. Even in the chemistry section, which had been viewed at the foundation of the Institut as a single science, advances in organic chemistry made it a discipline distinct from the traditional inorganic chemistry. Thus when Moissan, whose greatest claim to fame was the isolation of fluorine, came before the Académie in 1889 and again in 1891, the inorganic chemist Troost evaluated Moissan's work, leaving the organic chemists in the section to evaluate work in their own field.

Much work was put into these reports, and not only for the favoured candidate. Thus Brown-Séquard in 1882 obtained a 12-page report on his candidacy by Vulpian. By 1886 Marey thought it necessary to compile a 24-page report on Brown-Séquard's latest candidacy.

In 1885 Moissan's research was the subject of a report by Debray extending to 11 folio pages. Yet Moissan had to submit to three further candidacies, each with a separate report, before he was finally elected. Of course, subsequent reports did not have to start from the beginning. If a candidate had been the subject of a previous report by a colleague, it was considered legitimate to quote briefly from it. If a scientist repeated his candidacy within a few years, which was not an uncommon occurrence, a report might draw on what had been said earlier with an additional section reporting on recent work.

The *rapports* reveal that even for the experts within a section it was sometimes very difficult to reach agreement on the relative merits of rival candidates. The task of a section was to rank the candidates in order of merit. It did not matter too much if in second or third place several candidates were ranked equally, but the Académie did not like to be presented with more than one name at the top of the list. In rare cases where two first choices were proposed, this meant that the section had not been able to reach absolute agreement and was transferring its responsibility to the Académie as a whole. In 1876, the chemistry section took the unprecedented step of recommending three candidates for the first place, out of a field of eight.[21] In presenting its apologies to the Académie, the section explained that it had met on five different occasions with meetings often lasting two hours in order to resolve the problem. They were unanimous in wanting to present three candidates in first place, but they hoped that when they presented their detailed reports the chemists in other sections of the Académie would be able to choose the best candidate. This indicates incidentally that chemistry was able to have more representatives in the Académie than the six officially allowed. They included the secretary, J. B. Dumas, Berthelot in the physics section, Henri Sainte-Claire Deville in mineralogy, and Boussingault and Peligot in agriculture. The section claimed that within chemistry there was a wide diversity of scientific knowledge and it was difficult to arrive at a responsible decision over such a wide range of activities. Also it was difficult to compare work which had stood the test of time side by side with more recent work. In organic chemistry there was a great deal of very recent work which it was difficult to assess. The chemistry section made two very interesting claims: it said that many organic compounds were difficult to prepare—a reflection on the replicability of experiments—and that the language used by organic chemists was often obscure since they introduced terms without defining them. It was not only 90-year-old Chevreul who found it beyond him—it was too much

---

[21] *Presentation de la liste des candidats à la place vacante dans la section de chimie par le décès de M. Balard*, Académie des sciences archives, H. Debray dossier.

even for the other organic chemists, Wurtz and Cahours, let alone for the other members of the section, Fremy and Regnault, who were known for their work in inorganic and physical chemistry. The candidates represented different branches of chemistry. Debray (1827–88) was an inorganic and physical chemist, Friedel (1832–99) an organic chemist, and Cloëz another organic chemist and a protégé of Chevreul.

An Election in the Physics Section

Reports, where they survive at all, are scattered in the archives of the Académie and a complete set of reports for all candidates at most elections is not available. Nevertheless, one particular election illustrates the character of the contest. It also illustrates the problem raised by specialisation within the Académie. The death of Pouillet in 1868 created a vacancy in the physics section. Among the stronger candidates for whom Fizeau drew up written reports were Jamin, Janssen and Lissajous, who were all of approximately the same age and all of whom finally received some recognition from the Académie. Lissajous (1822–80), who is remembered today principally for the figures produced by vibrating bodies, was no more than a teacher at a *lycée*. The report on Lissajous gave him full credit for his work on accoustics, but the brevity of the report indicates that his candidacy was not taken very seriously by the physics section.[22] Janssen (1824–1907), one of the pioneers of astrophysics, was a much more serious candidate, although since his work was in more than one field, he was in danger of being considered less of a specialist in any one science, in this case physics. Fizeau commented that his work on spectral analysis had been the subject of several *notes* addressed to the Académie. He had also competed for the Bordin prize in 1865 and won an award. Singled out for special mention was a recent eclipse expedition in which Janssen had played a major part. His day, however, had not yet come. Janssen and other competitors had been bracketed together as the second choices of the physics section. In first place it presented Jamin (1818–86).

Fizeau's report on Jamin ran to 15 pages—three times as long as the report on Janssen. The report began by saying that several of Jamin's memoirs had been judged worthy of inclusion in the *Mémoires des savants étrangers*. Jamin's contributions to optics were outlined and particularly his work in the polarisation of light, which was related to the earlier work of Fresnel, Brewster, Malus, Cauchy, Arago and de Senarmont. It was obviously the task of the *rapporteur* to place the candidate's research within the appropriate tradition for the benefit of those members of the Académie who could not be expected to have a detailed knowledge of it. The fact that Fizeau's main research had been in optics enabled him to discuss it with a special expertise and enthusiasm.

---

[22] The actual report of 1868 on Lissajous is missing from his dossier but his position can be reconstructed from the very brief reports of his candidatures of 1863 and 1873. Lissajous never became a full member of the Académie but was elected as *correspondant* in 1879.

Optics was not, however, the only branch of physics to which Jamin had contributed. Both in his own name and in collaboration with his students, Jamin had presented to the Académie papers on electro-magnetic induction, electrical sparks, capillarity and the compressibility of liquids. This was all recent research and Fizeau pointed to it as "certain proof" that Jamin was not slowing down in his research although he was reaching the age of 50. Fizeau concluded that Jamin was the unanimous first choice of the physics section. He added that although his presentation was as a representative of the section, in a personal capacity he also supported Jamin and would cast his vote for him.

This particular election was complicated by the candidacy of an "outsider", Pierre Antoine Favre (1813–80), best known for his earlier collaborative work with Silbermann on thermochemistry.[23] Favre was a physical chemist but, because the competition for membership of the chemistry section was particularly intense and his work lay at the intersection of physics and chemistry, he thought he had a chance to be considered as a physicist. He accordingly compiled a *notice* in which most of his work was presented as falling within "physics",[24] with such sub-divisions as "heat", "electricity" and "electrodynamics". In fact, much of his work was connected with thermodynamics and he had been particularly concerned with the equivalence of heat and electricity as different forms of energy. In the discussion at the election, Favre was able to count on the powerful support of his former teacher, the chemist J. B. Dumas,[25] then permanent secretary of the Académie. Two documents provide evidence of the discussion. One consists of two pages of rough notes, drawn up by Fizeau on behalf of the physics section and containing the damning conclusion that "M. Favre is a chemist". The physics section, therefore, refused to include Favre's name on their list of recommended candidates, even at the bottom of the list. The section then had to present their recommendations to the whole Académie and here it was certain that Dumas would be the advocate of Favre. The second relevant document is marked, "*Notes prises pour répondre à M. Dumas dans la discussion des titres*", again by Fizeau. Although the physicists accepted that Favre had done work on the mechanical equivalent of heat, normally regarded as physics, they again insisted that most of his work did not belong to physics. Occasionally a strong case made before the whole Académie could override the recommendation of the specialists, but in this case when the vote was taken

---

[23] *Recherches sur les quantités de chaleur dégagées dans les actions chimiques et moleculaires* (Paris, 1853).

[24] It is interesting that the standard large French biographical dictionary, which gives nearly a full column to P. A. Favre, quotes his Académie *notice* as a principal source, not realising the special pleading involved. Balteau, J., *et al.* (eds), *Dictionnaire de biographie française* (Paris, 1933–), Vol. XIII (1971), cols 873–874.

[25] Klosterman, Leo J., "A Research School of Chemistry in the Nineteenth Century: Jean Baptiste Dumas and his Research Students", *Annals of Science*, XLIII (1985), pp. 1–80, esp. p. 18.

426

the candidate favoured by the physics section, Jamin, was elected by 37 votes to 13 for Favre out of 56 votes cast. No one at all voted for the *lycée* teacher, Lissajous, although today his name is probably known to many physicists who may never have heard of the other scientists in the contest.[26]

Successive Candidacies

It was very unusual for an *académicien* to be elected at his first candidacy. Repeated candidacies therefore became increasingly the rule. It was normal for a good candidate to increase his favour with the Académie on the successive occasions on which he stood for election. Thus as a young man he might be placed last in order of preference. He might later find himself rising from third choice to second, and finally his name might be put forward by the section as its first choice. The order of merit decided by the section, however, required full justification in the *rapports*.

As far as the public record is concerned, all the information available on elections is to be found in the *Comptes rendus*. But if one pursues research into the unpublished documents, there might occasionally be something in the record of the secret sessions of the Académie to increase one's knowledge of a particular election. The normally privately printed *notices* of the candidates set out their principal qualifications as they saw them. In practice, in the nineteenth century this increasingly meant a *catalogue raisonée* of their research publications. The *notices* by the candidates were, in principle, supplemented by manuscript reports on each candidate, but there is hardly any scientist for whom all the printed *notices* and manuscript reports survive. Nevertheless, it is possible to construct plausible accounts from the available documentation.

The documentation of the electoral fortunes of Grimaux (1835–1900) permits an account of his experiences as a candidate in the highly competitive chemistry section.

TABLE

Grimaux's Candidacies

Candidacy	Printed *notice*	Ms *rapport*	*Rapporteur*
1	—	1877	Cahours
2	—	1878	Cahours
3	1881	—	?
4	—	1884	Debray
5	1888	—	?
6	—	1889	Friedel
7	1891	(Supplement 1891)	Friedel
8	1894	—	?

SOURCE: Archives of the Académie des sciences.

[26] Lissajous was finally elected as a *correspondant* in 1879, but died the following year.

The lack of *notices* for Grimaux's candidacies in the 1870s when he was in his early forties could well mean that he did not have any printed. He was hardly a serious candidate compared to more senior figures in this period. Nevertheless, Cahours as *rapporteur* in 1877 and 1878 wrote a report filling most of five foolscap pages. The reports were very similar. It was reported that Grimaux was the author of some 27 papers, which were praised for their originality. Some of his work had practical applications. The report discussed some details of the contributions to organic chemistry, but ended with the statement that all Grimaux's experimental work had been closely related to theory, i.e. he was not guilty of blind empiricism.

By the time of the vacancy in the chemistry section, which occurred in 1881, the number of Grimaux's publications had risen to 58. By 1884 it was the turn of Debray to act as *rapporteur*. He presented Grimaux as a serious organic chemist with an interest in theory. By 1888 Grimaux listed 78 publications in his *notice*. In 1889 Friedel was the *rapporteur*. The ten quarto pages of his report began by praising Grimaux for his contributions to science over nearly 30 years. Friedel stressed Grimaux's interest in relating the properties of organic compounds to their structure. When a further vacancy occurred two years later, Friedel used the same basic text with a supplement. By now the chemistry section was unanimous in placing Grimaux first in order of merit. His *notice* listed 97 publications, but he was again rejected. On his final presentation in 1894, he produced a supplement to his *notice*, listing his most recent publications now amounting to 109. Grimaux, now approaching 60 years of age, was clearly the senior candidate. He had nearly a lifetime of solid research and publication behind him. He had earned his place.

Criteria of Assessment, Careers and Seniority

The death of the physical chemist Victor Regnault in 1878 came when there was a recrudescence of nationalism following the defeat of France in the Franco–Prussian war. The report by Debray, supporting the candidacy of Troost (1825–1911), emphasises the specifically French tradition. The positivist legacy of Regnault and many other French scientists of the time also affected the report, which concluded:

. . . M. Troost in his research has touched successfully on the most varied aspects of experimental chemistry in which he has made numerous discoveries of some importance. But what characterises his work in a special way is his research on physical chemistry which is related to the philosophy of science embodied in the work of Lavoisier and Laplace, and pursued with such success by many physicists and mathematicians of our period and in particular by M. Victor Regnault. Well prepared by a broad scientific education, he has been able to devote himself to the study of physical chemistry in which the scientists of our Académie have been successful in finding such fine laws, expressed with that wisdom, clarity and precision of which French science offers such fine models. M. Troost seemed to us to be worthy to be included among the candidates who can aspire to the place left by the death of M. V. Regnault.[27]

The character and style of the candidate was also referred to. In addition to scientific qualifications, personal qualities were sometimes brought out. Gaudry as *rapporteur* gave a useful contemporary view of Lapparent (1839–1908) at the time of his candidacy in 1895 for a vacancy in the mineralogy section: "By the vivacity of his mind, by the elegance of his speech he has made a considerable reputation for himself among scientists. In meetings of geologists whether in France or abroad, no one knows better than he how to make himself heard and to carry conviction."[28] The candidate's research was also mentioned as well as his work as a writer and editor, but the conclusion is somewhat slighting, saying merely that he was "to be included in the list of savants who aspired to the honour of being a member of the Institut". The implication is that Lapparent was a young man who knew how to make himself attractive, but was not yet ready for election. In fact Lapparent was 56 years old and only had to wait a further two years before being presented at the top of the list of candidates. The *rapporteur* in 1897 was again Gaudry, and he provided the Académie with a longer report and a constructive summary of the candidate's research:

When one looks at the whole of M. Lapparent's scientific work, one sees that his activity has been devoted for the last 34 years to an almost uninterrupted series of publications in which he has taken up in turn specialist problems of stratigraphy, general problems of the history of the earth, the didactic exposition of geology and mineralogy and finally the application of geology to geography. No one in France has contributed more than he has to publicise geology and to protest against its almost complete abandonment in university syllabuses. At the time when M. Lapparent's treatise on geology appeared, the teaching of this science in this country depended largely on translations of English and German works . . .[29]

This particular theme was taken up again in the concluding remarks of the report which spoke not only of Lapparent's charm but also said that he had done much to make French science appreciated. This time Lapparent was successful, receiving 48 votes out of 56. Indeed, Lapparent's personal qualities did not go unremarked by his colleagues and ten years later he was elected as one of the two permanent secretaries of the Académie, but he died within 12 months of that election.

When *rapporteurs* were faced with a number of good candidates they often had recourse to considerations of seniority. Thus after the unsuccessful candidacy in 1876 of Debray (1827–88) for a place in the chemistry section, he had the good fortune of a second candidacy within 12 months. The anonymous report referred pointedly to his "long career"—he was now 49—and the Académie elected him.[30] When several years later

[27] Académie des sciences archives, Troost dossier.
[28] *Ibid.*, Lapparent dossier.
[29] *Ibid.*
[30] Académie des sciences archives, Debray dossier.

Debray himself had the task of writing a report on the candidacy of Troost, placed first in the list by the chemistry section, he pointed out that Troost had been publishing the results of his research over a period of nearly 30 years.[31] At the same meeting of 30 June, 1884, Debray had to consider the candidacies of Gautier (1837–1920), Grimaux (1835–1900) and Jungfleisch (1839–1916).[32] He explained that the chemistry section had agreed to place these three names in the third rank since they were all younger than Schutzenberger (1829–97), the candidate placed in the second rank. In a telling phrase, Debray said that the research of these relatively young men "had not yet received the precious consecration of time", nor had they yet achieved "the same total of discoveries".[33] This represents an interesting cumulative view of science in which quantity was stressed at the expense of quality. Debray went on to say that "when their hour came" they would have added to their current qualifications. In due course, all three chemists were elected—in 1889, 1894 and 1909 respectively.

It always helped a scientist to emphasise his discoveries. Picard recalled that he had once heard Fizeau when presenting the merits of two candidates in a secret session speak of the "discoveries" (*découvertes*) of the one but the "lucky finds" (*trouvailles*) of the other.[34] It was clear to the Académie that Fizeau had little sympathy with "lucky finds". Here the vocabulary rather than the enthusiasm and length of presentation indicated the bias of the *rapporteur*.

The Assessment of Foreign Candidates

The Académie was mainly concerned with French scientists, but through its corresponding members and *associés étrangers* it also had contact with eminent scientists in other countries. A corresponding member was a person of some scientific eminence, who was not qualified for full membership because of his residence outside Paris. Most were French, although there were also an appreciable number of foreigners. But, whereas corresponding members (*correspondants*) were in a sense second-class members of the Académie, it was a special honour granted to a foreigner to be elected as *associé étranger*, since the number was limited to eight for the entire world. Indeed, a French *correspondant* of the Académie living in a foreign country might hope to be promoted to *associé étranger* if he were eminent enough.

By the late 1860s, the British Astronomer Royal, George Biddel Airy (1801–92), already a corresponding member since 1835, was considered for promotion. The manuscript report was drawn up by the astronomer Delaunay, who pointed out that Airy had been Astronomer Royal for over 30 years. As director of the Greenwich Observatory, he was responsible for

---

[31] *Ibid.*, Troost dossier.
[32] *Ibid.*, Gautier dossier.
[33] "*La même somme de découvertes.*"
[34] *Mémoires de l'Académie des sciences de l'institut*, LVIII (1926), p. xxxiv.

VIII

an annual publication of observations, which provided valuable data. He had carried out improvements at Greenwich and had also done research in a number of different fields—five are mentioned. Yet the size of the report—four small pages—indicates that Airy, although given a fair trial, was not yet a favourite candidate.

When he was considered again for a vacancy as *associé étranger* in 1872, a more detailed report of five folio pages was drawn up by Faye. Airy was then presented as one of the grand old men of science. His first publications, on the shape and dimension of the earth, went back to 1826. This was the period of Young and Fresnel, when the wave theory of light was necessitating a revision of physical theory. The report, after reviewing Airy's research, concluded:

Such, gentlemen, is the scientific career of Mr Airy, at least in so far as one can analyse in a simple report a series of publications which appeared over the last 45 years, at the rate of one or two quarto volumes per year. A profound physicist, an able mathematician, a careful observer, an eminent engineer and inventor, Mr Airy is considered in all civilised countries of the old and new world as representing the complete scientist [*comme le type du savant complet*], combining theory and practice at the same time and with the same power, always in the van of activity and progress. That is not all; this scientist, this professor, is also an administrator of the first order, whose upright and firm mind, stable character, perseverence and untiring activity have raised higher than ever one of the premier institutions of the scientific world [i.e. the Greenwich Observatory] . . .[35]

A final title to eminence was that Airy was currently president of the Royal Society.[36] With such qualifications, Airy was elected.

Private evaluations are always liable to bias and misrepresentation and the foreign scientist was probably particularly vulnerable. Of the scores of reports examined, only one stands out as an example of distortion of the most chauvinistic kind. It concerns the candidacy in 1892 of Joseph Lister (1827–1912) for the position of *associé étranger*, a position which Lord Lister was eventually to regard as one of his greatest honours. In 1892, however, the task of compiling the report fell to Bouchard, who was far from being an enthusiastic supporter of antiseptic surgery which was Lister's main claim to fame. In an unprecedented way Bouchard began his report on a strong negative note:

Mr Lister did not discover the animate nature of the agents of infection, nor the application of antiseptics to the treatment of wounds; he is not even the inventor [*l'inventeur*] of carbolic acid or of its use in operations. I do not hesitate to acknowledge that his theory of the source of serious accidents which complicate wounds is true only for a small number of cases and that the doctrine which we must accept today and which he himself adopts is that which Mr Le Fort proposed in this country in 1865, two years before the work of Lister . . .[37]

Lister is thus presented as an opportunist who had taken Pasteur's work and

[35] Académie des sciences archives, Airy dossier.
[36] November 1871–November 1873.
[37] Académie des sciences archives, Lister dossier.

applied it to the treatment of wounds. What was needed was an antiseptic, said Bouchard. He continued: "Carbolic acid was this substance which was able to kill germs while not damaging the tissues. In this country M. Déclat made a panacea of it from 1865 and since 1861 Lemaire had already shown its good effects on festering wounds and published on this medicament in 1863 and 1865 two editions of a book which has been largely overlooked". This report was unique both in minimising the contributions of the person it was supposed to be supporting and in finding French predecessors for all aspects of Lister's work. Having written such a maliciously condemnatory report, Bouchard did not attend the relevant meeting and it was left to his colleague in the section of medicine, Brown-Séquard, to present it. With such advocacy, Lister did not receive the necessary majority of votes.

Another vacancy arose in the very next year and this time the task of presenting the merits of Lister was given to Charcot, already famous for his studies of hysteria at the Salpêtrière. The fact that he was an expert in a very different branch of medicine did not prevent him from carrying out this task well. He was positive where Bouchard had been negative. He mentioned the innovations of Lister and pointed to his international reputation. Lister did owe a debt to Pasteur and had freely acknowledged it in his famous paper of 1867. What Charcot took special pains to emphasise, however, and this was characteristic of Académie reports, was Lister's claims to have used scientific methods. Lister was not to be honoured as a famous surgeon but as a scientist who did systematic experiments: "One cannot say too often", said Charcot, "that one of the fundamental characteristics of the researches of Mr Lister is that they have always been pursued simultaneously in the laboratory and in the hospital ward."[38] In other words Lister was presented as the representative of scientific medicine. Accordingly he was elected *associé étranger* on 6 March, 1893.

If an Englishman like Lister could suffer from chauvinism in the Académie, the chances of a German, against the background of the great bitterness towards Germany after the French defeat in 1870–71, were not at all good. The only nineteenth-century figure comparable to Pasteur and Lister was Robert Koch (1843–1910), often considered the founder of bacteriology. The occasion to elect Koch as *associé étranger* came in 1903 with the death of another German medical scientist, Virchow. Koch was the obvious choice for a replacement. Pierre Roux, the former assistant of Pasteur, would not admit in his report[39] that Koch was superior to Pasteur in any way, but he expressed unqualified admiration for his isolation of the tubercular bacillus 20 years earlier. The fact that it took so long for the Académie to recognise the important contributions of both Koch and Lister was only partly the consequence of chauvinism. It was also the result of the great conservatism of the French medical profession and the unwillingness of French scientists to allow the fame of Pasteur to be diminished by

[38] *Ibid.*
[39] Académie des sciences archives, Koch dossier.

432

comparison with others who had developed his work in important practical ways.

Conclusion

Despite very occasional problems of nationalism or—more important—personal favouritism, its reports show that on the whole the Académie des sciences went to great lengths to proceed objectively to debate the merits of rival candidates for election.[40] The systematic collection and presentation of relevant data are impressive. From these data an evaluation was made of a scientist's whole career with special emphasis on research. In most cases, the persons drawing up the reports were among the best qualified to make the judgement. But they were obliged to explain to *académiciens* from other disciplines what the research meant. The ideal report therefore, while mentioning some of the technical details of the research, also brought out general principles of scientific work. There are many cases where method is emphasised, often with a streak of positivism. Thus favourable attention is drawn to the patient collection of data and its correlation, the very antithesis of wild speculation. Reports for successful candidates also often call attention to a long period of devotion to a subject, marked by a succession of publications. Although the candidate could list his publications, it was for his peers in the Académie to evaluate them and in the reports of the Académie we have a record of such evaluation which, since it covers all branches of science and at the highest level, is not only important but unique. The reports made explicit criteria for the evaluation of scientific excellence which helps us understand better the development of the profession of science.

[40] Considering that the Académie des sciences with its judgemental role and major prizes was well known to Alfred Nobel, who lived in Paris for several years, one might see the Académie as providing a precedent for the establishment of the Nobel prizes. In particular, nineteenth-century Académie reports might be compared with the reports submitted after 1900 for Nobel prizes. A recent study of the Nobel archives in Sweden has revealed that the reports written for awarding committees were hardly reasonably objective assessments of the work of a range of outstanding candidates, but rather a subjective *post facto* justification of the choice of a particular candidate. History seems to have largely vindicated the objectivity of the French Académie as compared with the Nobel prize committee. Crawford, Elisabeth, *The Beginnings of the Nobel Institution: The Science Prizes, 1901–1915* (Cambridge: Cambridge University Press, 1984), reviewed by Pinch, Trevor, *Times Higher Educational Supplement*, 13 September, 1985, p. 19.

IX

The Emergence of Research Grants within the Prize System of the French Academy of Sciences, 1795–1914

Prizes, particularly if they have some official status, may constitute powerful symbols of excellence as well as financial rewards. By the mid-eighteenth century many academies had come to adopt the practice of offering prizes from time to time for both literary and scientific subjects; and no country possessed so many important literary and scientific academies as France.[1] Some provincial academies in France offered prizes on broad intellectual issues and it was largely as a result of winning such a prize, offered by the Dijon Academy in 1750, that Jean-Jacques Rousseau rose to fame. An ambitious man, whether pursuing a literary or a scientific career, might enter for the prize competitions of several academies in the hope of winning a name for himself. By the end of the *ancien régime* the idea of honorific prizes was well established in the public mind.

Yet, despite the vitality of several provincial academies in France in the eighteenth century, none could equal in importance the institutions of Paris. As far as science and technology are concerned, the most prestigious and the most numerous prizes were those offered during the eighteenth century by the Royal Academy of Sciences, in Paris. These prizes served as a precedent for the nineteenth-century Paris Academy, which became effectively the *national* Academy, serving the whole of France. The Academy continued to provide prizes which were largely honorific but, as science became more professionalized and research more sophisticated, men of science became increasingly concerned with the costs of scientific enquiry. Science was no longer simply a hobby. The new system of higher education, concentrated in Paris, provided salaries for men of science,[2] but the cost of apparatus and equipment constituted a barrier, restricting ambitious laboratory science as well as field work in foreign countries.

In nineteenth-century Britain, a *laissez-faire* philosophy encouraged personal initiative[3] and private patronage, but in France there was a contrasting tradition of state-financed science. From the eighteenth century, senior members (*pensionnaires*) of the Royal Academy of Sciences received modest salaries and there were several important state-sponsored scientific expeditions, notably those concerned with studying the shape of the Earth. There was also some state patronage of science in an industrial context. In the absence of a patent system in France before the Revolution of 1789, inventors were sometimes rewarded with monopoly rights, subsidies or pensions by way of recompense.[4]

Although there were no grants for science, there were prizes administered by the Paris Academy. These prizes were not awarded until the research had been completed or its success demonstrated. It will be the task of this paper to show how this prize system was transformed from an honorary reward system into a monetary system. Once monetarized, the prize system could lend itself to becoming partly a grant system. The Academy itself did not make an explicit distinction between honorific prizes (consisting of a medal)[5] on the one hand, and monetary prizes (including grants) on the other, and the very existence of grants has therefore been disguised from historians and sociologists of science.[6]

In 1980, two articles made important preliminary contributions to the study of Academy prizes in the nineteenth century. One, dealing with several aspects of the emergence of the grant system, untangled the complex finances of the Montyon legacy (to be explained below), and argued that this legacy had considerable (though hidden) influence on many activities of the Academy.[7] Another, by Elizabeth Crawford,[8] focused on the large number of private donations to the Academy from the 1880s onwards. Although the latter provided a valuable sociological analysis of the prize system, it was limited by being confined to the *second* half of the nineteenth century, and by being based exclusively on *printed* sources.

The present paper is able to take a longer term view, and makes use both of the published record and of archival sources, which often reveal what was going on behind the scenes. We wish to show that grants developed much earlier than has hitherto been recognized.[9] The fact that the prizes and grants were given by the Academy means that this precedent was of the greatest importance, since the Academy in Paris was at the centre of French science, and in a unique position to reward and encourage outstanding scientific achievement. We shall see that throughout the nineteenth century the Academy acquired an increasing variety of funds, which it was able to use for many different purposes. In the 1820s it was embarrassed by a large legacy from Baron Montyon, but it gradually learned to use these funds for a variety of purposes, including grants. It would be naive to assume that the story is an uncomplicated one, in which increasingly large sums of money were simply given to the Academy to provide grants. The traditional idea of prizes as rewards for past achievement was deeply ingrained in the Academy, and to make a more flexible system required several decades of negotiation both between the different interest groups within the Academy and between the Academy and potential donors.

In the eighteenth century, the Paris Academy provided a model for scientific societies in other countries.[10] The prize system became an important feature of the Academy which, by the end of the nineteenth century, was awarding prizes with a total value much greater than that of any other European scientific society.[11] Although the Royal Academy of Sciences (along with the other Royal Academies) was suppressed in 1793,[12] it was reborn as the First Class of the National Institute in 1795, and assumed its former title of Academy of Sciences in 1816 with the Restoration of the Bourbons.

According to its constitution, the First Class was to award an annual *grand prix*. One feature of the nineteenth-century Academy was the way in which this prestigious state-sponsored prize was complemented by a number of prizes offered by private donors. The most important single donor was Baron Jean-Baptiste Antoine Montyon (1733–1820), who left large sums of money mainly for philanthropic purposes. Later, several scientists left modest sums of money to the Academy to found prizes in order to further their own respective branches of science. There were also many wealthy donors in the latter part of the nineteenth century, who left money to the Academy in order to commemorate their family names. It is worthy of note that most of the donors favoured the traditional prize system and ceremonies associated with recognized achievement, and only a small minority considered anything approaching a grant for aspiring researchers. It was the members of the Academy rather than the donors who saw the value of grants, and sometimes tried to make subtle alterations in the terms of prize legacies.

To suggest a continuous and conscious desire throughout the nineteenth century to produce something approaching the modern grant system would be a historiographical error. What we in fact witness is a fascinating period of transition between a traditional view of prizes as honorific awards given after completion of work of a very high standard, and a nineteenth-century concept of the use of prize funds given *in advance* to promising younger scientists to enable research to be undertaken in the future. We must patiently follow the historical development of this transition, taking into account the wishes of donors of prizes and the negotiation of new principles between different groups of scientists within the Academy.

The Honorific System of Prizes in the Early Nineteenth Century

Two main features characterize Academy prizes in the early nineteenth century. First, prizes consisted of medals, honorific symbols of success, awarded *after* the contributions to science had been made. In cases where work submitted for the prize was not considered of a sufficiently high standard, the award of a prize would be postponed, a not infrequent occurrence. Secondly, the award of medals was intensely élitist. These main features of the honorific prize system can be easily characterized: (1) the number of medals awarded was small, which helped to increase their prestige; (2) prizes were awarded only for work that was outstanding; and (3) their conferment was intended not only to reward a few *savants* but also to stimulate others.

In the early nineteenth century the Academy regularly awarded one honorific prize, the *grand prix*, which had even greater importance than the common prizes (*prix*). The significance of the *grand prix* derived from the fact that, through it, the Academy guided research, by offering the *grand prix* for the best answer to a major scientific problem selected by a group of expert members. The prize drew the attention of first-rank men of science outside the Academy towards problems which were important in the eyes of Academicians. These scientific problems were called 'prize questions' ('*questions de prix*'). In the early nineteenth century, they were regularly selected every two years in each of the two broad Academy divisions, for the mathematical and for the natural sciences. Due to the extreme difficulty that the successful solution of prize questions implied, especially if they involved higher mathematics, only a few prize entries (usually less than half a dozen) were regularly submitted to the Academy, but they were normally of a very high standard. During the early nineteenth century the Academy fostered outstanding research through prize questions, especially in the mathematical sciences. Brilliant young scientists like Etienne Louis Malus, Joseph Fourier, Pierre Louis Dulong and Augustin Fresnel successfully solved Academy prize questions in that period.[13] Without considering the extraordinary importance that the Academy attached to the *grand prix* it would not be possible to understand fully the emergence of grants from the 1820s.

The honorific prize system, revolving around prize questions, was necessarily élitist and consequently useful only for a limited number of *savants*. This spirit was in marked contrast to the demands placed upon the Academy throughout the nineteenth century. Even before monetary prizes emerged, Academicians began to consider the desirability of a more open system, in which *savants* would be free to choose their own subjects of research. Prize questions came to be replaced by a different kind of honorary prize, awarded retrospectively for the best work in a general field. The change from prize questions to retrospective prizes provides a second axis in the story. From the perspective of the Academy the decline in prize questions in the second half of the nineteenth century meant a decline of its influence on the course of scientific research. But this was compensated by its grants to influence future research.

If we are to understand the emergence of grants, the vital question to consider is not the growth of retrospective prizes, but the conversion of honorary prizes to monetary prizes.[14] Before passing from honorific to monetary awards, however, it is necessary to consider certain differences in the administration of monetary awards. While a mechanism to produce a carefully evaluated order of merit in scientific research had been developed as essential for the conferment of honorific awards, this was inappropriate for the allocation of monetary awards, for several reasons. First, honorific prizes (medals) were symbols of success, and success could only be recognized after research had been completed. The conferment of honorific prizes thus involved an *a posteriori* assessment. Monetary awards, in contrast, were practical means to assist researchers, whose assessment centred on the potential of their research programmes.

The honorific mechanism was unsuitable also because it was governed by traditional ideas of merit. Honorific prizes were conferred on those who deserved recognition. Monetary awards were given to those men of science who needed financial support, although they would also have been judged as having suitable research potential. These two main criteria were very different, because *savants* who deserved public honour did not necessarily need financial support; others simply needed money. One such was the young medical doctor Nicolas Deleau (1797–1862), who received a total of Ffr 16,500 from the Academy in four different *encouragements* (1826, 1828, 1829 and 1832). Deleau needed the money to do research on the speech of deaf children. Financial need was stressed in the Academy not only by Deleau, when he solicited help, but also by the *rapporteur*, Etienne Geoffroy Saint-Hilaire, who considered his research programme promising, and strongly supported his request.[15]

While honour was a factor generated and controlled by the Academy itself, funds came largely from private donors. The monetarization of the reward system during the third quarter of the nineteenth century, which weakened the institutional *grand prix*, brought about intense competition among the Academy's individual sections. For example, after the valuable Jecker contest began to be organized by the chemistry section alone, other sections wanted similar valuable funds. The absence of a central authority in the reward system in that period was accompanied by the lack of a specific general policy for rewards.

A few words may be said about the mechanics of the prize system. Subjects for the *grand prix* were announced at the annual public meeting, which was widely reported in the press. The prizes were given further publicity in printed programmes and in the *Comptes rendus*. In the early nineteenth century, prize memoirs had to be submitted anonymously and before a deadline. For retrospective prizes, contestants would submit recent work, although for major prizes they might submit a bundle of off-prints and the prize would be awarded for half a lifetime's work (*'pour l'ensemble de ses travaux'*). There were some important exceptions to the direct submission of work by candidates (for example, the Lacaze prize, where members of the judging commission sometimes took the initiative and made their own nominations on a secret list). Such nominees had a greater chance of success than normal applicants. Yet on the whole the judging system seems to have been remarkably objective. Judging commissions drew on a wide range of expertise within the Academy and sometimes *savants* would be rewarded for work contrary to the prevailing orthodoxy. An example of this is the award of a *grand prix* to Fresnel in 1821 for a memoir on the wave theory of light, which proposed arguments contrary to the ideas of the influential Academician, Laplace.

For grants there was no formal applications procedure until the early twentieth century. The *encouragements* that were given depended on a system of private patronage based on informal contact between members of the Academy on the one hand and their junior associates, students and other active members of the scientific community on the other.[16] An Academician who held a position at one of the many institutions of higher education in Paris, might propose one of his own students who was particularly able. Alternatively, the bond of patronage might be forged by a common allegiance to one of the *grandes écoles*; thus a graduate of the *Ecole Polytechnique* might hope for support from an Academician who was also a *polytechnicien*. In any case, a young scientist hoping for financial support from the Academy would need to be well known to at least one Academician and convince him of his ability and the feasibility of a particular project. The Academician would then speak up for his protégé on a suitable occasion when the Academy was meeting in secret session. There are few records of the arguments used in support of early grants; the most that one can hope for is a record of the decision to give a specific sum of money, and even this was usually a *private* record. One suspects that the efficacy of the earliest 'grant applications' depended most of all on the political muscle of the patron.

In general we have no reason to doubt that money given by the Academy was actually used for the purpose stated. Let us take some examples from the mid-century. Exceptionally, when Fizeau was given Ffr 3700 in 1856 'for the construction of apparatus. . . required for his work on the velocity of light', a note was added to say that he was expected to provide the Academy with an invoice.[17] In other cases, such as when Quatrefages was given Ffr 2000 in 1859 in connection with a journey to the Midi to continue his research on silkworms,[18] it is less clear that the grant corresponded to the actual cost of the expedition. The sum in round figures suggests that the 'grant' was essentially a subsidy. On the other hand, when the young palaeontologist Gaudry was given the large sum of Ffr 8000 in 1860 to study certain fossil bones of quadrupeds in Greece, it is quite clear that he could never have undertaken this prolonged research in a foreign country without the money and, moreover, we learn that he was carrying out the research at the specific request of the sections of zoology and geology of the Academy.[19]

The Arrival of Monetary Prizes: The Montyon Fund

By his will of 1820, Baron Montyon left the Academy of Sciences a large sum of money to be used for prizes in medicine.[20] One prize was for any contribution to medicine or surgery and a second prize was specifically related to improving public health and hygiene in relation to unhealthy occupations. Discussion over the implementation of the will continued for several years and it was only in 1825 that the first Montyon medicine prizes were awarded. The capital sum which constituted the Montyon medicine fund was by now producing an annual interest of approximately Ffr 20,000 (almost £40,000 at today's values). The legacy was the largest in the history of the Academy and remained so until the late nineteenth century. We shall see how its very size unbalanced the whole prize system of the Academy. It became almost an embarrassment for the Academy to use this money in prizes for medicine. It produced an anomalous situation since the most prestigious prize awarded by the Academy, the *grand prix*, consisted of a medal to the value of only Ffr 3000.[21] The unexpected and unplanned large sums of money in the Montyon account guided the Academy towards the monetarization of the prize system. Whereas previously there had normally been a single prize winner, with perhaps some honorary mentions, the Montyon award became a multiple prize system with different sums of money awarded, graded according to merit.

We must point out that the medical interest in the Academy was confined to only one section ('medicine and surgery') of the eleven specialist sections.[22] Some of the Academicians even questioned the relevance of science to medicine, which had been favoured by Montyon mainly for philanthropic reasons. On the other hand, Paris in the early nineteenth century had become an international centre for hospital-based medicine.[23] There was a large pool of potential candidates for medical prizes, since medicine was a thriving profession which had recently improved in status through its claims to be scientific.

Although the terms of Montyon's will limited the prize fund to medicine, the Academy, overwhelmed by the size of the fund, was in the exceptional position of being unable to state in advance the value of the prizes:

> The sums of money which will be awarded to authors of discoveries or works rewarded [by the Academy] cannot be indicated exactly in advance, because the number of prizes has not been determined; but *it is possible for the sums of money available to exceed by a considerable amount the value of the largest prizes awarded up to the present.* The liberality of the founder and the order of the King have given the Academy the means of raising these prizes to a considerable value, so that *authors may be compensated for costly experiments or research which they have undertaken* and will receive recompense proportional to the service which they have rendered, either by preventing or considerably diminishing the unhealthy effects of certain trades or by contribution to the improvement of the medical sciences.[24]

The Montyon Surplus

In 1825, when Montyon prizes began to be awarded, the Academy was dominated by the idea implicit in the *grand prix* — that is, the idea of the prize-winner as a model to be emulated. The main awards were exclusively for outstanding research, and for research fully corroborated. The bestowal of a main award was a great responsibility, and meant the full approval by the Academy of a medical idea or technique that was to be used on human beings. At the beginning of the Montyon medicine contest, the Academy noticed that several candidates, in order to win awards, deliberately concealed aspects of their researches in which they had obtained negative results. This attitude increased the precautions taken by the Academy and slowed down or even halted the process of evaluation of competing research. In 1826 the Academy decided not to give any Montyon main award. In their report, the members of the commission emphasized:

> The Academy has considered it necessary to delay its final judgement on important work, because its prize commissions have not been completely convinced of [the veracity of] the results presented. Candidates have described in detail the cases in which they have thoroughly succeeded, but have said nothing about the cases in which they have failed.[25]

The Academy announced that it would give a main award the following year only 'if candidates become aware that they should give science an exact account not only of their positive results, but also of the obstacles, reverses and relapses observed'.[26] Because of this greater moral responsibility in the case of medicine, the Academy reduced the number of large awards.

Since the sum available every year from the Montyon medicine and surgery prize fund was nearly Ffr 20,000 and only Ffr 4000 had been expended in awards in 1825, there was consequently a surplus of about Ffr 16,000 in the Montyon account. Two factors led to the emergence of several valuable subsidiary rewards that lacked the great honour associated with a main award: the pressure to consume a surplus that would increase enormously in the near future, and also the reluctance of the Academy to give main awards. In 1826, the second year of the contest, the Academy announced:

> No main prize (*grand prix*) for original contributions to medicine or to surgery will be awarded [this year]..., but, from the surplus destined to that noble aim by M. Montyon, a sum of 16,000 francs will be given to candidates...as *encouragements.*[27]

In 1827, when the excess had increased to Ffr 40,000 (corresponding to the interest of 1826 and 1827), the Academy gave Ffr 16,500 in nine *encouragements*.

The new and very important change in the Montyon medicine contest, and indirectly in the whole of the reward system, was the emergence of dozens of comparatively small monetary awards. Their number and total financial value greatly surpassed those of the main awards, even though the latter maintained their higher honorific rank. Between 1825 and 1842, the Academy spent a total of Ffr 283,000 in Montyon medicine awards. Of this total, Ffr 68,000 (25%) corresponded to eight main awards and the rest — that is, Ffr 215,000 (75%) — to more than one hundred secondary awards.

In 1831, the Academy stated that not only promising research could be stimulated by prizes but also research that had not been begun. The commission, addressing the Academy, said that the confirmation of the awards

> would be a legitimate recognition of useful work already accomplished and a strong encouragement for the completion or for the beginning (*'pour faire entreprendre'*) of research, which was as important as the accomplished research originally proposed for reward.[28]

This conception of rewards as help given before research was carried out, and not afterwards, represents a major innovation in the Academy's reward system, and it brought about some other important changes.

First, because research had not begun when the Academy gave its blessing, the focus of the Academy's assessment shifted from the 'memoir' to the 'man'. This innovation in the Academy's system appeared for the first time in the Montyon contest for medicine. Second, the relationship of the Academy with the individual recipients of the Montyon medicine awards became very close. Recipients of cash rewards had an implicit obligation towards the Academy; and the Academy had to develop evaluating criteria for their competence and professional responsibility. By the early twentieth century, applicants for awards might be expected to be personally known to an Academician or senior professor and also to have some higher qualifications in science — for example, a doctorate (see Table 3 [p. 90, below], last column).

In the first quarter of the nineteenth century, it had been the exception for a scientist to win more than one prize. In the Montyon medicine competition, a new kind of winner appeared in the Academy's sweepstakes: the 'recurrent' winner — that is, the recipient of a number of awards over a period of several years. In the late 1820s and 1830s, there emerged the practice of competing for Academy rewards on a regular basis, very probably because of the great and permanent disproportion between the enormous sum of money available for medicine awards every year, and the relatively small number of medical doctors considered brilliant enough to win prizes outright.

The Montyon surplus funds gave the Academy a new sense of freedom. It was, for example, in the interest of most Academicians *not* to award Montyon prizes so as to build up its surplus. This surplus permitted the Academy to begin the publication of a weekly journal of its proceedings, the famous *Comptes rendus*, for which the government had made no initial financial provision.[29] However, the diversion of Montyon funds to cover publication costs had the effect of reducing the money available for other purposes, which in turn reduced the number of candidates for Montyon prizes, the number of which had been steadily increasing from 1825 to 1835. The graph in Figure 1 reveals a close relation between supply (availability of money) and demand, suggesting good informal communications between the Academy and the scientific community.

Another reason for the decline in the number of candidates in the late 1830s and 1840s,[30] was that the Academy deliberately raised their standard, thus discouraging less able candidates. But the number of candidates in the 1840s also declined because many non-medical Academicians, notably among the chemists, objected to a distortion of the prize system in a way which favoured so blatantly only one of the eleven subject areas represented. The Academy consequently agreed to restrict both the number of Montyon prizes awarded and their total value.[31]

It was only in the 1850s that the number of candidates for Montyon prizes suddenly increased. This feature, clearly shown in Figure 1, requires an explanation. The prize system of the Academy (and particularly the setting of prize questions) had from the beginning demanded some general consensus among Academicians. After a period of jealousy and restriction in the 1840s, we shall refer to a more harmonious state of affairs in the 1850s,[32] when a negotiated settlement between competing interests in the Academy brought widespread benefits from the use of the large accumulation of Montyon funds.

The Use of the Montyon Surplus as 'Encouragements'

By 1850, a large part of the printing costs of the *Comptes rendus* was covered by a regular government allocation, thus freeing the Montyon

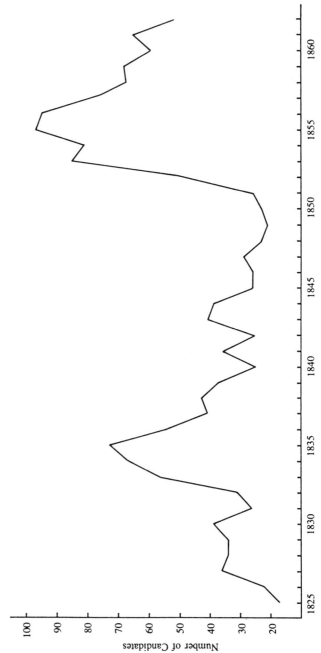

FIGURE 1
Number of candidates for Montyon prizes for medicine

fund surplus for other purposes. Moreover, in the 1840s, the chemistry section (whose members regularly collaborated with the medicine section in judging Montyon prizes) had effectively restricted the amount of money given in these prizes. It was the chemists, led by the future secretary J. B. Dumas, who insisted on a more equitable distribution of Montyon money for the benefit of all fields of science. By the early 1850s, after a decade of restrictions both on Montyon medicine prizes and the use of the Montyon surplus, an agreement was reached between the medicine section on the one hand, and the non-medical majority on the other. The latter would withdraw their objection to lavish medicine prizes publicly awarded if they were allowed to make generous use of the substantial Montyon surplus in the form of private *encouragements* for non-medical subjects.

In the 1850s, therefore, the sums of money awarded from the Montyon funds for prizes in medicine were considerably increased (from less than Ffr 10,000 annually to more than Ffr 25,000 in 1852, 1853 and 1854), and the number of medical candidates for Montyon prizes increased proportionately (see Figure 1). At the same time, generous grants were awarded privately in areas quite unconnected with medicine.[33] The sum awarded in *encouragements* in 1852 was Ffr 12,500; in 1855 it was more than Ffr 20,000, and in one year (1856) more than Ffr 30,000. These were really bonanza years for French researchers, who heard by word of mouth of the opportunities, usually through patrons within the Academy. Biological scientists received grants to pay for numerous plates in their respective publications. Those involved in field work abroad were given generous expenses,[34] the best example being the palaeontologist Albert Gaudry, who received Ffr 6000 in 1855 and a further Ffr 8000 in 1860 to enable him to spend several months in Greece, studying fossils. In 1868, the astronomer Jules Janssen was authorized to spend the large sum of Ffr 12,000 in three equal instalments on an astronomical expedition to India. Within physics, Hippolyte Fizeau received a grant in 1856 to pay for his apparatus to determine the velocity of light, and César Despretz benefited from a succession of grants for research on electric batteries. In chemistry, Charles Gerhardt received Ffr 2000 in 1853 for research on organic chemistry, and Henri Sainte-Claire Deville was allocated an equal sum in 1854 to continue his preparation of pure aluminium. Thus most subjects, except mathematics, benefited considerably from the new policy of liberality in the use of the Montyon surplus. When an impoverished descendant of the astronomer Lacaille received Ffr 1000 from the fund in 1857 it was noted simply as '*secours*' (that is, general financial help) — a reminder of the charitable connotations of grants at this time. On 20 February 1860, Louis Pasteur, who had recently been awarded a prize medal (worth Ffr 900) for experimental physiology was awarded a sum of Ffr 2000 *'en accroissement du prix'* — that is, in order to increase the value of the prize.[35] The Academy was in fact showing that it appreciated that an active young scientist like Pasteur needed money even more than honour, in order to be in a position to carry out further research.

Although the public and the medical profession thought of the Montyon fund as one benefiting medicine, for the non-medical majority of the Academy, Montyon meant a release from the tight financial control of state funds. From the point of view of the historian, the main prizes (*'prix'*) of this time were much less important and certainly less innovatory than the *encouragements*, even though historically the residual Montyon funds which paid for them were subsidiary to the Montyon prizes. Modern scholars have failed to appreciate the importance of *encouragements*, which have lain hidden in the shadow of prizes. Yet *encouragements* not only provided money which could be used for further research, but also provided a psychological push, suggested by the literal translation of their name — they *encouraged* scientists to continue their research. On the other hand, *savants* were often too proud to accept charity and few wished to admit their poverty publicly. Thus when Gegner founded an annual prize of Ffr 4000 in 1869 for 'poor scientists', many Academicians urged that this phrase should be omitted from the published offer of the prize (or grant), since they felt that it would discourage applicants.[36]

The Monetarization of the Prize System in the Second Half of the Nineteenth Century

We are now in a position to review the progressive monetarization of the prize system throughout the nineteenth century. It is clear that the Montyon legacy had widespread repercussions for the Academy. The first of these was the monetarization of the prize system. The decision to award actual money was to open up the prize system and eventually to transform it. Even in the most honorific of all Academy prizes, the *grand prix*, where medals continued to be awarded as first prize, 'runners up' were from 1825 sometimes given small monetary *encouragements* rather than honorary mentions. In the 1850s, the Academy began to accept an increasing number of new prize funds, notably the Jecker prize for chemistry. In the great majority of these new prize contests, money was to be awarded rather than traditional medals. By the end of the 1850s, the Academy annually awarded more than Ffr 40,000 in monetary prizes. By now the ten monetary prize contests had surpassed the number of honorific prizes (nine). (See Figure 2.) By the end of the century, the Academy awarded nearly Ffr 150,000 annually in monetary prizes and the number of monetary contests (fifty-eight) had risen to nearly four times that of honorific contests (twelve). If we consider the total sums of money awarded, instead of the number of contests, the disproportion is even greater, because of the establishment from the 1870s of several very large competitions involving monetary prizes.

FIGURE 2
Total Number of Contests Organized by the Academy

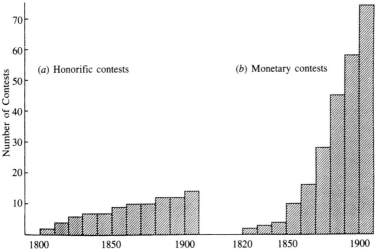

The early 1850s constituted a watershed in the history of Academy prize funds, not only because it was a period of great liberality in the use of Montyon funds as grants, but also because it was the beginning of a period when the Academy began to accept an increasing number of legacies (see Figure 2(b)). Some of these legacies came from people who had previously been awarded Montyon *encouragements*. The new funds, especially the valuable ones like the Jecker chemistry fund (1851), constituted a threat to the traditional honorific kind of reward system, characterized by the multidisciplinary *grand prix*. By accepting such gifts from the beginning of the Second Empire (1852–70), the Academy opened its doors to the full monetarization of its reward system.

The first cash awards from the Jecker Fund for the rapidly expanding science of organic chemistry were distributed in 1857. The Academy announced that the legacy was intended 'to accelerate the progress of organic chemistry',[37] which suggests that it was looking into the future rather than into the past. Nevertheless, in practice, the Jecker prize was most often given for past achievement, which was more easily judged, than future promise. Its importance lies in the fact that it was money, which allowed the Academy considerable flexibility in using the prize as an incentive system. The chemistry section had at its disposal an annual sum of Ffr 5000, to be divided at its discretion. It opted for giving several medium-sized monetary awards over a period of several years rather than fewer and larger sums of money only once. This policy permitted the Academicians to give financial support repeatedly to the same candidates. In the first two decades of the contest, for example, Marcellin Berthelot, Charles Friedel, Auguste Cahours and Adolphe Wurtz all received the Jecker award twice, and S. Cloez three times. Berthelot and Friedel were only thirty-three years old when they first won the prize.

As the prize system grew during the second half of the nineteenth century, and also as the terms of prize legacies by private donors became more open, the Academy was able to develop and expand the policy of rewarding the same scientist repeatedly — for instance, by awarding him several monetary prizes, but each from a different prize fund. Pierre and Marie Curie, for example, two of the scientists who most benefited from the Academy prize system at the end of the nineteenth and the beginning of the twentieth century, received half a dozen different prizes (see Table 1). The only award that they received several times was the Gegner 'prize', which we would now classify as a grant.

TABLE 1
Academy Prizes Awarded to Pierre and Marie Curie (1895–1906)[a]

Prize-winner	Year	Prize	Subject Restriction	Value (Ffr)
Pierre Curie	1895	Planté	Electricity	3000
Marie Curie	1898	Gegner	Any subject	4000
Marie Curie	1900	Gegner	Any subject	4000
Pierre Curie	1901	Lacaze	Physics	10,000
Marie Curie	1902	Gegner	Any subject	4000
Marie Curie ⎫ Pierre Curie ⎭	1902	Debrousse	Any subject	20,000
Pierre Curie	1906	Reynaud	Any subject	10,000
			Total	55,000

[a] Pierre Curie was elected to the Academy in 1901 and this precluded him from being awarded any further prizes. The Reynaud prize was a rare exception of a prize that (outstanding) Academicians could win. Pierre Curie died in 1906 at the age of fifty-six as the result of a road accident.

Resistance to Monetary Prizes and Their Final Adoption

It should not be assumed that the shift towards monetary prizes was equally acceptable to all Academicians. There was strong opposition, especially in the 1830s and 1840s, to the award of monetary prizes, mainly by mathematicians,[38] whose research, in contrast to that of the chemists, did not involve any great expense. The selection of mathematical prize questions for the *grand prix* in the early nineteenth century had continued a tradition among mathematical Academicians from the *ancien régime*. Honorific prizes involving prize questions relating to mathematics and physics had been particularly successful up to 1827, although admittedly there had been some decline subsequently. The opposition of the mathematicians is understandable in the light of several features of monetary prizes. For example, monetary prizes could be awarded to scientists who were not of the first rank.

Because of the resistance within the Academy to the monetarization of the prize system, the Academy preferred to award money privately — that is, without reporting it in the *Comptes rendus*. In this way, in the late 1850s, it distributed more than one-third of the total annual interest raised in the valuable Montyon medicine fund. One of the consequences of the Academy's policy of confidentiality was that it became easier to reward the same scientist more often than the usual maximum of two or three times. For example, the medical scientist and future Academician Charles Robin (elected 1866), who had won three Montyon medicine 'rewards' (*'récompenses'*) in 1853 (Ffr 2000), 1854 (Ffr 2000) and in 1856 (Ffr 1000), could be given privately three further *'encouragements pecuniaires'*, in 1859 (Ffr 1000), 1860 (Ffr 2000) and 1861 (Ffr 3000). Thus the Academy could not be publicly criticized for favouring one person disproportionately through its rewards.

Through monetary prizes, the Academy widened the circle of prize-winners, especially from the 1850s, when it began to organize a growing number of new monetary funds. But it was not until after the Franco-Prussian war (1870–71) that the monetary prize policy was generally accepted in the Academy. Previous élitist policies were now criticized, and there was a general feeling that Academy funds should be distributed more widely. At the public meeting of the Academy in 1874 the president, the astronomer Hervé Faye, spoke of the changing needs of science. The cost of apparatus had increased significantly and there had been a considerable growth in the size of the scientific community. The Academy had available for prizes a large sum of money (more than Ffr 100,000 annually) in interest on its capital investments. How was this to be used?

88

Faye went out of his way to criticize the traditional system of *grand prix* and championed the monetary alternative to prizes:

> The Academy prefers to help in advance all those [scientists] with talent, than to wait indifferently that they succeed exclusively through their own resources and reward them afterwards.[39]

He claimed that there existed 'an army of workers' (*'une armée de travailleurs'*) which needed encouragement, in contrast with the élitist groups that had formerly competed for a small number of honorific prizes (*'couronnes'*).[40] There is no doubt that the growth of the science profession justified greater resources to help young researchers, who had obtained a qualification in science but had not advanced very far in their careers.

Wealthy Donors and the Expansion of the Grant System

Although the Academy welcomed the increasing amount of private money available to the prize system in the second half of the nineteenth century, there was a constant tension between the wishes of wealthy donors and the needs of the Academy. After defeat in the Franco-Prussian war of 1870 to 1871, there were many manifestations of patriotic zeal. One of these was the decision of several wealthy Frenchmen to bequeath large sums of money to be used for science, now seen as a neglected source of national strength.[41] The Academy encouraged such donors but was seldom able to dictate the kind of prizes to be given. When a large and impressive sum of money was made available in prize funds, the last thing that most donors wanted was that these sums should be broken down into a number of smaller sums to reward several scientists. Table 2 gives examples of large donations, which were increasingly offered as *indivisible* cash awards. Donors saw division as reducing the prestige of the prize and, by association, the distinction attaching to their donations. Such prestige was thought to be proportional to the size of the prize and, consequently, it was arranged that large prizes were to be awarded not annually but at intervals ranging from two to six years in order to increase the sum available (see Table 2). Several wealthy donors insisted that the sum of money available for prizes should never be divided. This led to the establishment of a few huge monetary awards constituting the most conservative type of prize of a type comparable with the Nobel prize.[42] Examples of such prizes are the Leconte prize of Ffr 50,000,[43] first awarded in 1889, and the Osiris prize of Ffr 100,000, first awarded in 1903.

TABLE 2
The Larger Monetary Prizes

Contest	Subject	Value (in francs)	Frequency (in years)	Year of First Award
Montyon	Medicine	20,000 (variable)[a]	1	1825
Jecker	Organic chemistry	5000 (10,000 from 1877)[a]	1	1857
Bréant	(a) Cholera	100,000[b]		
	(b) Epidemic diseases	5000[a]	1	1858
Lacaze	Physiology	10,000[b]	2	1873
	Chemistry	10,000[b]	2	1873
	Physics	10,000[b]	2	1873
Reynaud	Any subject	10,000[b]	5	1881
Petit d'Ormoy	Any subject	10,000[b]	2	1883
Leconte	Any subject	(a) 50,000[b]		
		(b) 5000[a]	3	1889
Osiris	Any subject	100,000[b]	6	1903

[a] Prize money could be divided and therefore sometimes awarded in instalments to promising scientists.
[b] Prize money could not be divided.

These prizes were necessarily retrospective awards and it was sometimes difficult to find a single individual who really deserved such a large sum of money. Because most of these large prize funds were available to reward any subject, the greatest practical problem was for the Academicians to agree among themselves which field should be rewarded. The Academy appointed commissions with many members so that all sections of the Academy could be represented in the final decision.

Yet, as we have seen in the public speech by Faye in 1874 previously quoted, many members of the Academy believed that it should be using its funds to encourage future research rather than ignoring the financial needs of experimental scientists and giving lavish retrospective rewards to the most successful. We find a compromise agreed in the case of the Leconte legacy. There is evidence of a protracted negotiation between the wealthy business man Victor Eugène Leconte (d. 1886) and the secretary of the Academy, J. B. Dumas.[44] Although by the terms of the Leconte legacy an indivisible sum of Ffr 50,000 was to be offered every

TABLE 3
The Bonaparte Grant Fund

	Supply and Demand			Purpose of Grant						Origins of Grantees			Status of Grantees		
Year	Grant (Ffr)	Number of Applications	Number of Grants Awarded	Continuation Grants	Apparatus, Materials, and so on	Travel Field-work	Publication	Personnel	Special Purpose (Morocco Expedition)	Paris-based	Provincial	Foreign	Professors, Directors of Observatories	Some attachment in higher education	PhD Science
1908	25,000	107	10	—	9	0	1	0	—	6	4	0	4	6	3
1909	25,000	35	9	2	4	3	1	1	—	7	2	0	3	8	5
1910	30,000 (incl. 5000 supplement)	34	11	2	8	1	1	1	—	9	2	0	1	7	6
1911	30,000 (incl. 5000 supplement)	34	11	1	8	3	0	0	—	6	5	0	4	8	1
1912	41,000 (50,000 available)	87	16	—	6	3	1	2	4	5	11?	0	3	9	—
1913	55,000 (59,000 available)	63	21	—	8?	3	7	1	2	(Data not published)					
1914	54,500 (57,000 available)	60	18	—	7?	3	4	1	3	3	14??	1	7	9	7

three years for outstanding work done in the past, some interest from the fund (not exceeding one-eighth) could be awarded in *encouragements* for future work.[45] It is worthy of note that in the period 1905 to 1916, the Academy did not give any major Leconte award but managed to give a total of Ffr 30,000 as *encouragements*.[46] These large monetary prizes clearly did not meet the actual needs of science. Although the sums of money bequeathed by scientists to the Academy in the nineteenth century were understandably less than the sums given by businessmen, it is not surprising that scientific donors (who constituted the majority before 1870) were more sensitive to the needs of science.

While scientists, although understanding research needs, were not usually very wealthy, and wealthy donors did not understand the needs of science, the divide was bridged in the early twentieth century by Roland Bonaparte (1858–1924), who was both a wealthy man and a man of science, although hardly a specialist. As a wealthy man and a dilettante, he was happy to act as patron to science in general rather than one particular branch, which had been the limitation of most scientist donors. Since his family name was one of the most famous in the history of France,[47] it did not need to be associated with a large undivided sum of money in its commemoration by the Academy. When he was elected an *Académicien libre* in 1907, he showed his appreciation by founding a Bonaparte grant-fund. As a first step he gave a sum of Ffr 100,000[48] to cover grants for the next four years. He insisted that this should be divided between people with research potential in order to encourage further research (*'provoquer des découvertes'*).[49] From 1912 the sum available was Ffr 50,000 per annum. The foundation of the Bonaparte grant constituted a landmark in the history of Academy grants, since for the first time written applications were required, stating the purpose for which the money was wanted, and grantees were asked to report subsequently on the use of the money. There was no longer any question of charity.

The large sum of money involved and the publicity given to the new fund, which cannot have been lessened by its description as a 'Bonaparte fund', produced a flood of applicants. There were 107 applications in the first year but, as the Academy decided not to award anyone less than Ffr 2000, there was money available for only ten. The fund produced a large number of small grants, Ffr 6000 being the maximum. In later years supplementary funds were available and the number of successful candidates doubled by 1913 (see Table 3). A wide range of different fields of study was supported by Bonaparte grants but with a marked preference for research projects in spectroscopy and in geography, particularly in relation to Africa. Pressure was put on successful applicants to produce written reports and, in a few cases where reports were not submitted, subsequent requests were refused.[50] In 1915, Roland Bonaparte suggested that it should be the responsibility of any member of the Academy, who had supported a grant application, to report back within two years on the use of the money.[51] The fact that Bonaparte grants produced quality in applicants, as well as mere quantity, is suggested by the subsequent election to the Academy of more than one-third of the recipients of grants up to 1914.

Prizes and the Direction of Research

We have deliberately left to our final review the question of the direction of research, a major issue in science policy. Some of the traditional honorary prizes involved the Academy setting prize questions to be answered by competitors in the following two years. As long as the Academy set prize questions, it attempted to direct research, and had some remarkable successes in the first quarter of the nineteenth century. There were, however, many occasions when the prize questions set were too difficult or of little interest to potential competitors, and the tradition of setting prize questions consequently declined, to be replaced by retrospective prizes. On the surface the prize system continued, but the Academy risked losing its former influence on the direction of research.

In the late 1820s, however, the Academy introduced a new method of directing research by offering substantial sums of money. In the early years of the Montyon fund, the Academy announced major monetary awards for the new developments in lithotrity (the term used to describe the bloodless surgical procedure for pulverizing the stone in the bladder through a probe). The offer of money clearly acted as an incentive, since many surgeons, some without previous experience in the field, were drawn to the research and the Academy spent more than Ffr 50,000 over a decade in Montyon medicine awards on this subject. One of the beneficiaries was the impoverished Jean Civiale (1792–1867), who won a Montyon *encouragement* of Ffr 6000 in 1826, which he seems to have invested in further research, for which he was given the Montyon main award of Ffr 10,000 in 1827.[52] Another case of an Academy prize influencing the direction of research was through the Bréant prize (1858) of Ffr 100,000 for the cure for cholera. Although this sum was never awarded, the existence of the prize encouraged work on other infectious diseases, which the Academy rewarded by giving subsidiary prizes.

After Civiale, there were several other cases where prize-winners used their prizes as a grant in order to do further research. Thus Louis Pasteur, who was awarded the Academy prize for Physiology in 1859, used a part of the prize to buy a hundred flasks which he used for Alpine experiments at different altitudes in his controversy with Pouchet to disprove spontaneous generation.[53] Pasteur had previously received Ffr 2500 from the Academy in 1858 as an *encouragement* in order to continue his experiments on fermentation. We have also highlighted the case of Pierre and Marie Curie, who received a succession of different Academy monetary prizes (see Table 1), which they undoubtedly treated as research grants.

What can be said about the 'efficiency' of the old prize system and the emerging system of grants? As far as traditional prizes are concerned, the most spectacular results with a minimum financial investment were achieved by prize questions, in the period from about 1807 to 1827, but these involved only a tiny élite. In the following decades, when the efficacy of prize questions seriously declined, the Montyon surplus was used to provide proto-grants. Although their distribution was rather arbitrary, it included a much wider spectrum of recipients, and we can surmise a reasonable efficiency of expenditure. In the final decades of the nineteenth century, when there was almost a superfluity of wealthy donors wishing to bequeath money to the Academy, it is clear that the very size and complexity of the system decreased its efficiency. There were, finally, too many prizes with too specific terms of reference, so that it was sometimes difficult to find a deserving winner.

But we can approach the question of efficiency more directly. Given the innovatory decision of the Academy to provide money *in advance* for research, was that money really needed and was it well spent? Unfortunately the very *informality* of early grants provided a situation in which no systematic records exist.[54] More research would be needed in individual cases. In some well documented cases, like that of Pasteur, there is no doubt that he really depended on the money, that he used it for the purpose stated, and that nearly every grant given was a good investment. With his great genius in applying science to solve practical problems, such as those in agriculture, Pasteur could even argue that science paid for itself many times over in the benefits it conferred on the national economy.[55]

Yet it would be a mistake to focus exclusively on the economic dimension. The reaction against prize questions in the second quarter of the nineteenth century was partly a reflection of scientists' search for independence. Some undoubtedly did better work without following the directions imposed from above. Also there can be little doubt that throughout the nineteenth century the Academy, both by the traditional prize system and the gradual introduction of grants, stimulated hundreds of able scientists to carry out research of a high standard. Prizes were given considerable publicity and helped to raise the level of scientific awareness in France. Grants were, at first, less public. Drawing on funds which varied considerably, *encouragements* were sometimes awarded secretly, a situation which might lead one to suspect corruption. An examination of the names of some of the secret recipients, however, reveals the names of many who later achieved fame. One concludes that, in general, money was given to persons of ability and in many cases for worthwhile research. The secrecy must therefore be interpreted partly as discretion to avoid jealousy in the early years of a new system and partly as the implementation of patronage.

The Academy was capable of distorting the wishes of donors in order to produce a more flexible prize system, and it repeatedly tried to negotiate with future donors in order to widen terms of any bequest. All prizes and grants had to receive the formal approval of the Minister of Education, who was responsible for the Academy, but an examination of the private correspondence and meetings in secret sessions of the Academy shows that it was often less than frank in its dealings with the Minister. We have also mentioned that many of the monetary awards of the Academy, other than monetary prizes, were given in secret. Anyone reading the *Comptes rendus*, which was supposed to provide a full report on the work and decisions of the Academy, has no means of knowing about the system of private patronage operated by the Academy to supplement the official prize system. All of these factors provide a fascinating insight into the way in which a rigid institutional framework could be adapted to changing circumstances by the determination of members of the institution and the growing needs of experimental scientists. The Academy was remarkably successful, given its historical constraints, in adapting and transforming a system of honorary prizes into a system of monetary prizes and grants.

Whereas there is an obvious parallel between the transition from honorary to monetary prizes on the one hand and the transition from eighteenth-century amateur science to nineteenth-century professionalism on the other, it is misleading to take this connection further and assume a causal link between the growth in the number of Academy 'prizes' in the late nineteenth century and the growth of the scientific community. The number of donors who appeared in the 1880s and 1890s was out of all proportion to the demand for money, and even more out of touch with the needs of specific sciences at the time. Yet the eccentric growth of the system increased the mismatch between supply and demand, and by the end of the century much of the prize money was regularly unclaimed, since contestants had not met the necessary criteria. This chaotic situation was greatly relieved by the introduction of the Bonaparte fund. A system of formal grants open to all and with conditions clearly specified, opened a new chapter in the history of the organization of science. The Bonaparte fund provided a model for the modern research grant. By now there was a general feeling in the scientific community that the best experimental science deserved appropriate finance and should not depend on heroic sacrifices by experimental scientists.

Conclusion

This analysis of the nineteenth-century Academy prize system has ended with a discussion of the Bonaparte grant fund, which, it has been argued, brings us into the modern world.[56] Yet the Bonaparte fund was remarkably similar in many ways to the earlier Montyon fund. Both involved large sums of money awarded annually. In both, the principle of divisibility of the available money was important. Both attracted a large number of recurrent winners in both contests. Yet the Montyon legacy was unique in upsetting the equilibrium of the Academy. The Academy was to derive enormous benefits from it in a way which could hardly have been foreseen either by the philanthropic donor or by the early nineteenth-century Academy, tied as it was at the time to the restrictions of a tight state budget and a purely honorary prize system.

   The Bonaparte fund, on the other hand, had the effect of stabilizing the different interests within the Academy since it was open to all subjects, as opposed to previous legacies which were restrictive in scope and had caused some jealousy within certain sections of the Academy. The early success of the Bonaparte grant fund may well have encouraged the industrialist, Auguste-Tranquille Loutreuil, to bequeath to the Academy the enormous sum of Ffr 3,500,000 in his will of 1910. The interest from this capital sum was also to be used for grants rather than prizes, individual grants being as much as Ffr 15,000. As the Loutreuil fund only began in 1915, however, it lies outside the scope of this paper. Its story really belongs to some future analysis based on twentieth-century developments in which a prominent place would be given to the famous *Centre Nationale de la Recherche Scientifique*, founded in October 1939 and re-founded in 1945.[57] The scale of research grants has been greatly increased in recent times but the *principle* of giving grants for future research was established well before World War I, and in this the Academy of Sciences played a key role.

• NOTES

We must thank MM. les Secrétaires perpetuels of the Académie des Sciences in Paris for their permission to consult the archives of the Academy. We should also like to thank M. Pierre Berthon and Mme Claudine Pouret for their personal assistance. Published material relating to the Academy and, in particular, the *Comptes rendus* was regularly consulted in the library of the University of Kent. Finally, we should acknowledge the advice offered by several referees, and we should like to thank Roy MacLeod in particular for helpful editorial assistance.

1. D. Roche, *Le siècle des lumières en province. Académies et académiciens provinciaux, 1680–1789*, 2 Vols (Paris: Mouton, 1978). See especially Vol. 1, 324–55; see also A. F. Delandine, *Couronnes Académiques, ou recueil des prix proposés par les sociétés savantes*, 2 Vols (Paris, 1787). The emphasis of the book, as implied in the title, is on the *honour* attached to winning prizes.

2. M. P. Crosland, 'Salaries and Science', in his *The Emergence of Science in Western Europe* (London: Macmillan, 1978), 141ff; Crosland, 'Salaries and Sympathies', in his *Gay-Lussac, Scientist and Bourgeois* (London: Cambridge University Press, 1978), 228–34.

3. See, for example, J. B. Morell, 'Individualism and the Structure of British Science in 1830', *Historical Studies in the Physical Sciences*, Vol. 3 (1971), 183–204.

4. See R. Hahn, *The Anatomy of a Scientific Institution: The Paris Academy of Sciences, 1666–1803* (Berkeley, CA, and London: University of California Press, 1971); Harold T. Parker, 'French Administrators and Scientists during the Old Régime', in Richard Herr (ed.), *Ideas in History* (Durham, NC: Duke University Press, 1965), 85–109; Shelby T. McCloy, 'Patents and Encouragement', Chapter 12 of his *French Inventions of the Eighteenth Century* (Lexington, KT: University of Kentucky Press, 1952); C. C. Gillispie, 'Invention', in his *Science and Polity in France at the End of the Old Régime* (Princeton, NJ: Princeton University Press, 1980), 459–78.

5. For a British context of the award of medals by the Royal Society, see R. M. MacLeod, 'Of Medals and Men: A Reward System in Victorian Science, 1826–1914', *Notes and Records of the Royal Society*, Vol. 26 (1971), 81–105. For grants in Britain, see MacLeod, 'The Royal Society and the Government Grant: Notes on the Administration of Scientific Research, 1849–1914', *Historical Journal*, Vol. 14 (1971) 323–58. In 1850, the British government decided to give an annual grant of £1000 to the Royal Society to distribute to encourage research. This sum was later increased to £4000. Given a rate of exchange of Ffr 25 for £1 sterling, it should be possible to compare the individual grants, ranging from £50 to £300 for British men of science in the second half of the nineteenth century, with the sums of money made available through the Academy for French savants. Interestingly, in France, these were *not* government grants, but money made available to the Academy through private benefactions.

6. The two main published sources of data (mainly list of awards and prize-winners) are: E. Maindron, *Les Fondations de Prix à l'Académie des Sciences, 1714–1880* (Paris, 1881) and P. Gauja, *Les Fondations de L'Academie des Sciences, 1881–1915* (Hendaye, 1917). Because the terminology used by the Academy blurred the distinction between honorific and monetary prizes, this distinction is not made clearly by Maindron and some later scholars have been misled. It may be noted that, after the emergence of grants, the Academy continued to refer to all legacies as *'prix'*. This permitted it to award money from the legacies in any form as if it were a prize.

7. M. P. Crosland, 'From Prizes to Grants in the Support of Scientific Research in France in the Nineteenth Century: The Montyon Legacy', *Minerva*, Vol. 17, No. 3 (Autumn 1979), 355–80.

8. E. Crawford, 'The Prize System of the Academy of Sciences, 1850–1914', in R. Fox and G. Weisz (eds), *The Organisation of Science and Technology in France, 1808–1914* (London: Cambridge University Press, and Paris: Editions de la Maison des Sciences de l'Homme, 1980), 283–307. A more recent study, drawing substantially on Crawford's analysis, is: H. W. Paul, 'Science Funding in the Twentieth Century', Chapter 8 of his *From Knowledge to Power, The Rise of the Science Empire in France, 1860–1939* (London: Cambridge University Press, 1985). Paul emphasizes the weakness of the Academy retrospective prize system without apparently being aware of the previous evolution of grants.

9. Crawford (op.cit. note 8) associates the beginning of grants with the 1870s and 1880s, whereas the present paper shows that they originated half a century earlier.

10. J. E. McClellan, *Science Reorganized: Scientific Societies in the Eighteenth Century* (New York: Columbia University Press, 1985).

11. See P. Forman, J. L. Heilbron and S. Weart, 'Physics circa 1900: Personnel, Funding, and Productivity of the Academic Establishment', *Historical Studies in the Physical Sciences*, Vol. 5 (1975), 1–185, at 75–79.

12. Hahn, op.cit. note 4.

13. One of us (A. G.) has made a general study of the *grand prix* and is preparing a paper on this for publication.

14. The distinction made here between prizes in the Academy which were honorific and those which were monetary may be more difficult for people today to appreciate in a period when so much attention is paid to the Nobel prizes, which combine great honour with the award of a large sum of money.

15. Geoffroy Saint-Hilaire reported that 'the enthusiasm of M. Deleau was greater than his financial means', and suggested that the Academy should award Deleau a sufficient sum of money to provide board and lodging for three deaf and dumb children for three years: Institut de France, *Procès-verbaux de Séances de l'Académie [des Sciences] tenues depuis la Fondation de l'Institut jusqu'au mois d'Août 1835*, 10 Vols (Hendaye: Imprimerie de l'Observatoire d'Abbadia, 1910–22), Vol. 8, 446–48 (23 October 1826).

16. The idea that patronage is the key to understanding the social context of nineteenth-century French science was developed by M. P. Crosland, *The Society of Arcueil: A View of French Science at the Time of Napoleon I* (London: Heinemann, 1967), for example 'This is a book about patronage' (1).

17. Fizeau was not only exceptional in being asked for a receipt but also in being a man of independent financial means. He did not really need the money but the award of a proto-grant from the Academy helped to give him semi-professional status: Académie des Sciences, Archives (hereafter abbreviated as AS), *Commission Administrative, 1829–77*, 114.

18. AS, ibid., 128.

98

19. AS, ibid., 133.

20. Crosland, op.cit. note 7.

21. Although it is possible to give the monetary value of gold medals, they were obviously not intended to be exchanged for cash.

22. From 1803, the eleven subject areas were: (1) Mathematics, (2) Mechanics, (3) Astronomy, (4) Geography and Navigation, (5) Physics, (6) Chemistry, (7) Mineralogy, (8) Botany, (9) Agriculture, (10) Anatomy and Zoology and (11) Medicine and Surgery. Further information on the nineteenth-century Academy is given in M. P. Crosland, 'The French Academy of Sciences in the Nineteenth Century', *Minerva*, Vol. 16, No. 1 (Spring 1978), 73–102.

23. E. H. Ackerknecht, *Medicine at the Paris Hospital, 1794–1848* (Baltimore, MD: The Johns Hopkins Press, 1967).

24. Institut de France, op.cit. note 15, Vol. 8, 235, our italics.

25. Ibid., 387–88. Charles Heurteloup, for example, was criticized for having consciously concealed negative results in his prize memoirs, in order to win prizes, a criticism from which he strongly defended himself: see ibid., 392.

26. Ibid., 387–88. This reservation, although made in the context of surgery, was to apply equally to medicine.

27. Ibid., 385.

28. Ibid., Vol. 10, 662–25. In 1929, the Academy had awarded four Montyon medicine *encouragements* of Ffr 2000 each, 'either for results or else for tests that promise future useful results': ibid., Vol. 9, 259.

· 29. See Crosland, op.cit. note 7, 367–68.

30. *Comptes rendus des Séances de l'Académie des Sciences* (hereafter·abbreviated as *CR*), Vol. 15 (1842), 1140.

31. AS, *Comité Secret, 1837–44*, 70 (22 March 1841); *Comité Secret, 1845–56*, 19 (17 September 1845).

32. After the revolution of 1848 and the short-lived Second Republic, the year 1852 marks the beginning of a period of greater political stability and economic expansion, that of the Second Empire.

33. The following data, which the Academy regarded as confidential at the time, is to be found in AS, *Commission Administrative, 1829–77*, 95ff.

34. The monetary value of cash awards may be compared with salaries at the time. Academic salaries in mid-nineteenth-century Paris were of the order of Ffr 5000–10,000 for a full professor. Pasteur, as Director of the Ecole Normale in the 1860s, received Ffr 8000 per annum. A grant of Ffr 6000, therefore, was comparable with an annual salary of an academic scientist. By the 1880s, the salary of professors in the Paris Faculty of Science had been increased by a sympathetic government to Ffr 15,000, but cash awards also tended to be higher in the late nineteenth century.

35. A precedent has already been established in 1858, when Claude Bernard had commented on the small value of the physiology prize, and had proposed that several of the competitors who had not won a prize should be given an *encouragement* to help them continue their work, the money to be taken from the Montyon surplus: AS, *Comité Secret, 1857–69*, 78.

36. AS, *Comité Secret, 1870–81*, 23 (23 May 1870).

37. *CR*, Vol. 46 (1856), 317.

38. The mathematicians exerted great influence in the Academy, an influence out of all proportion to their number. Quite apart from their individual prestige and the prestige of mathematics in the French educational system, the term 'mathematical' was used to

describe one of the two broad divisions of the Academy, so that mechanics, astronomy, geography and physics were all classified as 'mathematical sciences'.

39. *CR*, Vol. 79 (1874), 1525–31, at 1530.

40. Now that France was again a Republic, the derogatory use of the term 'crown' may be noted.

41. As an example of nationalism, we may cite the Osiris prize, which the donor insisted should only be awarded to Frenchmen.

42. Nobel spent nearly twenty years of his life in Paris and was very probably inspired by the prize system of the Academy: see E. Crawford, *The Beginnings of the Nobel Institution: The Science Prizes, 1901–1915* (Cambridge: Cambridge University Press, and Paris: Editions de la Maison des Sciences de l'Homme, 1984), 10.

43. Leconte, however, did allow some use of interest for subsidiary awards.

44. AS, dossier donneur Leconte.

45. See text of the Leconte legacy in Gauja, op.cit. note 6, 301–02.

46. Ibid., 304–06.

47. Roland Bonaparte was a nephew of the former Emperor Louis Napoleon, who was himself a nephew of Napoleon Bonaparte.

48. That is, not capital to earn interest as was usual but an outright gift of money.

49. Gauja, op.cit. note 6, 486.

50. *CR*, Vol. 173 (1921), 1282.

51. AS, *Fonds Bonaparte*, Vol. 1, 112.

52. Institut de France, op.cit. note 15, Vol. 8, 544. There is a long report in Civiale's research in ibid., 43–51.

53. P. Vallery-Radot (ed.), *Correspondence de Pasteur, 1840–95*, 4 Vols (Paris, 1940–51), Vol. 2, 80 (letter of Pasteur to his father, 15 September 1860): see A. Gálvez, 'The Role of the French Academy of Sciences in the Clarification of the Issue of Spontaneous Generation in the Mid-Nineteenth Century', *Annals of Science*, Vol. 45 (1988), 345–65.

54. It was only after 1900 that recipients of grants were required to report on the use of money.

55. For example, Pasteur, writing to the Minister of Public Instruction on 13 December 1859, asking for a personal research grant ('*allocation*') of Ffr 1500 annually, concluded his letter by asking: 'Who would dare compare. . . the value of the new discoveries which I have just described on the composition of wines with the expenditure of a few hundred francs which provided the material means of making these discoveries?': see Vallery-Radot, op.cit. note 53, Vol. 2, 61.

56. A full study of the period after 1900 would also need to take into account the *Caisse des recherches scientifiques*, founded in 1901. This followed the precedent of the Academy in giving grants, and members of the Academy were strongly represented in it, but they did not have control.

57. See Frédéric Blancpain, 'La création du CNRS: histoire d'une decision, 1901–1939', *Bulletin de l'Institut Internationale de l'Administration Publique*, Vol. 32 (1974), 93–143, quoted by Crawford, op.cit. note 8, 291.

X

HISTORY OF SCIENCE IN A NATIONAL CONTEXT

THE history of science can be approached in several different ways. It may be studied, as in the classification once favoured in the long-established Department of History and Philosophy of Science at University College London, by considering separately the history of individual sciences: physics, chemistry, biology, etc.—Partington's monumental *History of chemistry* is a good example of the cross-section of history of science obtained by considering a single discipline. This approach is understandable when history of science is the work of retired specialists in a particular science. On the other hand, many of those who have approached the history of science from a training in general history have tended to favour a study of a particular period as an alternative to an orientation by subject. This is particularly valuable before the nineteenth century, when subject boundaries were not so tightly drawn as some of the old science historians tended to assume. A third possibility is area studies, usually the history of science within a particular country. Sometimes this is done unconsciously, as when historians claim that they are dealing with a general theme, such as science and religion or scientific institutions, but do so with special reference to their own country. French historians of 'the Enlightenment' often study French authors exclusively. Language as much as country is a limiting factor here.

There are advantages and disadvantages in all of these approaches. Let me make a brief criticism of the subject approach. Partington too easily rejected as irrelevant, ideas in what he saw as 'physics' or 'biology' even though such ideas might have been pertinent to chemistry. The subject classification itself is something of an anachronism in, say, the sixteenth century, when vitalistic ideas permeated proto-chemistry, and even astronomy or astrology had connexions with it. But if I mention Partington, it is not to deride him. His massive though hardly imaginative scholarship is of permanent value to us, his successors, whose linguistic ability and sheer *Sitzfleisch* may hardly be on the same level. The period approach may be less open to objections than the subject orientation, though if the period chosen is very restricted, or is very early, or very modern, other historians might feel the work too narrow.

If we think of history of science in relation to a country, we may consider the evidence for national characteristics. The British, for example, have often been associated with empiricism.[1] I quote the nineteenth-century astronomer G. B. Airy, who claimed that

> In England an observer conceives that he has done everything when he has made an observation . . . In the foreign observatories on the contrary, an observation is considered as a lump of ore . . . and without value until it has been smelted.[2]

It can happen that in one country astronomy is seen as largely an observational science, whereas in a neighbouring country the major advances may be in mathematical astronomy: it is interesting to remember that William Herschel and Laplace were contemporaries. Was it by chance that all the basic experimental work in pneumatic chemistry was done in Britain? One thinks of Boyle, Hales, Black, Cavendish, and Priestley, who might collectively be described as a British school. However, we should not exaggerate the concept of a national style. There is always the danger of caricature, and I prefer to discuss wider and less intangible questions of a social and institutional context. This approach can be valuable, particularly if one can manage eventually to look at more than one country.

Most contributions to the understanding of the natural world from the seventeenth to the early twentieth century were made within a local or national context. Most men of science wrote in the vernacular, primarily for their friends and compatriots, and it is only comparatively recently, with wonderful improvements of communication, that we have been able to think of science on an international stage. After the collapse of the medieval world, where Latin had provided a medium of communication that was understood by educated men from one side of Europe to the other, linguistic barriers were added to other barriers between communities: geographical, political, and religious. The English Channel and the Alps were probably two of the most effective physical barriers in western Europe. Of course, it was always possible for any *savant* to strive to reach beyond his local context. The publication of Volta's famous paper on the pile in the *Philosophical transactions* and of Avogadro's memoir in the *Journal de physique* were attempts by the authors to obtain wider recognition than would have been possible in their own country in the early nineteenth century. Nevertheless, Volta and Avogadro must be understood in the context of the Italian states in which they lived. Similarly, many Swedish men of science felt that they lived on the edge of the civilized world and made great efforts to overcome isolation. The European tour of Berzelius, and particularly his year in Paris, brought him out of that isolation. Yet he remains a Swedish scientist, and we must make some attempts to understand the Swedish scene if we are to understand Berzelius. We must know something of the structure of intellectual life within a country. In brief, we must study institutions.

Science has an important institutional dimension. This is most obvious in the case of experimental science, but even the theoretician usually has institutional support. The isolated thinker may work outside any university or academy, but, even for him, publication and the reception of his ideas involve institutional factors. The formulation of ideas is usually influenced by education, reading, and discusssion, and these in turn presuppose schools, channels of publication, and associations whether informal or in a scientific society.

The institutional dimension can, of course, be examined in relation to a specific branch of science. It is possible to consider the provision of university posts and laboratories for the study of one science without considering others. In some cases, however, this will impose a rather artificial division. Again, it is possible to consider institutions in a particular period. Martha Ornstein's book *The role of scientific societies in the seventeenth century* has shown that such a general survey on a European scale is possible for that period; but, as the number of scientific societies has grown, it becomes increasingly difficult to make such a survey at more than a superficial level. Finally, one can consider institutions within a particular city or country. In so far as different scientific subjects are often studied within the same institution, it is convenient to look at science in general. But if the overlap of source materials is one reason why one might study science in a national context, it is hardly the ultimate justification. This must be an appreciation that science is a part of the intellectual life of a country and cannot be divorced from social, political, and religious history. One may need to examine attitudes towards science—for example, an evaluation of its utility in the economic life of the country or its implications for established religion. Government policy has sometimes encouraged particular kinds of science, such as astronomy as an aid to navigation, or mineralogy as a guide to the exploitation of natural resources. Appropriate facilities have then been provided. The absence of financial support, on the other hand, may rule out the pursuit of certain types of experimental work involving expensive instruments, and may encourage the development of a cheaper branch of experimental science or even desk science. Local conditions have also had a decisive influence on specimens available for the study of the various branches of natural history. Thus even when scientists in different countries have begun with similar interests, their achievements have often been significantly different.

The idea of studying the science of a particular country has sometimes been criticized on the grounds that science is international. There is a sense, of course, in which science does transcend both time and place. But the historian of science, being concerned with the development of ideas about the natural world, is committed to the study of science in a perspective that does not transcend time. Equally he might consider those aspects of science that do not transcend place.

In the reception of scientific theories, national factors are of major importance. Thus German reactions to Lavoisier's oxygen-centred theory and nomenclature were influenced by factors that were frankly nationalistic. Some German chemists took a patriotic pride in the large part played by their fellow-countrymen Becher and Stahl in the early formulation of the phlogiston theory which was now threatened by the new oxygen theory. The theory was sometimes known in Britain and Germany as the French theory—much to the annoyance of Lavoisier, who sought credit for himself. During the Revolutionary and Napoleonic wars the use of such a phrase as 'the French theory' hardly helped objective study, and there can be little doubt that Humphry Davy was stimulated to criticize the new theory as an act of patriotism as well as of science.

But the thesis that there is a national context for scientific work does not always depend on the crudest feelings of nationalism. The reception of Darwin's theories in France, for example, did not depend simply on a patriotic preference for Lamarck. Yet Lamarck and Geoffroy Saint-Hilaire had discussed evolution and the evidence for it half a century before Darwin published his *Origin of species*, and French biologists could not help but see evolution through Lamarckian spectacles. If the publication of the *Origin of species* in 1859 began a drama that was to dominate British thought for several decades, the situation in France was different. As a recent writer has remarked, the French stage was not empty and waiting;[3] the evolutionary plot had already been explored and driven off the stage, for example in a number of skirmishes which took place in the Académie des Sciences in the spring of 1830, when attempts by Etienne Geoffroy Saint-Hilaire to defend the mutability of species had been attacked by Cuvier. Hence discussion of evolution in France in the nineteenth century was a much longer-drawn-out affair, in which, for the first two acts of the play, Darwin did not even come on the stage. When he did appear to the French public, it was in the translation of Clémence Royer, who gave her own twist to Darwin's ideas. If some French biologists saw Darwin as the successor of Lamarck, she saw him as the successor of Condorcet and even incorporated the idea of progress into the French title.[4] Darwin's disappointment that his book caused little excitement in France is suggested by his comment about 'horrid unbelieving Frenchmen'.[5] In considering the reception of the ideas of a scientist in another country, therefore, the prejudices of his translators may be all-important. Heinrich Bronn, the Heidelberg palaeontologist who translated Darwin's *Origin* into German, did not accept the theory;[6] he added notes criticizing the text and omitted the controversial sentence of Darwin that 'light will be thrown on the origin of man'. I think I have said enough to suggest some of the differences between the reception of Darwinism in Britain, where strong traditions of natural theology made Darwin's ideas particularly controversial, and in different European countries, where a different religious, social, and intellectual history, not to mention the accident of the translator, could make a major difference to the reception of the idea. Darwin himself mused: 'It is curious how nationality influences opinion.'[7]

Before leaving the general discussion of science in a national context, I should refer to one or two complexities. But if they are difficulties, they do not invalidate the concept of national scientific patterns. The first objection is the complication caused by migration. In the understanding of American science in the twentieth century, one often has a picture that is unusually complex in so far as it involves people from different European countries crossing the Atlantic to begin a new life. A full biography of any such scientist must obviously examine both the old world and the new. It is only those concerned with scoring in the Nobel prize game who feel they need to assign scientists to one country only. Nor is a brain drain a purely modern phenomenon. Although Huygens came from the Netherlands and Roemer was a Dane, some of their work comes within the orbit of French science in so far as it was done in Louis XIV's Académie Royale des Sciences. Similarly a large part of the work of Lagrange comes within the context of French science despite his birth in Turin as the son of Guiseppe Francesco Lodovico Lagrangia. Most scientists, however, do not change their names, nor do they move from one wealthy patron to another.

Nor does the existence of border areas invalidate the approach to a national context for science. Obviously the north of England has been more influenced by the products of Scottish education than, say, the west of England. On the Continent, Alsace provided a fertile meeting ground for French and German ideas. Often a language helped both to erect an external barrier and to provide some internal coherence before political unity existed; Germany is an obvious case in point. We do not have to wait until Bismarck before we make generalizations about German science or German universities.

Early colonial or expeditionary science may, in another sense, be a border area in so far as it combines the education and ethos of the mother country with the influence of the local environment. In Bonaparte's expedition to Egypt in 1798 and the establishment of the Institut d'Egypte,[8] Berthollet's study of chemical reaction suddenly found a new meaning in the trona deposits, Malus was encouraged to consider optical phenomena, and it has been suggested that Fourier became obsessed with the problem of heat when he returned to the colder climate of France.[9] Despite such strong environmental influence, the Institut d'Egypte was strongly French in character, a microcosm of the French

National Institute in Paris. Similarly the British transported their civilization to India. In the early nineteenth century the Asiatic Society of Bengal made some attempts to study science; a specifically scientific periodical was published in Calcutta from 1829 called *Gleanings in science*. The editor spoke of 'the scientific community of India',[10] but in his list of subscribers, all but one were British names. The editor felt that science would help to counteract 'the apathy and indolence which are the bane of our Indian clime'.[11] The pursuit of science, therefore, had a moral dimension, and the journal was printed, significantly, in Calcutta by the Baptist Mission Press. It not only kept expatriates in touch with work carried out in Europe but it encouraged local research: the growing of indigo, the analysis of Indian woods, meteorology.

In north America British cultural patterns persisted in the nineteenth century in Canada even more than in the United States. A Canadian writer in 1852 deplored the meagre contributions which had been made to science in North America, attributing their meagreness to 'the great vice of Society in America, that eternal sabbathless pursuit of a man's fortune . . . which leaves to the mind neither leisure, taste or capacity' for the cultivation of such pursuits as science.[12] He considered that Canadian science should not try simply to imitate science in Britain but should try to harness the 'fund of practical knowledge and thought, the wisdom of the workshop, the field and the loom' which was present in every community. Here, then, is the view that colonial science should be more practical in character or at least in inspiration.

In discussing science in a regional way, the country is not, of course, the only unit. In some cases one needs to take a larger cultural area and consider, say, science in Western Europe or Islamic science. At the other extreme, useful information may sometimes be provided by the local historian, and the study of a city, or of one institution within it, may produce valuable results. Finally, it hardly needs to be said that the study of science within a particular area does not preclude a study of a period or of a subject; for a short-term study, they can often be combined most effectively.

French science

For a case-history of a national context of science, I shall turn to France. A uniform national structure of education makes it a clearer case for study than Britain, with its traditions of local initiative and independence. But although in some ways it is easier to speak of French science than British science, there is a sense in which it is perhaps even more urgent to grasp the character of the British activity. In so far as scientists in France were trained as such and fulfilled the role of scientists, they came closer to an ideal trans-national science. The British man of science, who was not at all a specialist, was to a greater extent immersed in the culture of his country.[13] There were many features of British science at the time of the Industrial Revolution that were not shared by other countries. It might be misleading if the student of science in this country were to generalize and to assume that what was true of the England of George III was true for the Austria of Joseph II or the Russia of Catherine II.

So I turn to France as a country which for the past 350 years at least has been a major contributor to the scientific endeavour. My aim will be neither to bestow extravagant praise on French science nor to attempt a systematic exposure of its weaknesses. I want simply to suggest that science in France had certain features not present in science in other countries and that national educational patterns and institutions in France provided a general encouragement for the pursuit of science, but in certain directions rather than others. Prizes offered by the Académie des Sciences for research on particular topics were only one way in which latent genius was encouraged to express itself. Within such a system valuable scientific work was done. The intellectual, religious, and political environment of the French was different from other countries. Methods of teaching,[14] social support, and economic stimulus were different. Science in France was highly structured, with career patterns[15] marked out and membership of the Académie an unbelievably important goal.

The founding of the two major seventeenth-century scientific societies, the Royal Society and the Académie des Sciences, represented two contrasting approaches to the patronage of science by the state, and since that time science has flourished or languished in Britain and France in different ways, for different reasons, and at different times. Science in France, of course, was not only different from her northern neighbour, but also from those on the other side of the Rhine and of the Alps.

The absolutism of the French monarchy after Louis XIV remained unrestrained by anything which in England would be called a Parliament. In the eighteenth century the nobility continued to enjoy privileges at the expense of the bourgeois and the peasants. Paris was the only large city. France remained an essentially agricultural country, influenced very little by the industrial changes taking place in Britain. In so far as science was thought of at all, it was studied by a handful of people as an intellectual exercise, and by rather more, as amusement. A smattering of science was a part of the conversation of the enlightened man. Voltaire and the Encylopaedists had more influence on the educated than the Catholic church, although nearly all education was in Church hands. A series of mounting crises in the 1780s finally led to the Revolution, which is a watershed in French history, education, and science.

In the creation of the new society, education was thought to be fundamental, and science had an important part in that education, partly for ideological reasons and partly because of the utility it had shown in the Revolutionary wars. After the Terror a constructive period followed, which saw the establishment of the first Ecole Normale and the Ecole Polytechnique.[16] The Jardin du Roi was transformed and expanded into the Museum of Natural History, and under Napoleon the first faculties of science were established. Science could now become not a hobby but a full-time job, after training and the acquisition of the appropriate qualifications. Positions in academic life or the civil service depended on the *concours*. If we are to understand the conditions under which science was done in France in the nineteenth century, we must take into account the examination-orientated system, in which certain skills were at a premium.

Although some of my general remarks about French science will range more widely, I shall concentrate on the period of the early nineteenth century, not only as a period of great achievement, but as a key to understanding all of subsequent science in France. The foundations laid in 1794-5 and in the succeeding quarter of a century coincided with the greatest period in the history of science in France. They represented a national investment in science unparalleled in any other country at the time.

Two important features of the French educational system and the organization of science were centralization and integration. They are both perhaps symbolized by the establishment of the Museum of Natural History,[17] formed from the Royal botanical garden and the transfer of animals from the Royal menagerie at Versailles to Paris. The Museum was a national institution with the duty of supplying specimens for the whole of the country when required. The menagerie was not simply in juxtaposition to the botanical garden but was integrated as a department of the Museum. To give an example: on the death of an animal it was immediately taken to the anatomy laboratory, where the skin was removed and given to a taxidermist to prepare for exhibition. The skeleton was preserved and became invaluable material for the study of comparative anatomy. Sceptics who doubt the contribution of institutions to the pursuit of science might ponder the context of Cuvier's work on comparative anatomy.

The centralization of French science struck visitors forcibly. Charles Lyell, writing on a visit to France in 1823, commented: 'If a man is thought to display talent, he is hurried to Paris, as the only soil where it can be nourished or admired.'[18] The claim that French science was centralized requires some qualification according to the period under discussion. Although centralization was part of the policy of Louis XIV, the eighteenth century witnessed much important intellectual activity in the provinces, particularly in the Academies. With the Revolution, however, centralization was greatly increased. The Académie des Sciences in Paris was replaced by a National Institute, all full members of which were required to reside in the capital. When the university system was set up, no provincial universities were established, but only provincial faculties, which were local branches of a system based on Paris, where the decisions were made. It was only at the end of the nineteenth century that real provincial universities were re-established, although they naturally suffered from lack of prestige. With this concept of centralization I associate integration. The Napoleonic University of France was really a Ministry of Education concerned with national education from infancy onwards. All examinations and grades were carefully specified, and science had to fit within this structure. Although Church schools were re-established in the nineteenth century to form a parallel system of education, their teaching was very much influenced by state examinations.

The centralization of French science was largely achieved by a concentration of major institutions in Paris. The Collège de France was itself a university in miniature. The Museum of Natural History was a centre both for teaching and research, the latter encouraged by the living-quarters provided for the staff and their families, who lived in proximity to their collections. With its many galleries and gardens, therefore, the Museum constituted an enclave of major importance in the history of the biological sciences. Then there was the Ecole Polytechnique, which impressed foreign visitors not only because of the standard of the students and the eminence of the staff but also because of the facilities provided for practical work. There were also higher educational establishments concerned with civil engineering, mining, and, of course, medicine. Finally, there was the Paris Faculty of Science and the Ecole Normale, which later in the nineteenth century was to rival the Ecole Polytechnique as a national establishment training mathematicians and scientists. All these institutions attracted the leading French scientists, who nearly all worked and lived in Paris. To list the staff of the respective schools would be to give a roll-call not only of the most distinguished French men of science but, in many fields in the early nineteenth century, of many of the most distinguished anywhere in the world. Even after the founding of the Universities of Berlin and London, no other capital city could provide such a concentration of leading scientists. Perhaps centralization tended to impose a certain uniformity of attitude towards any particular issue. Perhaps there would have been greater opportunity for critical examination of scientific theories if there had been competing centres of excellence.

Another characteristic of French science was specialization. In the late eighteenth century one might have expected to find in a country which took science seriously, chairs in mathematics, natural philosophy, and natural history. But in the Ecole Polytechnique there were chairs not merely of mathematics but of analysis, mechanics, descriptive geometry, physics, and chemistry. The Paris Faculty of Science, established in 1808, had similar subjects, though astronomy replaced descriptive geometry, and mineralogy and zoology were added. Mathematics was represented by three chairs: calculus, higher algebra, and mechanics. It was understandably at the Museum of Natural History that the biological sciences received their most specialized treatment. Thus zoology was subdivided, having three chairs, the first concerned with mammals and birds, the second with reptiles and fish, and the third with insects, worms, and microscopic creatures. Lamarck held this third chair, and at his death in 1829 a further subdivision was introduced. To many a British naturalist in the early nineteenth century such specialization must have seemed very narrow; but it did enable the professors at the Museum to make significant advances in their respective fields, and daily contact with colleagues in related studies prevented the sterility of isolation.

Research in depth rather than in breadth was also encouraged by specialized journals. The *Annales de chimie*, first published in 1789, is the oldest surviving journal devoted specifically to chemistry. The first journal in the world specially concerned with mathematics was Gergonne's *Annales de mathématiques*, which first appeared in 1810; and when it ceased publication in 1831, it was soon succeeded by Liouville's *Journal de mathématiques pures et appliquées*. The new institutions also had their journals. Although here the glory of the establishment took precedence over any one subject, the journals did tend to cover a particular area. The *Journal de l'Ecole Polytechnique* was very mathematical, and the *Annales* of the Museum of Natural History covered the whole range of subjects studied in that institution.

The charge of extreme conservatism has been levelled against science in nineteenth-century France. There may have been a rigidity in French institutions which, with the passage of time, came increasingly to impose constraints on new developments. I want to consider particularly the flexibility of the institutional framework to enable it to incorporate new branches of science. The establishment of a chair in a new subject could have major implications for the development of that subject. The recognition of organic chemistry by the establishment of a chair in 1837[19] marks an important advance in one of the major sciences and is typical of the specialization which is a necessary part of the growth of science.

What were the possibilities for innovation within French science? In the eighteenth century most developments were of a comparatively minor nature, and the same may be true for much of the nineteenth century; but between the two there was one epoch-making generation in which recent changes in science could be encapsulated in an institutional framework. It is almost as if the French Revolution not only had to make up for the slow rate of development under the Ancien Régime but had also to anticipate the emergence of new sciences, since there were to be few opportunities in the nineteenth century. As schools such as the Ecole

Polytechnique had been far in advance of other countries and had achieved so much in the early years, it was easy for the French to be complacent and to assume in the mid-nineteenth century that they still led the field. Research was given a boost under the Second Empire by the foundation of the Ecole Pratique des Hautes Etudes, but it took the major catastrophe of defeat in the Franco-Prussian war to bring the French government to a fundamental reappraisal of the organization and financial support of science.

The Académie des Sciences, founded in 1666, was given a detailed list of regulations in 1699, which governed the major body of French science until its suppression in 1793. There were minor changes of title and sections, but the hierarchical organization and the clear delineation of the rights and duties of academicians remained. The Revolution threw all this into the melting pot, and the Académie that emerged under the title of the First Class of the Institute was quite different in its approach. Its recognition of a section for mineralogy was a reflexion of the work of the late eighteenth century, in particular of the crystallography of the abbé Haüy. The Académie in 1795 could not have foreseen the emergence of geology as a major branch of science. Because of the historical accident of the date of foundation, geology remained unrecognized throughout the nineteenth century in the premier scientific body of France as a major branch of science. One or two geologists like Elie de Beaumont were eventually rewarded by election to the mineralogy section of the Académie but, as long as the section was called mineralogy, it could largely ignore geology. Fortunately the inflexibility of the Académie did not prevent the introduction of geology in the Museum nor the later foundation of chairs of geology at the Ecole des Mines and at the Paris Faculty of Science.

The Collège de France is of some importance in the process of innovation, since it combined prestige with a certain independence of action.[20] Its status was such that the participation of one of its professors in a new activity constituted a stimulus to the legitimation of that activity. Founded in the sixteenth century as the Collège Royal and a bastion of Renaissance culture independent of the Sorbonne, it took pride in its independence. It pioneered many new subjects by the agreement of its professors to change the subject of chairs as they fell vacant. It was in this way that several science chairs were introduced in the 1770s: for example, a chair of physics was established to replace a chair of Greek and Latin philosophy. A century later, in 1888, a chair of experimental psychology was created by the transformation of a chair of law.

How was the actual science done in France affected by the social and institutional context in which the scientists worked? After all, it could be argued that, if the difference between science in Britain and France was that scientists had to carry out their research in their spare time in Britain, whereas many were able to do so as part of their employment in France, the results would be similar but more rapid on a full-time basis. But the differences between countries are usually more complex and subtle than this. In this discussion I shall touch on three specific areas: the question of standards, education, and the general intellectual climate. In fact, all three are related, but it is convenient to look at them in turn.

If French science reached a high standard in the early nineteenth century, it was not purely through native genius. One must think rather of ability guided in certain directions and benefiting not only from the education received early in a career but from the stimulus and direction of a highly competent and expert body, the Académie des Sciences. The Académie was the inner circle of the scientific community, but although its membership was severely restricted, its professional expertise was not. Any literate person could submit a memoir to the Académie for its judgement, and, if it was competent, the memoir would probably receive encouragement. John Herschel, in 1830, considered that the reports drawn up by commissions of the Académie 'contributed, perhaps more than anything, to the high scientific tone of the French savans'.[21] Pursuing this theme with some enthusiasm, Herschel continued:

> What author indeed but will write his best, when he knows that his work, if it have merit, will immediately be reported on by a committee, who will enter into all its meaning; understand it, however profound: and, not content with *merely* understanding it, pursue the trains of thought to which it leads; place its discoveries and principles in new and unexpected lights; and bring the whole of their knowledge of collateral subjects to bear upon it.

Of course, Herschel is too uncritical. Reports in the Académie were by no means always immediate; nor were they all works of art. However, they did constitute a kind of superior referee's report and one that authors were proud to have and that they sometimes published as an appendix to their own work. However, a system in which the approval of the Académie was so important encouraged the perpetuation of current orthodoxies rather than the introduction of new ideas.

Turning to education, one observes mathematics becoming an increasingly prominent part of school education in the eighteenth century (by 'mathematics' I mean rather more than just arithmetic and simple geometry.) The tradition may be found in the Ancien Régime in the schools conducted by the Jesuits and Oratorians, but it found its most advanced expression in the military academies. When, with the Revolution, education became the responsibility of the state, any special feature of education was likely to have the widest possible influence. In the écoles centrales, planned in 1795, mathematics and science had a prominent place and several future scientists were to benefit from this training. In 1802 the écoles centrales were replaced by the lycées and, because these reverted to the traditional classical education and drastically reduced the time spent on science, it has sometimes been thought that this marked a serious reversal for the whole cause of science. In fact, it marked a decision to postpone the study of all but elementary science to university level.[22] But the main point I want to make is that in the lycées, mathematics, as opposed to science, was by no means restricted. Mathematics constituted a prominent part of the later classes of the lycée curriculum, and, from the very beginning, the lycées had two distinct mathematics appointments, often with assistants, in elementary mathematics and special mathematics.

So for one *professeur* who was responsible for the teaching of physics, chemistry, and natural history, there would be two or three to teach mathematics. When the Napoleonic lycées became the collèges of the Bourbon Restoration, mathematics continued to play a prominent part. Historians hardly seem to have appreciated the significance of the special *baccalauréat* in sciences mathématiques, which encouraged boys to specialize in mathematics. This was introduced in 1821 when Poisson joined the Council of Public Instruction. Those boys interested primarily in *experimental* science (or who were not good enough at mathematics to specialize) might take the *baccalauréat* in sciences physiques, which still involved a mathematics examination but only at the same level as that for students who had opted not to study science. After following courses in the faculties, a student could take a *licence* or first degree in sciences mathématiques, or sciences physiques, or sciences naturelles, and there was a similar division for the doctorate. There was a steady stream of students taking the *licence* in mathematics from 1811 onwards, and throughout the nineteenth century there were as many students taking this *licence* as those in the physical and biological sciences together.[23] The majority of those graduating would go on to teach, so strengthening the French mathematical tradition in the schools. This growing tradition was not without its critics among experimental scientists: in 1847, for example, J. B. Dumas as Dean of the Paris Faculty of Science presented a report in which he attacked 'the minute and sterile study of pure mathematics' and called for a broader scientific curriculum.[24]

If mathematics had a prominent place in the faculties, it had an even more important place in the prestigeous Ecole Polytechnique. Entry to the Polytechnique was on a national basis in which the applicants' ability and knowledge of mathematics were the controlling factors. As the Polytechnique in its early years provided free education, and was the gateway to a wide variety of careers, the requirements of the entrance examination had an important effect on French education. The ambition to enter was widespread enough to extend numeracy in the population as well as to stimulate mathematical talent at the highest level.

The importance of entry to the Polytechnique soon led to the establishment of special preparatory schools in the capital which gave a good grounding in mathematics. By 1800, candidates for entrance were examined on quadratic equations, progressions, logarithms, trigonometry, conic sections, and elementary statics, and further requirements were added in later years. The lycées and collèges sometimes gave special training in mathematics to senior pupils. Richard, professor of 'special mathematics' at the Collège Louis-le-Grand from 1821, built up a reputation for his preparation of young men for entry to the Polytechnique, and he could boast among his former pupils Galois, Le Verrier, and Hermite.[25]

At the beginning of the nineteenth century the encouragement of mathematical education and the establishment of teaching posts was of great benefit to physics, as is evident from the biographies and the work of Ampère and Fresnel. Over-specialization, however, had a drastic effect on physics. To the extent that mathematics provided both a training and a career parallel to but separate from physical science, this valuable dimension was lost to physics. The very separation of mathematics from the practical aspects of experimental science tended to increase its prestige, and the mathematics class was something of an élite. Nor was the pursuit of mathematics at a disadvantage in the later nineteenth century, when governments failed to provide adequate laboratory facilities.

The encouragement of mathematical talent and its syphoning off into a specialized group had, therefore, two important effects. First, it fostered the study of mathematics on a national scale, so that any boy with mathematical ability, wherever he lived in France, would be likely to find encouragement within the educational system. On the debit side, however, was an impoverishment of physics. Those who were experimentally minded often followed the French tradition of chemistry, which looked back to Lavoisier and included such men of the first rank as Gay-Lussac, Dumas, Wurtz, and Berthelot.[26]

While paying tribute to French scientific education, it is tempting to ask if it was not sometimes over-intensive. The problem of what constitutes an ideal scientific education and how to avoid indoctrination is a subject in itself, and I shall content myself with a passing reference to the problem and to one instance when the accusation was made. The fact that the testimony comes from a student who had been expelled from the Ecole Normale for political activity may make us view it with some caution, but as the student was Evariste Galois, who was to show in his short and tragic life that he did have important and original contributions to make, his remarks can be given more weight. In an article on the teaching of science,[27] Galois argued that mathematics was taught in a way that stifled the creative talents of students. He likened the teaching of mathematics to the teaching of Latin. Perhaps the problems set, which helped consolidate a mathematical technique for ordinary students, were too many for a mathematical genius. He felt that scientific education had become a game in which one prepared to please the particular prejudices of the examiners.

Here, then, was an attack against the system. It was partly political, in so far as it represented the antagonism of a young republican to a system supported by a monarchy; but it also had serious educational implications. The implications of Galois's remarks are that within a generation the examination system, intended to create a meritocracy, had ossified as a test of conformity. Inevitably the French educational system did produce a certain uniformity. There was less room for the eccentric or the rare genius. This was one of the weaknesses of French science in the nineteenth century when it is compared with the diversity of the German or the British universities. In this way, the comparative uniformity of the French system and the stress that it laid on conformity had important implications for science.

Those who, in the nineteenth century, argued against government support for science were often exercised by misgivings about political control. There were a few cases of political interference—for example, the expulsion of Lazare Carnot from the Académie under the Directory and the expulsion of Monge under the Restoration—but such events were very much the exception. Holders of state appointments, including members of the Académie, were required to take an oath of loyalty to the government. Such a requirement can, perhaps, be defended, but for those who lacked the flexibility of a Laplace it could make life difficult in a fluctuating political situation. After the July Revolution of 1830 the legitimist Cauchy left France, but it was a self-imposed exile. Under the government of Napoleon III both the royalist Cauchy and the republican Arago were exempted from the oath of allegiance.

Government support of science could take many forms. Most obviously it provided institutions in which scientists could work, and it paid their salaries and some expenses. One area which has not been examined by historians is that of publication, something that was particularly costly in the case of the biological sciences where many illustrations were required. The state was able to help by taking out multiple subscriptions. To give a specific example: in 1805 the government was subscribing for 36 copies of the *Annales* of the Museum of Natural History.[28] It was also taking out a few subscriptions to the *Annales de chimie*, but in the case of the latter publication the Minister of the Interior intervened in a far more effective way: in 1806, at the behest of Fourcroy, he sent a circular to the prefects in all the departments of France urging them to encourage local subscriptions to the journal as one of public utility.[29] Not many scientific journals have been able to benefit from such government pressure. Indeed, the incident reminds us of the political dangers of a close association of government and science. It would have been sad if support had led to control of publication. A long-standing accusation that the Napoleonic government intervened to prevent the publication of an alternative scientific theory has been refuted.[30]

In the last twenty years the history of French science has gradually gained greater recognition as a field of study. With greater professional support in the United States and Britain than in France for history of science, more has in fact been done in this field by 'les Anglo-Saxons' than the French,[31] but I do not think this is necessarily a bad thing. Since the 1950s, studies have widened from exclusive concentration on the Revolutionary period, and scholars have looked back to the Ancien Régime as well as forward to the early twentieth century. The greatest task ahead for historians probably lies in the nineteenth century. There is still much to learn about scientific education and scientific societies as well as about the science actually done. For many institutions, the principal source available is a centenary study, possibly published in the 1890s and with the editorial or financial assistance of alumni. The personal reminiscences contained in some of these histories hardly qualify as history, and we urgently need more scholarly studies.

Even the general public now recognizes that scientists are human beings and that we need to know more about them. Biographical studies of scientists can provide insight into the whole scientific endeavour in a national context. The French tradition of *éloges* has hardly fulfilled this purpose. It is seemly to praise the dead, but the historian usually wants greater freedom to analyse and criticize. Government support of science must be studied by looking at the records of the different ministries which contributed in some way to the support of science, as well as by studying parliamentary debates and statements of government policy. Influential scientists such as Laplace, Cuvier, Dumas, Pasteur, and Berthelot were able to exert important influence on government support of science. We are still awaiting a detailed study of laboratory facilities in nineteenth-century France.[32] We know that by twentieth-century standards they were inadequate, but we should beware of exaggeration—the shed with the leaking roof where Marie Curie toiled over her extraction of radium from pitchblende (as in the film) can hardly have been typical for scientists after they had gained recognition.

Apart from language, there are certain technical difficulties in pursuing the study of French science. For the benefit of foreign scholars there is now a specialized American company based in Paris, which in its advertisements claims to 'cut through the lack of co-ordination and bureaucratic difficulties which plague French research facilities'.[33] I think such remarks are a little unkind. In many instances the researcher is met with friendly co-operation, and there is a French tradition of documentation which goes back to the Revolution and which means that valuable information is recorded in archives. All this should encourage the potential student.

I am well aware that in this general talk I have been guilty of indulging in generalizations, some of which are only partly true—that is inevitable if one is to cover a broad canvas and not continually to insert tedious qualifications. I have also omitted many aspects of French science in this rapid and impressionistic survey. I should like to have had time to discuss other peculiarly French contributions to the history of ideas, such as positivism, or aspects of the organization of science, such as the *cumul*, but these would need separate articles; and to give even a superficial survey of French science over one century would require a book. But it seems to me that a prior need is for detailed studies of certain key figures in science, of the scientific community, and of institutions. When these have been published, we shall have a clearer idea of what is meant by French science.

I have tried to convey something of the atmosphere of science in France. I spoke earlier about some of the difficulties and limitations, but I cannot end without acknowledging a real danger in the study of science in a national context. Such history could easily degenerate into flag-waving, perhaps even encouraged by government grants. If governments became obsessed with the history (including history of science) of their own countries in order to bolster nationalism, a dangerous situation would result. Some past history of science has been too nationalistic. I think one achieves a greater objectivity if one decides to study a country other than one's own. The multi-volume *History of the English people in the nineteenth century* by the French historian Halévy is an example of a history written from the outside. Such a history can gain in perspective what it might lose in intimate local knowledge. But inevitably most history in a national context is likely to be written by natives and so with the dangers I have suggested.

I conclude, therefore, with the suggestion that a study of history of science in a national context should not preclude a wider frame of reference. In such history one is always implicitly making comparisons; but an explicit comparative study should be encouraged. Not only are we being less parochial if we know something about science in more than one country but we can learn more about science. It is so easy to take particular attitudes and institutions for granted. If we look at these in more than one country, we may gain a valuable perspective. We are finding at Canterbury that a study of the transmission of scientific ideas from one country to another can be an enlightening and, I hope, useful exercise.[34] The diffusion of science can often be profitably studied in terms of country, and studies of the reception of the ideas of Descartes, Newton, Lavoisier, and Darwin have been followed in this framework. But although nations have provided geographical, political, and linguistic constraints, they have never managed to imprison science. Much experimental, observational, and even theoretical science can be usefully thought of as arising within a national context, but it finally becomes international. International co-operation in science has been growing in the past 300 years. In the end we must understand the development of science not only in its national but also its international context.

NOTES

[1] See M. B. Hall, 'Science in the early Royal Society', in M. P. Crosland (ed.), *The emergence of science in western Europe* (London, 1975), pp. 57–77.

[2] *Second Report of the British Association for the Advancement of Science; Cambridge, 1832* (London, 1833), p. 184.

[3] R. E. Stebbins, in Thomas F. Glick (ed.), *The comparative reception of Darwinism* (Austin, Texas, 1974), p. 122.

[4] *De l'origine des espèces ou des lois du progrès chez les êtres organisés* . . ., traduit . . . sur la troisième édition . . . par Mlle Clémence Auguste Royer. Avec une préface et notes du traducteur (Paris, 1862).

[5] Darwin to Lyell, 22 August [1867], in *Life and letters of Charles Darwin*, ed. F. Darwin (3 vols., London, 1888), iii. 72.

[6] O. Chadwick, *The secularisation of the European mind in the nineteenth century* (Cambridge, 1975), p. 176.

[7] Darwin, op. cit. (5), iii. 118.

[8] M. P. Crosland, *The Society of Arcueil. A view of French science at the time of Napoleon I* (London, 1967), pp. 14–16.

[9] J. W. Herivel, *Joseph Fourier. The man and the physicist* (Oxford, 1975), p. 235.

[10] *Gleanings in science*, i (1829), pp. xi–xii.

[11] Ibid., p. vii.

[12] *The Canadian journal. A repertory of industry, science and art and a record of the proceedings of the Canadian Institute*, ed. H. Y. Hind, i (1852–3), Introduction, pp. 2–3.

[13] S. A. Shapin and A. W. Thackray, 'Prosopography as a research tool in history of science: the British scientific community, 1700–1900', *History of science*, xii (1974), 4.

[14] For an excellent characterization of a certain style of teaching science in Paris in the nineteenth century, see R. Fox, 'Scientific enterprise and the patronage of research in France, 1800–70', *Minerva*, xi (1973), 442–73.

[15] Some information on nineteenth-century career structures is given in M. P. Crosland, 'The development of a professional career in science in France', op. cit. (1), pp. 139–59.

[16] An introduction to Revolutionary institutions is provided in M. P. Crosland (ed.), *Science in France in the Revolutionary era described by Thomas Bugge* (Cambridge, Mass., 1969).

[17] For the history of the Museum in the Revolutionary period, see M. Deleuze, *Histoire et description du Muséum royal d'histoire naturelle* (2 vols., Paris, 1823), i. 67–158.

[18] *Life, letters and journals of Sir Charles Lyell, Bart.*, ed. [Mrs] K. M. Lyell (2 vols., London, 1881), i. 150.

[19] A ministerial decree of 14 July 1837 retitled the chair of pharmacology at the Paris Faculty of Medicine: it was henceforth the 'chaire de chimie organique et de pharmacie'. It was given to J. B. Dumas. A further chair of organic chemistry was established in 1865 at the Collège de France for Marcellin Berthelot.

[20] *Le Collège de France 1530–1930. Livre jubilaire* (Paris, 1932). See also T. N. Clark, *Prophets and patrons. The French university and the emergence of the social sciences* (Cambridge, Mass., 1973).

[21] J. F. W. Herschel, in *Encyclopaedia metropolitana* (London, 1845), iv. art. 'Sound' [1830], 810 n.

[22] The time devoted to science in the lycées was a contentious issue; there were frequent changes which it would be tedious to list.

[23] O. Gréard. *Éducation et instruction. Enseignement supérieur* (2nd edn., Paris, 1889).

[24] Report quoted by R. D. Anderson, *Education in France, 1848–1870* (Oxford, 1975), p. 61.

[25] A. Dalmas, *Evariste Galois, révolutionnaire et géomètre* (Paris, 1956), p. 27.

[26] Although the major work of all four was in chemistry, Gay-Lussac and Berthelot had gained election to the Académie in the less fiercely contested physics section.

[27] 'Lettre sur l'enseignement des sciences', *Gazette des écoles* (2 January 1831).

[28] The cost was 3,120 francs for the year, the equivalent of a modest salary. Archives nationales, F^{17} 1023, dossier 12.

[29] Ibid., dossier 13. The full text of the letter will be given in a forthcoming article by the present author on the history of the *Annales de chimie*.

[30] M. P. Crosland, 'Humphry Davy—an alleged case of suppressed publication', *The British journal for the history of science*, vi (1973), 304–10.

³¹ See, for example, R. Taton, 'Sur quelques ouvrages récents concernant l'histoire de la science française', *Revue d'histoire des sciences*, xxvi (1973), 69–90.

³² For a recent contribution in this area, see, however, Margaret Bradley, 'The facilities for practical instruction in science during the early years of the *École Polytechnique*', *Annals of science*, xxxiii (1976), 425–46.

³³ *History of science society newsletter*, vol. v, no. 1 (January 1976), p. 14.

³⁴ For example, R. G. A. Dolby, 'The transmission of science', *History of science*, xv (1977), 1-43, and M. P. Crosland and C. W. Smith, 'The transmission of physics from France to Britain: 1800–1840', *Historical studies in the physical sciences* (in press).

XI

Science and the Franco-Prussian War

The historian of science is interested in factors which encourage the development of science and also those which inhibit it. In addition he must concern himself with influences which, a priori, are less clearly positive or negative. One of these is the influence of war.[1] This essay examines one particular case of science in relation to war, a war which came to the very doorstep of a city which had been a Mecca for the study of science and medicine. War came to Paris in 1870 in a way which did not happen to London until the 1940s. The involvement of a civilian population and of scientists were also foretastes of war in the twentieth century.

In considering the relation between science and war it is useful to distinguish two questions: (i) the influence of war conditions on the local development of science; and (ii) the influence of scientific

An earlier version of this paper was presented to a staff seminar at the University of Kent and I should like to thank Mr W. M. L. Bell, Mr R. G. A. Dolby and Dr M. J. M. Larkin for their criticisms. For publication the original paper has been drastically abbreviated and partly rewritten with the benefit of helpful editorial suggestions from Dr R. M. MacLeod. For the manuscript sources on which this study draws I am indebted to the Archives Nationales, M. Langlois-Berthelot, the Bibliothèque historique de la ville de Paris and the Société chimique de France. A final acknowledgement is due to the Nuffield Foundation for support in this research.

knowledge upon the outcome of war.

The Franco-Prussian war had a direct and immediate influence upon the Academy of Sciences, and a more long-term influence on the state of science in France. However, one should not begin by assuming a necessary connection between the development of science and success in war. Victory or defeat obviously depends on many factors quite outside science: morale, training, number of combatants, leadership, to mention a few of the more obvious. Although after the French defeat some French scientists argued that the Prussian victory had been prepared in the German universities and depended on superior scientific investment the argument is suspect. It was, however, a useful line for scientists to take in France after 1870 to increase the financial support of science by the French government.[2]

It is not my contention that, unknown to historians, this was really a scientists' war and not a war between armies. On the contrary, military historians have related victory to Prussian organization and preparedness in contrast to French over-confidence at the beginning of the war and inefficiency. Much of the outcome of the war can be explained in purely political and military terms. However in the later stages of the war science came increasingly to be involved and some of its applications might well have affected the outcome when Paris came under siege.

One of the most remarkable features of the siege was the extent to which science was called upon. The invocation of science (and technology under the name of science) depended on a number of factors implicit in the social and intellectual history of France. It is possible to characterize at least three factors: (a) a memory of the wars of the First Republic when Berthollet, Monge, Guyton, Fourcroy and other men of science had advised on the manufacture of munitions. An explicit comparison was frequently made with the crisis of 1793, the last time there had been a cry of 'the motherland in danger'; (b) a general idea that France was endowed with some of the world's best scientists; and (c) a feeling for the boundless possibilities of science and, among republicans, an ideological belief that a republic was peculiarly fitted to exploit science successfully.

Science was never held in greater expectation than during the siege of Paris in 1870. It was expected not only to provide (or supplement) the daily bread of the Parisians, it was also expected to provide ultimate salvation. Kranzberg, one of the few historians to consider the scientific element[3] in the siege, attributes a class basis to this belief in science as the provider of miracles. He comments that, although middle-class audiences were sceptical of some of the more fantastic plans,

> the lower and less well-educated classes were not so sceptical of the many notions advanced before their clubs. The illusion that science could accomplish all never ceased to give hope that perhaps some one invention might be discovered which would surely and promptly give them victory, or that some marvellous new food might be invented which would save them from starvation and make that victory possible.[4]

Some popular ideas emanating from the Paris clubs represent an extreme of fantasy. Schemes, like the idea of killing Prussian soldiers by arming women with needles dipped in prussic acid, seem no more than a joke in bad taste. Others, like the idea of using balloons to lift an enormous hammer 6 km in diameter and weighing ten million tons, and then cutting the strings to drop it on the Prussians, seem child-like in their naivety. In this essay attention will be concentrated on the contributions of men with scientific training, and ideas that were possible, at least in principle.

One leading French scientist who was most involved in the application of science to defence was Marcellin Berthelot. He later acknowledged that an account of the scientific work of the period would have been of interest to posterity, but, writing some thirty-five years after the event he felt that:

> It would be too tied up with an account of the troubles and the failings of this period for it to be appropriate to do it now or perhaps ever.[5]

Thus, more than a generation after the Franco-Prussian war, memories in France were too bitter to make an objective historical account possible. Apart from several descriptions of balloons and two very personal descriptions[6] of aspects of scientific work during the siege of Paris, accounts published in the rush of memoirs which appeared in 1871, immediately after the event, little attention has been given to the role of science and technology in the war.

THE EARLY STAGES OF THE WAR AND THE BEGINNING OF THE SIEGE OF PARIS

War was declared on 19 July 1870. The early successes of the Prussians, derived largely from their good training, organization and communications, enabled them to pour troops over the border before the French had recovered from the surprise of their first defeats. Six weeks later, on 1 September, Napoleon III, at the head of an army of 80,000 men, surrendered to the Prussians at Sedan. This marked the end of the Second Empire, but the fight was continued from Paris under a new republican government which gave itself the significant title of 'government of national defence'. The establishment of the new government on 4 September marks a turning point in the war. This date is of some importance in understanding the establishment of scientific commissions which will be discussed in the course of this paper.

The countryside round Paris was transformed into a theatre of war when two Prussian armies converged on the capital. Paris itself was defended by a line of fortifications which prevented a simple frontal assault. However, by 19 September the Prussian armies had managed to encircle the French capital completely and for the two million inhabitants there began a siege which was to last into the bitter winter of 1870-71. Paris, which had been the host at the glittering International Exhibition of 1867, was to become a very different place.

It was particularly during the siege that science (or what was regarded as science) became of importance. The centralization of the French state was reflected in the geographical distribution of scientists and many leading French scientists found themselves in the besieged city.

THE EFFECT OF THE WAR ON SCIENTIFIC ACTIVITY

One immediate effect of the war may have been to reduce normal scientific research and publication. The reduced size of the *Comptes Rendus*, the official publication of the Paris Academy of Sciences, is a first indication of this. Another French scientific periodical claimed retrospectively that because of the military and political situation 'science had been suspended in France for a whole year'.[7] It claimed that 'the second half of 1870 and the first half of 1871 have been empty as far as scientific productivity is concerned'. A new journal, the *Archives de zoologie,* was due to appear in July 1870 when war was declared and publication was postponed until January 1872. On the other hand, the leading French chemical journal, the *Annales de chimie,* continued more or less regular publication on the same scale apparently regardless of the turmoil outside.[8]

There were certainly some scientists for whom the war was simply a source of interference with their routine work. Thus, Charles Sainte-Claire Deville complained at a meeting of the Academy of Sciences on 12 September that his meteorological experiments in the suburb of Montsouris were being disturbed by French troops responsible for the defence of the capital.[9] A British correspondent, remarking on the continuation of meetings of the Academy throughout the siege, suggested that

> the members may have felt it a relief to have to deal with immutable and indestructable facts [sic] while everything around them was so mutable and perishing.[10]

For Pasteur, probably France's most famous scientist of the time, and a great admirer of Napoleon III, the collapse of the Empire was a nightmare. The man who had so brilliantly applied science to practical problems in the 1850s and the 1860s was paralysed by the Prussian invasion. He was persuaded to leave Paris at the beginning of September for the safety of the provinces. He later declared: 'The war sent my brain to grass.'[11] His most notable act during the war was to return to the University of Bonn an honorary doctorate he had received two years previously, and to make a verbal attack on the King of Prussia.

After Pasteur, France's most famous living scientist was probably the physiologist Claude Bernard. Bernard was 57 in 1870 and had suffered from poor health for several years. One biographer tells us that he was considerably agitated by the outbreak of war in July and found he could not work in his laboratory,[12] although he asssiduously attended the meetings of the Senate and the Academy of Science. In mid-September he managed to leave Paris for his home village, explaining to a friend that he had encountered obstacles which prevented him playing a part fot the public good. Bernard did not return to Paris until the following June, the intervening months having passed with little constructive work accomplished.

However, if Pasteur and Bernard did not find in the war inspiration for work, others did. Pride of place must go to the organic and physical chemist Marcellin Berthelot, a scientist not previously noted for his concern with applications of science. This is how Berthelot later explained the effect of the siege:

> Dedicated as I have been since the beginning of my life to the search for pure truth, I have never been mixed up in the struggle of practical interests

which divide men; I have lived in my lonely laboratory, surrounded by a few students, my friends. But in the supreme crisis which France experienced no-one was allowed to remain indifferent; everyone had an obligation to contribute his help, however humble it may have been. This is how I was torn away from my abstract studies and I had to concern myself with the manufacture of cannon, gunpowder and explosives.[13]

Individual scientists, however, could do little. It was through official institutions and commissions that they were able to contribute effectively, sometimes helped by government finance. The scientific institutions which helped in the French war effort can be classified into four groups. In the first place there was the venerable Academy of Sciences, proud, austere, but not altogether aloof. It maintained its tradition of weekly meetings but could not ignore the crisis at its doors. Its discussions reflected the problems of the moment and it did its best to offer solutions. Its direct military contributions were probably less than a second group of institutions, the specialist scientific societies which provided a cadre of experts who were prepared to volunteer their services. Both the Chemical Society and the Society of Civil Engineers made useful contributions.

Thirdly, and of supreme importance, there were the new ad hoc scientific commissions, although I think it would be a mistake to classify them together. On the one hand there were the top-level commissions, close to the government, which were able to make recommendations and see them carried out. These commissions, notably that chaired by Berthelot, must be distinguished from committees established from political rather than military motives. By encouraging popular participation, the scientific committees of the twenty Paris 'arrondissements' (districts) probably contributed more to boosting morale than to any direct military endeavour.

THE ACADEMY OF SCIENCES

The Academy of Sciences was involved in the war in a way completely new in its history.[14] During the Revolution it had been under suspicion as a monarchical and elitist institution, and the scientists who had helped in the war effort in 1793 had done so by virtue of their technical expertise rather than as Academicians. Yet the existence of the Academy in the besieged city of Paris in 1870 provided an obvious source of talent. Consisting of sections ranging from pure mathematics to rural science and medicine, it was prepared to take a broad view of science and its applications.

Since its foundation in the seventeenth century, the Academy had constituted a useful source of advice to the government on scientific matters. In the crisis of 1870 it was not the government which approached the Academy, but the Academy which took the initiative. However, the Academy had important deficiencies. It was very restricted in membership. The average age of Academicians was well over sixty so that many scientists of undisputed merit were excluded on grounds of seniority. Bound by its traditions, the Academy was quite unable to co-ordinate the pool of scientific talent in the nation's capital.

One thing the Academy was able to do was to maintain its regular Monday meetings. At the meeting of 5 September, with the Prussians within sight of Paris, attendance at the Academy was 42, or nearly two-thirds of its membership.[15] During the siege the average attendance was 36, and never fell below 30. There may have been something of the ethos of the theatre in this — despite the emergency the show must go on — but even at its worst the Academy was never merely a show. Although it did not provide facilities for research, it was always an important forum for scientific discussion.

The first sign of war impinging on the work of the Academy came on 22 August when a note on military projectiles was submitted.[16] From the beginning of September, none of the weekly meetings of the Academy was to pass without some discussion relating to the war. Academicians were able to contribute a variety of memoirs on matters relating to the war and siege, but when non-members sent in suggestions, discussion was often hindered by adherence to the tradition that memoirs from outsiders should not be discussed by the Academy before being considered by specially appointed committees.[17]

The war placed the Academy under great strain. Although the Academy established sub-committees on 'military arts' and 'culinary arts', it was quite unable to deal effectively with the flood of ideas canvassed at the time. It postponed indefinitely the consideration of memoirs on balloons which had been submitted by outsiders, rather than expose the ignorance of their authors both in practical ballooning and in the elementary principles of science:[18] yet it gave strong support to the impractical ideas of one of its own members.[19]

There is no doubt that the war, and particularly the siege, completely transformed the normal academic science customarily discussed at the Academy. A few members were inspired by the emergency to supreme efforts. But individuals were not likely to achieve much unless their ideas could be channelled through appropriate institutions. In the event, new institutions, specifically created in the early months of the war, allowed men of science to make their most effective contribution.

SCIENTIFIC COMMISSIONS

The first commission to organize the defence of Paris was set up in early August, but the president of the commission, Marshal Baraguey d'Hilliers soon left the capital and no action was taken to replace him or reconvene the commission. On 17 August the Empress, then in Paris, took the initiative and established by decree a 'committee of defence for the fortifications of Paris'.[20] Again this was thought of in military terms and consisted largely of high-ranking army officers and senators, but it also included the Minister of Public Works, Baron Jerome David. At the first meeting of the committee on 18 August, David reported that he had set up what he called 'a commission of engineers'[21] to report on particular aspects of defence.[22] Thus, a month before the siege, a commission was sitting on a number of specific technological and scientific problems relating to the war. The commission was asked for advice on the advisability of using any form of artificial lighting in powder magazines. On the question of using electric light to illuminate enemy positions, it recommended a particular portable generator.[23] At a meeting of the defence committee the next day, 19 August, the Minister of Works reported his engineers' advice that under siege conditions magnesium could provide a further source of illumination. The defence committee accepted this advice and immediately placed an order for 100 magnesium torches and fifty 100 kg barrels of magnesium, costing 27,000 francs.[24] Nothing more was heard from the engineers until September, when there was a new government, a new Minister of Works and a more urgent situation. Meanwhile, however, there had been a strong movement for direct parliamentary membership of the committee of defence. To meet this demand the government at the end of August had nominated four additional members to the committee, including Thiers and Dupuy de Lôme, a naval engineer and a member of the Academy of Sciences but also nominally a parliamentarian, since he had been elected as a deputy in 1869 with government support. Dupuy de Lôme continued as a member of

SCIENTIFIC COMMISSIONS, 1870

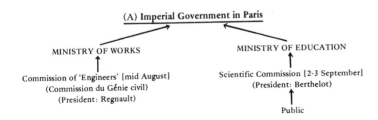

(A) Imperial Government in Paris

MINISTRY OF WORKS

Commission of 'Engineers' [mid August]
(Commission du Génie civil)
(President: Regnault)

MINISTRY OF EDUCATION

Scientific Commission [2-3 September]
(President: Berthelot)

Public

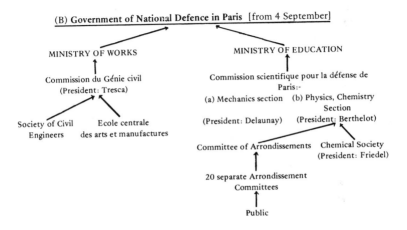

(B) Government of National Defence in Paris [from 4 September]

MINISTRY OF WORKS

Commission du Génie civil
(President: Tresca)

Society of Civil   Ecole centrale
Engineers   des arts et manufactures

MINISTRY OF EDUCATION

Commission scientifique pour la défense de
Paris:-
(a) Mechanics section   (b) Physics, Chemistry
Section
(President: Delaunay)   (President: Berthelot)

Committee of Arrondissements   Chemical Society
(President: Friedel)

20 separate Arrondissement
Committees

Public

(C) Government Delegation at Tours [October]

MINISTRY OF WAR

Commission d'étude de moyens de défense
(President: Serret)

Public

the committee of defence when it was reconstituted under the new government in September. On 6 September, he undertook to organize the mining of the outer perimeter of the Paris defences, making use of 180,000 metres of copper wire supplied by the navy so that the mines could be electrically detonated.[25]

The commission of engineers not only served a useful direct purpose in the early stages of the war but provided a definite precedent for a commission of scientists. Science in nineteenth-century France usually came under the Ministry of Education, and on 2 September the Minister, Jules Brame, established a commission with Berthelot as president to consider how science could be applied to defence.[26] The commission consisted of seven scientists (mainly chemists and physicists) and we may note that two members of the chamber of deputies were also included. It is one of the ironies of our story that the imperial commission had the briefest possible active life, since the government fell the day after its first meeting.

However, as such a commission seemed to have a useful function not only in giving advice to the government but also in acting as a clearing house for suggestions from the public,[27] it was reconstituted under the new Minister of Education, Jules Simon as 'la commission scientifique pour la défense de Paris' and given a budget of 40,000 francs.[28] The commission was divided into two sections: Berthelot continued as president of the physics and chemistry section,[29] while a parallel smaller committee was set up, called the 'mechanics section', with Delaunay, the new Director of the Observatory, as its chairman and an industrialist, an arms manufacturer and an administrator among its members.[30]

The physics and chemistry section was not only the most clearly scientific, but was also the most active. From its first meeting on 9 September it met daily, with possibly some exceptions towards the end of the siege. It either met in a room at the Ministry of Education or, if practical trials were to be carried out at a suitable place, it met there. It often visited forts on the perimeter of Paris and was not above visiting a municipal rubbish dump in the interests of public health.

An urgent problem considered on 11 September[31] was the supply of chemicals which might be desirable either for defence or public health before the Prussian army completed its encircling movement and cut off all supplies. On 15 September various methods of producing smoke screens were discussed (including the reaction between ammonia and hydrogen chloride) but a question which cropped up time and time again at the meetings was fire; both fire raising and fire extinguishing.

THE ORGANIZATION OF CIVIL ENGINEERS

In the early days of the new government of national defence several groups of engineers offered their professional expertise to the Minister of Works. One was the Society of Civil Engineers and another a group of former students from the Ecole centrale des arts et manufactures. On 13 September the potential value of engineers was recognized by a government decree establishing a 'Commission du génie civil' under the Ministry of Works.[32] This was intended to draw on the expertise of all the main groups of civil engineers in Paris and centralize them at the Conservatoire des arts et métiers, the deputy director of which, Tresca, was appointed president. The Commission du génie civil was asked to advise the government on the manufacture in Paris of the 'mitrailleuse', a primitive machine gun.

The Society of Civil Engineers under its president, Vuillemin, felt that they should do something to help and they decided to meet weekly to discuss how science and technology could be applied to the emergency. Memoirs were presented on dynamite, cannon and fire engines, but probably their greatest contribution was what they called 'mobile ramparts' or 'mobile redoubts'.[33] These were fortified waggons carrying riflemen and may be seen as a forerunner of the armoured car, if not the tank.

THE CHEMICAL SOCIETY

The other specialist scientific society which made a major contribution to the defence of Paris was the Paris Chemical Society. Feeling that the best hope of ending the siege lay in the possession of modern artillery, the Society on 10 October voted a sum of 5000 francs to pay for the manufacture of one breech-loading cannon to supplement the muzzle-loading cannon available.[34] Obviously one cannon would have made little difference to the outcome of the siege and on the following day, 11 October, Berthelot proposed that a public subscription should be opened for the manufacture of cannon in considerable numbers.

It so happened that on 12 October there was to be an assembly of the scientific committees of the districts of Paris (an organization discussed below). The steps taken by the Chemical Society were reported to it. The assembly agreed that the majority of the muzzle-loading cannon available to the French forces were inferior both in range and in rapidity of fire to breech-loading cannon, of which Paris had only about 500. However, if the manufacture of a further 1500 cannon could be organized – and this was well within the capacity of private industry in Paris – Paris might be saved.[35] Accordingly, the assembly agreed to open a public subscription for 1500 cannon and on 25 October, Friedel, the 38 year-old president of the Chemical Society, and Ribeaucours, the president of the assembly of scientific committees, had a meeting with government representatives who agreed to accept this help. Some young chemists turned their hands to the casting of cannon but they met many practical problems. They were better able to help in the manufacture of explosives, an activity discussed later in this paper.

POPULAR COMMITTEES

In addition to the scientific commissions and the professional societies, there were also more popular scientific committees, one for each arrondissement of Paris. The formation of such district committees concerned with defence was a logical extension of the idea of a government of national defence. The fact that they were explicitly *scientific* committees, with some pretensions to scientific knowledge, is a reflection of what science was thought to promise. It was Dorian, the Minister of Works, who suggested that each district should set up a committee consisting of a chemist, someone qualified in mechanics and an engineer or architect, the committee to be chaired by the local mayor. The minutes of the committees[36] append the occupation of most of their members and we can confirm that in accordance with the original intention a fair number of professional men were recruited, including civil engineers and architects. The 7th arrondissement was exceptionally well qualified in science, with three science teachers and a doctor of medicine, Ribeaucours, who served as provisional mayor of his district and was elected president of an assembly of the committees from all the districts. The proposal that there should be a general assembly of all the district committees came from the 7th arrondissement. It made obvious sense to try to co-ordinate their efforts and compare notes.

The arrondissement committees were directly in touch with the public. They passed on to the general assembly the more sensible proposals involving science. The general assembly had some dealings with Berthelot's commission but their most effective contact was with the Chemical Society, whose idea of manufacturing cannon they took up eagerly, using their popular base to launch a public subscription. The cannon manufactured under these unusual circumstances became of great importance in 1871 to the men who set themselves up in Montmartre as the true government of France. However, in the months of the siege politics were explicitly excluded from the work of the commission.[37]

SCIENTIFIC COMMISSIONS
IN THE PROVINCES

Finally, two commissions must be mentioned which were set up in the provinces with the object of using scientific expertise in the interests of war. The first step was taken on 28 September when the delegation of the government of national defence established at Tours set up a commission to examine the use of balloons in defence.[38] It was not long before the terms of reference of this explicitly 'scientific commission' were widened to include an examination of all aspects of defence to which science might be applied. The establishment of the commission was largely the work of Silbermann, a demonstrator in physics at the Collège de France. Silbermann was also vice-president of the French Meteorological Society, meteorology being a science of particular relevance to ballooning. The president of the commission was the mathematician Serret (1819-85), a member of the Academy of Sciences, and it included one or two staff of the Paris Observatory and two local lycée professors of physics.

The widening of terms of reference in mid-October was followed by a decree of 23 October, specifying more precisely the function of the commission.[39] It was now an advisory committee, which the government could consult on theoretical problems related to defence. This was in a tradition going back to the Revolutionary period of setting up committees of experts in a 'comité consultatif'. For its brief existence the commission had an unusually complex history and with the possible exception of maps, it seems to have produced very little in the way of science applied to defence. Its title of 'commission scientifique' represents an early ideal rather than actual achievement.

There was a second provincial government commission established at Tours in October 1870 by the Ministry of War, the 'Commission d'étude des moyens de défense'.[40] Its members were mostly military and civil engineers together with the chemist Naquet. It was concerned with the application of scientific, technological and other ideas to the urgent military situation. The Ministry of War had been overwhelmed with suggestions from patriotic citizens and the commission was intended specifically to examine these suggestions. De Freycinet, who helped to reorganize the Ministry of War, said that one of the problems facing the Ministry was to free the administration from

> a crowd of inventors, who encumbered the technical services and whom it was not easy to repulse because at this time public feeling was running very high; people were very much against negligence and the bureaucracy of the imperial administration and each inventor whose ideas were not discussed at length was almost ready to cry treason.[41]

This commission originated from the political situation of a 'nation in arms', when every citizen could claim the right to be heard, regardless of his education or formal qualifications. From the point of view of the Ministry it acted as a safety valve, granting civilians their democratic rights but allowing the Ministry to continue to function and to take as much or as little notice as it wanted of the suggestions passed on to it by the commission. The commission acted as a filter, passing on only a small proportion of the ideas submitted to it by the public.[42]

EXPLOSIVES

The explosive nitro-glycerine is very sensitive to shock but it had been made comparatively safe by the recent discovery by Alfred Nobel that it could be stabilized with the absorbent earth, 'kieselguhr'. The publication of his patent for the resulting 'dynamite' (1867) led to its use as a blasting explosive in several countries, including Sweden, Britain, Prussia and the USA before 1870 but it was almost unknown in France. Two pamphlets[43] on the subject were published in Paris in 1870 before the outbreak of war, but little notice had been taken of them.

Copies of these pamphlets were sent to Berthelot's committee and on 16 September the author of one of them, Achille Brühl, attended a meeting of the committee to speak about dynamite.[44] Berthelot delegated research to the Chemical Society, of which he was a past president. This emergency research, undertaken in the period of the vacation, is unfortunately not minuted in the proceedings of the Chemical Society[45] but a later account reads as follows:

> In order to achieve its ends the Commission [i.e. the scientific Commission for the defence of Paris] made use of the good offices of young scientists. Members of the Chemical Society who had remained in Paris, together with technicians with different specializations, did their utmost to contribute to the work of defence by patriotic devotion in the hours left free by their military service; they had meetings daily, which Berthelot normally attended.[46]

A number of young chemists, including C. Girard, A. Millot and G. Vogt, undertook to find a suitable absorbent material to replace keiselguhr which was not available in the besieged city. Their research, carried out in Wurtz's laboratory at the Faculty of Medicine was reported to the Academy of Sciences two months later.[47]

The minutes of Berthelot's commission, his diary and correspondence, all show his deep involvement in the manufacture of nitro-glycerine and dynamite during the months of September, October and November. By the end of November two factories had been established in Paris producing 300 kg of dynamite a day, and there is some evidence of its use by French troops.[48] Further independent research on dynamite under the aegis of the Ministry of Works was carried out by P. Champion, a demonstrator in chemistry at the Conservatoire des arts et métiers.[49] This, too, was reported to the Academy.

Other explosives were manufactured in Paris during the siege. Two small factories on the Quai de Javel manufactured gun-cotton until, after a serious explosion, production was stopped.[50] The manufacture of explosives with a potassium-chlorate base was also briefly revived. Berthelot himself devoted much time to exploring possibilities for the manufacture of gunpowder. Unfortunately the chief constituent, potassium nitrate, was not available. Berthelot therefore consulted historical works and reverted to primitive methods of extracting the saltpetre from natural sources such as stables and cellars where decaying organic matter and excrescences contained nitrates.[51] He enrolled the collaboration of the Paris district committee which were flattered to be asked for help by him.

FOOD

There had been confident predictions from the German side of capitulation of the French capital within a few weeks because of starvation.[52] However, at the beginning of the siege 25,000 head of cattle and 150,000 sheep had been brought into Paris. With a population of two million to consider, this was reckoned to be sufficient for about two months. As the autumn drew on and cattle and sheep were slaughtered and not replaced, people began to look elsewhere for meat. From horses they turned to domestic pets and in their extremity even a rat became a marketable item. Nor were the possibilities of the Paris zoo overlooked.

A number of attempts were made to apply scientific knowledge and principles to the urgent problem of subsistence. The first was the problem of preserving food, particularly meat. In addition to the traditional method of salting, the use of sulphur dioxide and immersion in a dilute solution of phenol were proposed.[53] The method of canning, introduced by a Frenchman, Appert, during the Napoleonic wars and further developed during the American Civil War, was not properly exploited in the conditions of the Paris siege. Only near the beginning was there really enough good meat for a far-sighted Parisian to have taken some action to preserve it. Throughout most of the siege the emphasis was on new or substitute foods rather than the preservation of existing stocks.

A new foodstuff from bones was introduced by Fremy under the name 'osseine'.[54] He explained that its chemical composition was similar to gelatine but he supported its use as a food not on theoretical grounds but in purely practical terms; that is, that osseine was edible and produced no harmful effects. He did not claim it as a satisfactory diet in itself, a claim which had rashly been made earlier for gelatine. The food crisis of 1870 caused a revival in the old gelatine debate[55] but more usefully it forced a general appraisal of the chemistry of nutrition. The only previous social incentive had been in charitable institutions where the problem had been to provide sufficient nutrition at a low cost.

Fats previously used in candles were investigated as sources of food. Colza (rape-seed) oil was in principle edible but it had an objectionable smell. Wurtz found that if steam at 116-120°C were passed through the oil, it would remove the objectionable impurity without decomposing the oil.[56]

Whatever the needs of adults, Parisians soon came to appreciate that milk was necessary for babies and perhaps for young children. Yet no provision was made for extra cows among the cattle brought into Paris at the beginning of the siege. The Prussian commander-in-chief, Moltke, regarded milk as so essential that he expected Paris to capitulate as soon as the milk supplies failed. With an increasing shortage of milk the Paris municipal council appealed to scientists for artificial milk.[57] Berthelot had preached in the 1860s the almost limitless possibilities of chemical synthesis;[58] now the problem was transferred quickly from a philosophical plane to one of immediate practicality.

Cows' milk had been analysed quantitatively by Boussingault and others in terms of nitrogenous matter, butter, sugar, salt and water. Using this analysis as a recipe, various 'artificial milks' were produced. The recipe suggested by Dubrunfaut was obtained by dissolving in a litre of water 100 g of sugar, 60 g of dry albumen, 4 g sodium carbonate and adding 120 g olive oil to give an emulsion which could hardly have resembled milk in consistency, colour or taste. The French Academy of Medicine could do no better than suggest an egg/sugar emulsion as a substitute and this was hardly helpful considering eggs were in short supply. The ubiquitous gelatine was used in further recipes, and other concoctions were made more attractive by the incorporation of nuts, particularly almonds. At best these 'artificial milks' were pleasant to taste and reasonably inert, but there must have been some which were harmful, particularly to young babies. The search for a substitute for milk in 1870 did not have therefore the same success as, for example, the search for a substitute for cane sugar in the Napoleonic wars.[59] It was one thing to synthesize a simple hydrocarbon, quite another to attempt the synthesis of something as complex as milk.

However, it was not meat nor milk but bread which was the staple diet of the people of Paris. Several memoirs on bread and flour were discussed in the Academy on 3 October.[60] Yet of the many articles of ordinary diet, bread proved the greatest challenge to science. The primary need was for administrative machinery to requisition stocks of flour and ration food. Steps were taken by the government and the amount allowed per person decreased during the winter until Parisians were literally starving. With agricultural resources and a suitable period of time, scientists might have been able to make an effective contribution, but in the confines of the city of Paris they concentrated their efforts in other directions.

MEDICINE

At the beginning of the nineteenth century French medicine, by concentrating medical training in the hospitals and bringing together surgeons and physicians, attained an international eminence second to none.[61] But by 1870 French medical practice no longer led the world and the country which produced Pasteur did not apply his work to medicine. It was left to Joseph Lister in Britain to lay the foundations of modern antiseptic surgery. In the late 1860s occasional use was made in France of Lister's discovery of the antiseptic properties of phenol. On 19 August 1870 an article published in the *Moniteur* recommended this treatment and pointed it out to the general public as an example of the constructive use of science in war. The use of phenol as an antiseptic was again advocated in a paper presented to the Academy of Medicine on 23 August.[62] However, the advice was largely ignored by the medical profession and one writer sums up the question of the use of antiseptic surgery in the Franco-Prussian war by saying: 'The Germans did their best, the French very little.'[63] There was a case of a French doctor, Lucas-Championniere, who took a bottle of phenol to a field hospital as a surgical aid but was prevented by his superior from using it.

The high mortality rate of the wounded was brought to the attention of the Academy of Sciences.[64] However, there were still Academicians like Chevreul who thought that diseases were caused by bad smells. It was J.B. Dumas who explained to his colleagues the use of phenol in terms of Pasteur's germ theory. It was soon appreciated that in a city under siege, public health was a matter of some importance and municipal health services were centralized at the Hôtel de ville as early as 10 September. The proceedings of the Academy of Medicine suggest that its main contribution to public health lay in the vaccination of troops against smallpox.[65] However, there were delays in carrying out the vaccination among the 'gardes mobiles' recruited from the provinces and the death-rate from this disease alone exceeded 400 per week by December. This was comparable with the death-rate from pneumonia and bronchitis which carried away hundreds of Parisians in that bitter winter.[66]

COMMUNICATIONS

Early in the siege the telegraph cable on the bed of the Seine was cut by the Prussians, thus severing Paris's last link with the outside world. Several proposals for a substitute 'telegraph' were put forward. Lissajous, professor of physics at the lyćee Saint Louis, experimented with an 'optical telegraph' or heliograph.[67] Bourbouze, a demonstrator at the Paris Faculty of Science proposed using the Seine itself as a conductor.[68] Preliminary secret trials suggested this would work and with the support of Berthelot's commission, Almeida, professor of physics at the lyćee Henri IV, was given a grant of 50,000 francs as well as a precious balloon in order to escape from Paris and re-establish communications from outside by this method.

One of the best-known features of the Paris siege was the use of balloons.[69] Their primary importance lay in the means they provided for communication with the outside world; they also enabled a small number of men, including the Minister of the Interior, Gambetta, to escape from the besieged city. It was hoped at one time that it might be possible to launch balloons from outside (e.g. from Rouen in the north or Tours to the south) to bring news to the beleaguered inhabitants, but this hope was never realized because of the difficulty of guiding balloons in a given direction. The way that balloons helped in reverse communication was by taking out carrier pigeons, which, when released in the provinces, would return to Paris. The balloon was above all a symbol of escape, and possibly of superiority. Sailing over the heads of the enemy tended to produce a feeling of superiority, all the more so as attempts to shoot down or intercept the balloons were nearly all unsuccessful. The balloon was all the more glorious as a means of escape because, in a period of intense nationalism, it was clearly a French invention.[70]

The idea of using balloons in the war was first suggested to the Minister of Works on 16 July, but no action was taken. In August, David's commission of engineers did discuss briefly their use as observation posts. The task of studying their potential value was given to a colonel of engineers, Usquin, who was appointed president of a commission specifically concerned with captive balloons.[71] This commission recommended making use of the balloons in the possession of the balloonist brothers, Eugène and Louis Godard. Although these captive balloons may have boosted popular morale, they never proved of any military value. Usquin did, however, start negotiations between balloonists and the government which were later to prove fruitful, and he also proposed the use of naval personnel which later provided the main source of manpower both for the manufacture and piloting of the balloons. It was only when the siege began that balloons became important and then it was no longer as captive balloons under the control of the army. Because they were used for communication with the provinces, they became the concern of the Director of Posts.

The construction and varnishing of balloons should be regarded as technology rather than science, although the approach was not purely empirical. Balloons were estimated to require some 2000 cubic metres of coal gas per day and a large volume of gas was also needed in the cannon foundries. A careful watch was therefore kept on supplies, and street lighting was first reduced to half and then dispensed with completely in the interests of defence. Balloons, when they left the factory, were subjected to a standard test. After filling with gas they were left for ten hours and were then required to be capable of lifting a weight of 500 kg. The basic advance in balloon technology was the ability to steer in a given direction. A contribution towards the construction of a dirigible balloon was offered by the Academician Dupuy de Lôme. He was given a grant of 40,000 francs by the Minister of the Interior but was unable to make a practical advance.[72]

The construction of balloons was as much a matter of organization as of the application of science and technology. The government gave orders for their manufacture to the Godard brothers who perfected a varnish which kept the gas within the balloons for several days — long enough for the balloon to be launched and reach a place of safety beyond the Prussian lines. Since railway stations no longer served any useful purpose in the besieged city, they were requisitioned for other purposes. The Godard brothers set up a balloon factory at the Gare d'Orléans where they gave basic training in the use of balloons to a company of sailors. By day the sailors helped make nets for the balloons and in the evenings they had practical instruction in ballooning. It was due to such preparation and training that some 66 balloons left Paris during the siege (an average of almost three a week), nearly all landing safely.

MICROPHOTOGRAPHY

The fact that news of the outside world reaching Paris depended largely on carrier pigeons imposed a severe limitation on the number of messages which could be carried at any one time. Obviously the thinnest possible paper was used to keep down the weight. The idea of using photography in communication[73] came from the chemist Barreswil, the author of a book on the chemical principles of photography. Gambetta had set up a provincial centre of resistance at Tours, and Barreswil was able to offer his advice for improved communication between Tours and Paris. Meanwhile in Paris a professional photographer, Dagron, was sent from Paris by balloon to organize the use of microphotography in communications from the provinces. Dagron set up a base first at Tours, and then at Bordeaux.

In the first stage of the development of microphotography at Tours in October 1870, telegrams were copied in capital letters and then arranged on a board so as to use all the available space. The board was then photographed and the resulting film gave a reduction of about one three-hundredth. It was soon appreciated that if words could be set up in small type, a greater number of messages could be included on one film. On receipt of the film, messages were read with the help of a powerful magnifying glass. Microphotography was taken further by Dagron who first produced a large negative of the printed page of messages and then, using a lens of short focal length, reduced this to a tiny collodion negative giving a reduction of approximately one nine-thousandth. Completed films were rolled into a tube taking up to a dozen, and in this way a single pigeon could carry up to 30,000 telegrams. To transcribe the telegrams, what was basically a magic lantern was adapted using suitable lenses to project the image onto a wall in a darkened room where four clerks were kept busy transcribing. The messages were finally delivered to various parts of Paris in the same way as orthodox telegrams. The weakest part of the system was the homing ability of the pigeons in the adverse conditions of the winter of 1870-71. Only about one in six of the pigeons that left Paris in balloons returned safely with messages and consequently messages were usually repeated several times to increase the chance of being received.

AFTER THE WAR

Extreme hunger, the failure of attempts by French troops in Paris to break the siege, and the abandonment of hope that Paris would be relieved by French provincial forces, finally forced the city to capitulate. An armistice was signed on 28 January 1871. The French defeat was compounded by further civil strife; but the great post-mortem on science in war began in the Academy of Sciences on 6 March, when Henri Sainte-Claire Deville opened a discussion which raised general questions of education and government support.[74] In the revolutionary wars of 1793 the scientists who had contributed to the success of the French armies were able to capitalize on this achievement in claiming support from the government for new scientific institutions.[75] It is ironical that the scientists of 1871 were able to use the argument that they had been *unsuccessful* in winning the war to claim additional government support. They argued that the science used on the French side had been too little and too late. The Prussians, on the other hand, had been able to use science or at least the 'scientific spirit' cultivated in the German universities.[76] It followed that nothing less than a complete rethinking of French scientific education was required. Deville suggested to his colleagues that they should introduce general discussions about scientific education into their meetings. Scientific education must become the concern of working scientists and could no longer be left to politicians and administrators. He argued that scientists could no longer afford the luxury of living studious lives in their laboratories:

> It is now our duty to intervene actively and directly in the affairs of the country and to contribute with all the force at our command to a regeneration by knowledge.

Under the Third Republic, scientists became increasingly involved in politics, and Berthelot was even a minister (for a short time) in two governments. The physiologist Paul Bert, who had helped in the hopeless task of organizing resistance in the provinces to the invading forces, was elected in 1872 to the Chamber of Deputies and later became Minister of Education, arguing for free compulsory secular elementary education which included science.[77] The chemist, Nacquet, who had had a chequered career under the Second Empire as an extreme left-wing republican, did no further science after the war, but devoted himself entirely to politics.

Berthelot's work during the siege was not forgotten, and in the elections of February 1871 the people of Paris gave him over 30,000 votes, although he was not officially a candidate. When, in 1881, he was elected a Senator for life, it was on the reputation he had won during the war with Prussia. From his influential position, Berthelot argued that France needed science more than ever in the economic rivalry between nations[78] and in the 1880s he could invoke a return to the spirit of determination of the siege. In 1885, after a move to cut spending on higher education, Berthelot, now inspector general of public instruction, intervened, taking up strongly the cause of science as a national asset. The pressure exerted by Berthelot and his friends resulted in the government agreeing to a long-term building programme, in higher education, including the building of several laboratories at a total cost of 49 million francs.[79]

The war also had some effect on the organization of science, particularly on its centralization. The existence of an almost independent provincial government when Paris had been cut off by the siege gave a new life to provincial pride, and a group of scientists met in 1872 to propose a national scientific organization with meetings in different cities, on the model of the British Association. They included Claude Bernard, Delaunay, Friedel, Quatrefages and Wurtz. Their 'Association Francaise pour l'Avancement des Sciences' owed something to Le Verrier's plan of 1864, but it was only after the war that the Association flourished with successive annual meetings in Bordeaux, Lyons, Lille and so on.[80] It was the outcome of the war as much as the state of science which prompted Almeida to found the first French journal of physics in 1872.[81] It was typical of the years after defeat that the editor, in the preface to the first volume, should have spoken not only about physics but about patriotism.

In 1870 there was already in existence a committee appointed by the Ministry of War to advise on technical and scientific matters relating to the artillery, but in practice this was usually concerned with purely routine problems like overseeing the quality of gunpowder. After the defeat, because it was widely felt in France that the country and the army had not kept up to date and in particular that she had not sufficiently benefited from the advice of her scientists, a new committee was set up to examine all kinds of explosives.[82] It is significant, after the events of 1870, that this new committee was given the official name of *scientific* committee of explosive substances, and that Berthelot was not only a member of this committee but its chairman, the other members being mainly high-ranking army officers. For a civilian to be placed above a general on a War Ministry committee is one indication of the increasing recognition given to science under the Third Republic.

Berthelot did much work in the 1870s on the theory of explosives. He wanted to turn the study of explosives into a science, and to this end he spent a major part of his time in the years 1871-83 collecting data relating chemical composition of different explosives to their explosive power.[83] His most important work however was that carried out in collaboration with Vieille. They discovered that explosions were propagated rather like sound waves, and they accordingly introduced the concept of a 'detonation wave'. The chronelectric method they introduced to measure the velocity is still used today.

CONCLUSION

Evidence has been provided of the considerable activity of French scientists in the Franco-Prussian war in which Parisian scientists were understandably most involved. The asymmetry of the situation, in which French scientists were involved in a way that German scientists were not, was a result of the presence of German troops on the outskirts of Paris. There the Prussian army was opposed by French regular soldiers, volunteers and civilians (including scientists). The scientists who contributed to defence represented the whole political spectrum, although one might have expected that those with republican sympathies, like Berthelot, should have been particularly prominent. A wide range of scientific expertise was used, but chemists and chemistry played a leading part in research on food and explosives, while physicists contributed most to communications.

The war brought out the worst on both sides, with Pasteur's cry of hatred and revenge on the one hand,[84] and the deliberate and systematic destruction by the Prussians of Regnault's laboratory on the other. The anthropologist Quatrefages, who had previously argued against the application of this new science to politics, was influenced by the war to write a paper undermining the claim of the Prussians to German leadership by claiming evidence that they were not of true German stock.[85] An unfortunate development of the bitterness of the war was the intensification in the late nineteenth century of nationalistic claims for science on both sides of the Rhine.[86]

In France the constructive effects of the war can be seen in the general rethinking of French scientific education. Since the 1830s proposals had been put forward for the extension of the Paris Faculty of Sciences, but it took the defeat of 1870-71 to justify the necessary funding.[87] Often proposed reforms followed a German model.[88] One can also see, as a direct result of the war, the development of particular branches of scientific research, such as Berthelot's work on the theory of explosions. The Franco-Prussian war might have marked the beginning not only of chemical[89] but even of germ warfare, since both ideas were proposed to Berthelot's commission. Its decision is worth quoting from the minutes:

> Propositions for the use of phosphorus, sulphuric acid, ammonia, wild animals, poisons and viruses as means of defence were discussed but the commission dismissed them all either as useless or as too cruel if not contrary to morality and humanity.[90]

If the war encouraged the development of various branches of applied science, it had also encouraged scientists to look back to the history of science. Suggestions for the use of observation balloons and saltpetre from natural sources were revivals of ideas from the wars of the French Revolution.

But these scientific activities were not more important than the organization of a whole range of scientific commissions which gave them prominence.[91] There was some understandable secrecy in the application of science and technology to war, and this, together with a more prosaic loss of interest, has meant that several of the sources used in this study have never before reached the public domain. The existence of archival barriers between the various Ministries involved, and the knowledge that many documents went up in flames in the Commune in 1871, has tended to discourage historical research. Had the scientific commissions achieved some spectacular success in the few months (or days) of their existence, or had the French won the war, greater publicity would have been given to them. They might then have provided a guide, or at least an established precedent, for the use of scientists in World War I.

NOTES

1. Most of the literature on science and war focuses on the twentieth century. There is however some brief historical background in e.g. J.D. Bernal, *The Social Function of Science* (London: Routledge & Kegan Paul, 1939), Chapter 7. A good general coverage is given by Louis Morton, in 'War, Science and Social Change', in K.H. Silvert (ed.), *The Social Reality of Scientific Myth* (New York: American Universities Field Staff, Inc., 1969), 22-57. H. and S. Rose, *Science and Society* (London: Penguin Books, 1970), Chapters 3 and 4, discuss science and war in World War I and II. However, their idea (p.38) that

science advisory committees originated in 1914 cannot be accepted if the French experience is considered. For science in the wars of the French Revolution, see C. Richard, *Le comité de salut public et les fabrications de guerre* (Paris, 1922); W. A. Smeaton, *Fourcroy, Chemist and Revolutionary, 1755-1809* (Cambridge: W. Heffer & Sons Ltd, 1962); M. P. Crosland (ed.), *Science in France in the Revolutionary Era Described by Thomas Bugge* (Cambridge, Mass.: MIT Press, 1969). On war and scientific productivity see D. de S. Price, *Science since Babylon* (New Haven, Conn.: Yale University Press, 1961), 102.

2. H. W. Paul, *The Sorcerer's Apprentice. The French Scientists's Image of German Science, 1840-1919* (Gainsville, Fla.: University of Florida Press, 1972). R. Fox, 'Scientific Enterprise and the Patronage of Research in France, 1800-70', *Minerva*, Vol. 11 (1973), 442-73.

3. Melvin Kranzberg's book, *The Siege of Paris, 1870-1871* (Ithaca, N.Y.: Cornell University Press, 1950), can be recommended as a social history introducing the scientific dimension, but unaware of the various scientific organizations. Recent popular works include Alistair Horne's *The Fall of Paris, The Siege and the Commune, 1870-71* (London: St Martin's Press, 1966), and Robert Baldick's *The Siege of Paris* (London: Batsford, 1964); both refer to well-known aspects such as balloons and microphotography, without mentioning scientists or scientific organizations. The existence of scientific commissions organized by the *provincial* government has been noted by Michael Howard, *The Franco-Prussian War. The German Invasion of France, 1870-1871* (London: Rupert Hart-Davis, 1960), 243. Howard urges their further study. Yet it was other commissions which made more significant contributions in the war.

4. Kranzberg, op.cit. note 3.

5. *Science et philosophie* (Paris, 1905), Preface, xiv-xv.

6. E. Saint-Edme, ex-secretaire du comité scientifique de défense des arrondissements de Paris, professeur des sciences physiques à l'école superieur de commerce, *La science pendant le siège de Paris* (Paris, 1871). G. Grimaux de Caux, *L'Académie des Sciences pendant le siège de Paris* (written in April 1871). The British Museum copy of the latter remained with pages uncut until 1974.

7. L. Figuier (ed.), *L'Année scientifique et industrielle*, Vol. 15 (1870-71), published 1872, Avertissement.

8. The *Annales de chimie* included descriptions of some of Berthelot's research on saltpetre and explosives, work inspired by the war.

9. *Comptes Rendus de l'Académie des Sciences de Paris*, Vol. 71 (1870), 425-26. The *Comptes Rendus* is hereafter cited as *C.R.*

10. G. F. Rodwell, 'Science in Paris during the Siege', *Nature*, Vol. 3 (20 April 1871), 490.

11. René Vallery-Radot, *The Life of Pasteur* (1906, New York: Dover reprint 1960), 180, 198.

12. J.M.D. and E.H. Olmsted, *Claude Bernard and the Experimental Method in Medicine* (New York: Schuman, 1952), 178, 180.

13. M. Berthelot, *Sur la force des matières explosives d'après la thermochimie* (3rd edn, Paris, 1883), Préface de la deuxième édition, xiv.

14. R. Hahn, *The Anatomy of a Scientific Institution. The Paris Academy of Sciences, 1666-1803* (Berkeley and Los Angeles, Calif.: University of California Press, 1971).

XI

15. G. Grimaux de Caux, op.cit. note 6, 1.

16. *C.R.*, Vol. 71 (1870), 383.

17. Ibid., 443-45, 451.

18. *Moniteur scientifique*, Vol. 1 (1871), 28-29.

19. Dupuy de Lôme.

20. France, *La guerre de 1870-1871. L'Investissement de Paris*, Vol. I (Paris, 1908-09, 4 vols.), 68-69. This valuable documentary source is henceforth denoted simply as *L'Investissement*.

21. The term 'engineer' should not be taken too literally. It was the appropriate name for an expert with scientific training employed by the Ministry of Works and avoided conflict with the Ministry of Education which was most concerned with the employment of scientists.

22. *L'Investissement, Documents*, Vol. I, 57-158. 'Procès-verbaux du comité de défense, 18 août-19 septembre'. See especially 57-58. David had only been appointed on 10 August, and this would seem to suggest that the commission of engineers was set up during his first week of office. The chairman of the commission of 'engineers' was the physicist Regnault, wrongly reported as 'M. Reynaud de l'Institut' – *l'Investissement*, Vol. I, 281.

23. On 8 September the same committee sent for two powerful electric lights with a reported range of 5-6 km which could be used to warn the defenders of an impending attack by the enemy. Ibid., *Documents*, Vol. I, 133. After Faraday's discovery of electromagnetic induction, one of the first to construct a magneto-electric machine was the Frenchman, Hippolyte Pixii (1832). The French lighthouse administration provided a market for the new electric generators. In 1870 Zenobe Gramme constructed 'the first dynamo of practical dimensions capable of producing a continuous current', but I have been unable to relate this to the war. The Prussians also made use of electric lighting in the siege, using the dynamo of Werner von Siemens.

24. Ibid., 61.

25. Ibid., 127. On 4 October Dupuy de Lôme supplied an impressive list of the mines he had planted (ibid., 238-40).

26. *Rapport de la commission scientifique pour la défense de Paris (Section de physique et de chimie)*, Vol. 2 (Archives Langlois-Berthelot). Hereafter called *Rapport*.

27. Berthelot had the services of a senior employee of the Ministry of Education to sort out the more extravagant suggestions submitted to his committee. He was grateful for this, as he claimed to have been receiving 260 letters a day from the public. E. and J. de Goncourt, *Journal*, Vol. IX (Monaco, 1957), 32-33 (6 September 1870).

28. G. D'Heylli, *Journal du siège de Paris* (n.d., Paris, 3 vols), 85.

29. The '*Section de physique et de chimie*' consisted of: Berthelot, Breguet, H. Sainte-Claire Deville, d'Almeida, Jamin, Fremy, Ruggieri, and Schutzenberger. Breguet, d'Almeida and Jamin were physicists; the remainder were chemists, apart from Ruggieri, who was a pyrotechnics expert.

30. The '*Section de Mécanique*' consisted of: Delaunay, Cail, Claparède, Gévelot and Rolland.

31. *Rapport*, op.cit. note 26, 4.

32. *L'Investissement*, Vol. I, 281. Sappers were also involved in defence work (e.g. constructing barricades), but this was less clearly 'science'.

33. *Mémoires et compte rendu des travaux de la société des ingénieurs civils* (1870), 492-95.

34. *Assemblée générale des comités scientifiques de défense des arrondissements de Paris,* Bibliothèque historique de la ville de Paris, M.S. 1085, 55: '12 octobre.' Dr Curie annonce à l'Assemblée que la Société Chimique a voté dans ses séances des 10 et 11 octobre une somme de 5,000f. pour la construction d'un canon se chargeant par la culasse, dont M. Claparède a accepté la commande.'

35. Ibid., 57. This initiative overcame the traditional refusal of the army authorities to accept the work of private contractors.

36. Op.cit. note 34.

37. Thus, when a demand was made for the summary execution of any man fleeing from the enemy, the reply was that this was 'a purely scientific commission' and such discussion was out of order (ibid., 38).

38. France, *La guerre de 1870-1871. Defense en Province* (Paris, 1911), 54. This is cited hereafter as *Défense en Province.*

39. Ibid., 56. F. F. Steenackers, *Les Télégraphes et les postes pendant la guerre de 1870-1871* (Paris, 1883), 389-90, gives the membership of the committee.

40. Ibid., 55.

41. Charles de Freycinet, *La guerre en Province pendant le siège de Paris, 1870-71* (1871), 18-19.

42. Altogether 1762 submissions were made to the commission, of which 1333 (or 75%) were rejected outright; 222 suggestions were considered of interest but no more, and only 207 (or 12%) were examined in detail or forwarded to an appropriate branch of the government or army. *Défense en Province,* 58.

43. Francois-Paul Barbe, *La dynamite, substance explosive inventée par M. A. Nobel, ingénieur suedois. Collection de documents rassemblée par Paul Barbe* (Paris, 1870), 77pp. Achille Brüll, *Notes sur la dynamite, sa composition et ses propriétés explosives* (Paris, 1870, 35pp. – Extrait des *Annales industrielles).*

44. *Rapport,* op.cit. note 26, 5 (16 September).

45. *Bulletin de la société chimique de Paris,* Vol. 14 (1870), 353, makes only laconic mention of activities between the reported meetings of 22 July and 4 November 1870: 'Dans plusieurs séances extraordinaires, qui ont eu lieu pendant la durée de ses vacances la Société s'est occupé de diverses questions relatives a l'approvisionnement et a la défense de Paris.'

46. E. Jungfleisch, *Notice sur la vie et les travaux de Marcellin Berthelot* (Paris, 1913), 33.

47. 14 November 1870, *C.R.,* Vol. 71 (1870), 688-92.

48. According to F. P. Barbe, 'diverse applications militaires furent faites au plateau d'Avron, au Drancy, à Buzenval et en divers autres lieux', 'Notes sur les usages de la dynamite', *Annales de chimie et de physique,* [4], 23 (1871), 300-01.

49. *Le Journal du siège de Paris,* (24 November 1870), 1871, 257-8, *C.R.,* Vol. 71 (1870), 728.

50. L. Figuier (ed.), *L'Année scientifique et industrielle,* Vol. 15 (1870-71), 143-49.

51. 'Rapport sur les Salpêtres par le Comité Scientifique de la Défense de

Paris', in Quesneville (ed.), *Moniteur scientifique*, [3] , 1 (1871), 455-58. The methods used in the eighteenth century for extracting saltpetre had virtually disappeared in Western Europe since the resumption of trade with India, the importation of 'Chile saltpetre' (sodium nitrate) and the exploitation of the Stassfurt deposits.

52. For a general review of food problems and contributions by scientists to their solution, see Payen, 'Des Subsistances pendant le Siège de Paris en 1870', *Moniteur scientifique*, [3] , 1 (1871), 460-69, and G. de Molinari, 'L'Alimentation de Paris pendant le Siège', *Revue des Deux Mondes*, Vol. 91 (1871), 112-23.

53. *C.R.*, Vol. 71 (1870), 479-86; Vol. 72 (1871), 61-63.

54. *C.R.*, Vol. 71 (1870), 559-62, 747-54. Osseine was obtained as a solid residue in a yield of about 35 per cent by boiling up macerated bones. In the current spirit of economy Fremy suggested that the filtrate containing phosphoric acid could be used to make fertilizer.

55. *C.R.*, Vol. 71 (1870), 796-98, 855-72, 912-27. For an account of the long drawn-out debate in the Paris Academy of Sciences on the merits of gelatine, see F.L. Holmes, *Claude Bernard and Animal Chemistry* (Cambridge, Mass.: Harvard University Press, 1974), 7-10.

56. E. Saint-Edme, op.cit. note 6, 187.

57. Ibid., 176-78.

58. *La Chimie organique fondée sur la Synthèse*, 2 vols (Paris, 1860).

59. M.P. Crosland, *The Society of Arcueil, A View of French Science at the Time of Napoleon I* (London: Heinemann Educational, 1967), 32-36.

60. *C.R.*, Vol. 71 (1870), 472-75, 'Sur l'emploi de la farine d'avoine dans l'alimentation'.

61. E.H. Ackerknecht, *Medicine at the Paris Hospital, 1794-1848* (Baltimore, Md.: Johns Hopkins University Press, 1967).

62. *Bulletin de l'Académie Impériale de Médecine*, Vol. 35 (1870), 714-27.

63. Sir R.J. Godlee, *Lord Lister* (2nd edn, London, 1918), 359.

64. *C.R.*, Vol. 71 (1870), 435 (19 September).

65. On 2 November Depaul reported to the Academy of Medicine that he had vaccinated 15,000 soldiers in the past two weeks. *Bulletin de l'Académie de Médecine*, Vol. 35 (1870), 813.

66. M.E. Decaisne, 'La Santé publique pendant le Siège de Paris', *C.R.*, Vol. 72 (1871).

67. This led to the creation of another commission: 'Commission de telegraphie optique', with Colonel Laussedat as president. *C.R.*, Vol. 72 (1871), 329. See also Berthelot, op.cit. note 5, 445-47.

68. Berthelot, ibid., 449-90. Berthelot's account was based partly on an unpublished report by Almeida.

69. There is extensive literature on ballooning and this is swelled by memoirs of balloonists who escaped during the siege. See e.g. L' Simonin, 'L'Aérostation pendant le Siège de Paris', *Revue des Deux Mondes*, Vol. 90 (1870), 612-30. The latest book on the subject is Victor Debuchy, *Les Ballons du Siège de Paris* (Paris: Editions France-Empire, 1973): this provides a convenient list of all balloons which left Paris during the siege. The present account draws extensively from E. Saint-Edme, op.cit. note 6.

70. As such names as Montgolfier, Charles and Pilatre de Rozier remind us.

71. *l'Investissement*, Vol. I, 214; *Documents*, Vol. I, 87.

72. For the text of the government decree making the grant award, see *Journal du Siège de Paris* (28 October 1870). Dupuy de Lôme's idea of using a steam engine, abandoned in favour of muscle power, *C.R.*, Vol. 71 (1870), 477, had not reached the trial stage when the war ended. His achievement compares unfavourably with Giffard's dirigible balloon of 1852.

73. A useful source on microphotography in the siege is F. F. Steenackers, op.cit. note 39, especially 199-219, 249-51.

74. *C.R.*, Vol. 72 (1871), 237-39. Among other Academicians who took part in the debate were J. B. Dumas and Quatrefages.

75. M. P. Crosland (ed.), op.cit. note 1, 6.

76. This was also the claim of the editor of the *Revue scientifique*, Vol. 8, No. 1 (1 July 1871).

77. N. Mani, art. 'Paul Bert', in C. C. Gillispie (ed.), *Dictionary of Scientific Biography*, Vol. II (New York: Scribners, 1970-), 59-63.

78. M. Berthelot, 'L'Enseignement supérieur', in *Science et Philosophie* (Paris, 1886), 275.

79. 'La Caisse des Ecoles et l'Enseignement supérieur', ibid., 277-79.

80. *Association Francaise pour l'Avancement des Sciences*, 'Notice Historique' (Paris: Secretariat de l'Association, etc., n.d.).

81. *Journal de physique*.

82. Decree of 14 June 1878: 'Considérant qu'il importe de donner au Comité special consultatif des Poudres et Salpêtres le moyen de se prononcer sur les questions relatives à la fabrication et l'emploi des substances explosives de toute nature dont l'usage peut être adopté ou essayé Decrète Art. 1. Il est institué près le Ministre de la Guerre une Commission scientifique dite des substances explosives.'

83. *Sur la force des matières explosives d'après la thermochimie*, 3rd edn (Paris, 1883).

84. Vallery-Radot (ed.), *Correspondance de Pasteur*, Vol. II (Paris: Flammarion, 1940-51), 492.

85. *Revue des Deux Mondes*, Vol. 91, 647-69 (issue dated 15 February 1871). *La Race Prussienne* was also published separately as a small book (12mo, Paris: Hachette, 1871).

86. *Moniteur Scientifique*, [3] , 1 (1871), 272-74.

87. Historians under the Third Republic liked to contrast their generous provision for scientific research with earlier indigence; O. Gréard, *Education et Instruction. Enseignement supérieur* (2nd edn, Paris, 1889), 51.

88. Claude Digeon, *La Crise Allemande de la Pensée Francaise* (Paris: Presses Universitaires de France, 1959), 370ff.

89. The possibility of chemical warfare had been discussed earlier, e.g. in the Crimean war. J.G. Crowther, *Statesmen of Science* (London: Cresset Press, 1965), 149. If one includes Greek fire as 'chemical warfare', the antecedents are much more ancient. One feature of the popular contribution to the war was suggestions for using Greek fire, and this was updated by chemists who were able to improve on the bituminous-based incendiaries used in former times by proposing preparations of phosphorus and carbon disulphide — F. Papillon, 'La Science et la Defense Nationale', *Revue des Deux Mondes*, Vol.6.(1870),374-61.

90. *Rapport*, op.cit. note 26, 8.

91. It should be emphasized that this essay has sought to draw attention to the main scientific commissions, and is by no means exhaustive.

XII

ASPECTS OF INTERNATIONAL SCIENTIFIC
COLLABORATION
AND ORGANISATION BEFORE 1900

INTRODUCTION

Most of the work that has been done on international science has been focused on the twentieth century by scholars whose primary interests have been political and sociological. Historians of science have not shown much general interest in this area, although occasionally a historian, concerned with the history of an individual science, has looked at some international collaboration within that particular discipline. Unfortunately the separate study of the history of individual sciences precludes the appreciation of general patterns. There are more similarities between problems of quite different sciences than one would at first imagine. I shall be referring in the course of this paper to the general question of agreement on a common language. I could also point out that expeditions concerned with one science, say astronomy, have benefited other areas of science like natural history, and vice-versa. No science, therefore, stands in isolation, whether considered methodologically or from the point of view of organisation.

Because this early period is not an area in which much research has been done, one of my principal objectives will be to present basic information and a few perspectives. What follows, therefore, is of the nature of a general survey. There is, of course, a vast documentation, but it may be possible later to explore more fully some aspects which I only touch upon here or even omit entirely. This paper presents some preliminary thoughts on a general problem which I think deserves attention in any comprehensive study of the organisation of science.

In the literature of scientific conferences, largely drawn up by librarians, compilers have delighted in searching out some particularly early date. Thus a recent compilation of scientific periodicals carries the title *Scientific Conference Proceedings, 1644-1972* . I might mention that the date 1644, predating the foundation of the Royal Society of London, refers to a meeting of falconers at Rouen - hardly a serious challenge to the established views on early scientific societies! It could be argued that such antiquarian tit-bits do no harm. My own view is that a title which implies an activity of an unchanging nature over three centuries as well as comprehensive treatment of that activity is grossly misleading. Another publication is called: *International Congresses ... Full List* . Here the claim for completeness is explicit. A third compilation claims an exhaustive enumeration of 1,978 international organisations founded since 1815. Any such claim to be complete is very dangerous, particularly in the earlier period. The scientific conferences of the twentieth century are well documented and can therefore be discussed without challenging the data. Before 1900 the data requires careful historical study. In superficial research it is all too easy to be misled by titles. The *Actes du 1er Congrès International de Botanique*, which refers to a meeting held in Paris in 1900, provides an example of such a title. It represents the vanity of its secretary, a certain Emile Perrot, and ignores the fact of international botanical meetings over the previous 40 years, some of which had made significant contributions to international collaboration in botany.

We may distinguish three kinds of international activity, each with its own development, although having implications for the others. The first and most basic kind of contact is the free flow of scientific information by correspondence. In the seventeenth century those indefatigable correspondents, Mersenne and Oldenburg, exchanged letters on a large scale and they passed on information to natural philosophers in all parts of Europe. This aspect of international collaboration continued to be important in science, and from the nineteenth century had spread from Europe to all continents of the world, although of course by then scientific journals had to a large extent superceded private correspondence. Translations, reviews and abstracts have all contributed to the dissemination of information on an international scale.

Travel is a second kind of transnational activity, often allowing an individual to discuss his work with other interested parties far from home. But scientists and particularly astronomers have also needed to make investigations beyond their national frontiers and they have sometimes required a guarantee of safe passage from the government through whose territory they were to pass. From the beginning the claim was made that the study of nature transcended all political and national considerations, except perhaps the pursuit of glory. This level of co-operation obviously involved a certain amount of negotiation at government level. I shall be focussing attention in this paper on the collaboration of scientists rather than the collaboration of governments but this is obviously a dimension that requires detailed study.

Whereas correspondence and travel were aspects of international scientific activity well established in the seventeenth and eighteenth centuries, the third aspect, the international meeting, involved a greater infrastructure and was slower to emerge. Nevertheless, by the early nineteenth century meetings, which had begun on a local or national scale, gradually came to take on an international aspect. The presence of foreign visitors at national meetings helped prepare the way for a multi-national conference, in which representatives of different countries met on equal terms. Arising out of international meetings, there came into being in the second half of the nineteenth century a number of permanent international scientific organisations. Thus all the main features of modern international science had appeared before 1900.

NATIONAL SOCIETIES

One might tend to think of a national meeting as being the antithesis of an international meeting. In fact the growth of scientific societies organised on national lines greatly helped the development of international contacts and prepared the way for international gatherings. Already in the seventeenth century the Royal Society had accepted foreign members, and in the eighteenth century foreign members represented a large fraction of the total membership. The Paris Academy of Sciences and the Berlin Academy each had a number of foreign associates. Although such a position was largely honorary, it did help to promote international awareness. In the nineteenth century the Gesellschaft Deutscher Naturforscher und Ärzte (founded in 1822), although explicitly a German organisation, attracted attention outside the German states and its meetings came to be attended by an increasing number of foreign scientists. The multinational character of the meetings became a source of pride to the organisers so that at the 1828 meeting in Berlin a map of Europe was printed showing the towns from which participants came, and a table of countries was attached showing that, in addition to the German-speaking states, there were several representatives from each of the following countries: Sweden, Denmark, Poland, Holland, as well as one each from Britain, France and Russia. The Scandinavian participation in the German meetings was particularly striking and this encouraged the Scandinavian scientists to organise their own congresses. From 1839 there were regular meetings of the Congress of Scandinavian scientists (*Scandinaviske Naturforskeres Møde*), which met every two or three years in the larger cities of Sweden, Denmark and Norway. These meetings attracted a limited amount of interest outside Scandinavia but a few visitors from neighbouring countries (notably Prussia, Finland and Russia) attended later meetings. The German society had important implications for the organisation of science in Britain and France as well as in Scandinavia, and it was not long before the question arose of coordinating meetings of the various national societies.

The best known influence of the Deutsche Naturforscher is that in Britain, where it served as a model for the British Association for the Advancement of Science. The British Association played a significant part in international relations, thus challenging the authority of the Royal Society to represent British science on the international scene. The first international congress on terrestrial magnetism was held in Cambridge in 1845 in conjunction with the meeting of the British Association there. The idea of such a congress in Britain had come from Kupffer, director of the Russian program of magnetic and meteorological observations. Although British physicists constituted the majority at the meeting, there were half a dozen important representatives from the Continent, including Erman from Berlin and Kreil from Prague. Weber, Gauss and Humboldt did not attend but sent letters to the organisers of the congress. Another magnetic congress in conjunction with the British Association took place in 1898. Meanwhile, however, there had been many proposals at meetings of the British Association which had implications for international collaboration. Joseph Henry wrote to point out to the 1855 meeting in Glasgow the desirability of an international catalogue of scientific literature, and in the 1860s there were several discussions on the rationalisation of electrical units.

On the political front the Congress of Vienna in 1815 was a landmark in the history of international collaboration and it helped to establish a system of international consultation. States were brought together in another way by the growth of railways in the nineteenth century and the introduction of the electric telegraph. Improving communications also brought new problems, some associated with the growth of national pride. However, there were obvious mutual benefits to be obtained by agreement on collective measures to improve trade and to check the spread of disease. International River Commissions were established in the early nineteenth century. The spread of epidemics brought about diplomatic conferences, although the first international health conference in 1851 in Paris achieved very little. The coming of the railway required collaboration at national frontiers and it was not long before there were demands to send telegraphic messages across national boundaries. In 1865 twenty European states met in Paris to work out an international convention on the telegraph. The Austro-German Postal Union of 1850 showed the advantages of a unified postal administration but the Universal Postal Union came into being only in 1874.

All of these international conferences involved minor concessions of national sovereignty. They paved the way for international collaboration in technological areas. The interaction of scientists for technological purposes in turn provided a good precedent for the organisation of pure science on international lines. Political and economic collaboration were not, therefore, the *cause* of scientific collaboration but they reinforced the trend and made it easier for scientists to cross frontiers. The political suspicion with which, for example, some of the earlier meetings of the Deutsche Naturforscher had been met came to be replaced by an acceptance and, by the end of the nineteenth century, a positive fashion for international conferences of all kinds.

It is inevitable that most of the discussion on international collaboration before 1900 should be focused on Europe because that was the centre of activity. The great powers in the eighteenth century were clearly Britain and France, whose political influence was rivalled only by the major contributions to science emanating from these two countries. Although French political power declined after the defeat of Napoleon, the prestige and high quality of French science ensured a continuing French authority in the nineteenth century. After 1800 German scientists too began to make some contribution: Humboldt and Gauss in geomagnetism, Bessel in astronomy and Weltzien in chemistry. After several important French contributions to geodesy it was the Prussian government which took the main initiative in the mid-nineteenth century in earth measurement. The first meeting of the Geodesic Association of Central Europe was held in Berlin in 1864. It was from this association that the International Geodesic Association sprang.

An important international venture, which began as a German collaboration, was the foundation of the International Association of Academies. In 1893 the Academies of Science of Vienna, Munich, Leipzig and Goettingen founded a cartel of German Academies. On the initiative of the Vienna Academy overtures were made to the Royal Society of London and the Paris Academy to suggest wider co-operation. This co-operation fitted in well with a plan of the Royal Society of London for collaboration to compile an international catalogue of scientific literature. The International Association of Academies was thus founded quite quickly after a meeting at Wiesbaden in 1899.

INDIVIDUAL AND COLLECTIVE AUTHORITY

A few outstanding individuals played a major role in the calling of international conferences. Although no single scientist in the second half of the nineteenth century could assume he had the authority to make decisions affecting his particular science, it was often the initiative taken by an individual that finally brought men together in an international congress. A good example of such authority and initiative is provided by the Belgian, Adolphe Quetelet. Although Quetelet is remembered for his pioneering work in statistics, he had begun as a geometer and astronomer. His appointment in 1828 as astronomer at the Brussels Royal Observatory and in 1834 as permanent secretary to the Brussels Academy helped to extend his contacts with scientists in other countries, and in the 1830s he was actively engaged in international collaboration in astronomy, meteorology and geophysics. His monograph of 1835 on social statistics gave him an international reputation and, if anyone was to call an international conference of statistics, he was the obvious candidate and Brussels was the obvious place.

Yet, although an individual might help to launch a congress, he could not usually expect to dictate to his colleagues. The growth of science and the parallel between the development of political democracy and the working of the scientific community was expressed as follows by a group of British zoologists writing in 1842:

> "The world of science is no longer a monarchy, obedient to the ordinances, however just, of an Aristotle or a Linnaeus. She has now assumed the form of a republic, and although this revolution may have increased the vigour and zeal of her followers, yet it has destroyed much of her former order and regularity of government."

This then was a nineteenth-century conservative British view of an indirect influence of the French revolution on the organisation of science. The zoologists went on to suggest that order could be imposed by general consent, which was more likely to be obtained by basing any rules of nomenclature on clear rational principles.

It is interesting that the group, which included Richard Owen and Charles Darwin, did not go as far as to propose an international meeting. Starting from the premiss that Britain was a world power, they felt that the British Association would have sufficient international authority. What they claimed publicly, however, was the authority of rationality:

> "All that is wanted then is that some plain and simple regulation, founded on justice and sound reason should be drawn up by a competent body of persons [i.e. themselves] and then be distributed throughout the zoological world."

Thus, despite the metaphor of science as a republic, they did not yet see it as an international community of equal partners. The idea of equal participation of many countries developed only slowly; a definite change can be seen in the international botanical congress held in Paris in 1867, a quarter of a century after the meeting of British zoologists. Alphonse de Candolle remarked ruefully [perhaps he was thinking of himself] that "Nowadays no-one likes to submit to the will even of a man of genius ..." He thought that many scientists would rather accept a decision arrived at by a majority of their colleagues, as expressed, for example, at an international scientific congress. One had to be careful, however, that any such vote really was representative of the growing international community of that particular branch of science. In so many international congresses the host country was so well represented that any majority decision might have effectively been a decision agreeable only to the scientists of one country.

In a subject in which national and local advantages were constantly sought it is pleasing to record at least one example of national abnegation in favour of a truly international spirit. The idea of an international congress of physiology came from the British Society of Physiology but, far from insisting on London as the venue, they proposed Switzerland. It was thus that a group of physiologists, representing Britain, France, Germany, Italy and Switzerland, planned the first international physiological congress, which took place in

Basle in 1889. No doubt arguments for location at the centre of Europe were advanced for the Swiss meeting, arguments which competed with considerations of ease of communications. Perhaps one should also remember that in the nineteenth century there was a growing appreciation of Alpine scenery, not to mention the physiological problems posed by high mountains.

Paris was chosen for the location of a large number of nineteenth-century scientific congresses for reasons geographical, political, linguistic and economic. In so far as it was the scene of several major international exhibitions which involved science and technology, it was appropriate to extend the occasion to bring together men in one particular scientific discipline. The supreme example of this is the way the International Exhibition of 1900 was used to mount international scientific congresses in mathematics, physics, meteorology, chemistry, botany, geology, psychology and many branches of applied science. It was also the occasion of the first international congress of the history of science.

THE STANDARDISATION OF LANGUAGE

One of the most important motives for calling an international congress was to obtain authority to introduce certain conventions in a particular science. Most basic was language, particularly in the biological sciences, geology and chemistry. Generally in physical science agreement was necessary on units of measurement, with the metric system receiving increasing support throughout the nineteenth century. The branch of physics in which there was most discussion about units was electricity, a science which in the nineteenth century, with the development of current electricity, the telegraph and electric power, had a growing industrial relevance.

In the eighteenth century Linnaeus had introduced into botany a binomial nomenclature, which provided a welcome solution to the confusion in the naming of plants at the time. Because it had been well thought out, because it made use of an international language (Latin) and because of the increasing stature of Linnaeus in natural history in the mid-eighteenth century, the nomenclature was widely accepted. The circumstances of the reform of chemical nomenclature in 1787 were rather different in so far as the main authority came from Lavoisier but the ground work had been done by Guyton de Morveau. A collaboration between these two and two other leading French chemists led to the publication of a joint work which came to be accepted not only within France but also in other countries. This can be explained by the dominant position of French chemistry at the time and by the good sense of the proposals. Although the historian may give these purely practical reasons for the acceptance of the new chemical language, Lavoisier would have claimed that, in following the sensationalist philosophy of Condillac, he was proposing a system related to nature. In so far as the terms derived from the Greek, he was tapping a universally recognised spring of western civilization.

In the early nineteenth century the Swedish chemist Berzelius was able to introduce the symbols used in modern chemistry based on the initial letters of the Latin names of the elements. Once again a Scandinavian scientist had been able to gain acceptance by using an international language. Yet we are now arriving at the end of a tradition. By the nineteenty century both the growth of national consciousness and the growth of science made it unthinkable that an individual could legislate for scientists in other countries. At the most he might name a new substance or define a unit and leave to a representative group of his colleagues the task of rationalisation.

A major achievement in the second half of the nineteenth century was the holding of international conferences in many different disciplines with the object of agreeing on a common language. The fact that they did not take place until the second half of the nineteenth century suggests that language was not the first reason for international *collaboration*. However, it was an important motive for holding international *meetings*. It may surprise some historians to learn how wide a range of international conferences were inspired by the goal of rationalisation of nomenclature. They range from statistics to geology and from chemistry to meteorology. Several subjects will be briefly reviewed in the chronological order of their first official international conference as examples of the importance of language in promoting international meetings.

It is worth citing the remarks of Quetelet at the first international statistical congress to show that the philosophy of the eighteenth-century French philosopher Condillac was not dead in 1855:

> "Le moyen le plus sur de faire progresser les sciences, c'est d'en perfectionner le langage et d'adopter des notations uniformes qui permettent de résumer plus facilement un grand nombre d'idées et de rapprocher plus de faits pour en saisir les rapports et les lois."

In statistics the problem was to persuade different governments to produce data on population, resources, mortality, etc., in ways which were strictly comparable. In chemistry a major problem was to persuade chemists, whether in the same country or not, to agree on a common system of notation in which the atomic symbols of Berzelius would stand for an agreed atomic weight. Some accepted atomic weights twice the value accepted by others, while others again had taken refuge in equivalents. The Karlsruhe congress, called by Kekulé, Weltzien and Wurtz in 1860 and attended by many leading chemists from most European countries, discussed the problem of nomenclature among other questions. Nothing was settled, but a valuable precedent was established for further international meetings.

The subject of nomenclature was not the first problem of the botanical congresses of the 1860s but it was a problem that did not take long to emerge. An important part of the 1867 congress was devoted to the question which had recently become more pressing because of the use by horticulturists for simple cultivated varieties of the Latin names used by botanists for the genuine species. Alphonse de Candolle headed a commission to examine the problem and make recommendations.

Turning to the subject of meteorology, which held its first international congress in Vienna in 1873, there is an obvious parallel with statistics in so far as the problem of language is concerned. Meteorology, with its agricultural and military implications, was also of interest to governments. But in so far as meteorology was a science like geomagnetism, transcending national frontiers, it is surprising that it had to wait until 1873 for its first international congress. Perhaps this merely illustrates the difference between international cooperation and the calling of a formal meeting. The conference was concerned to arrange for observations to be made in different countries at the same local time and with instruments so constructed and placed as to give results strictly comparable with each other. But the concept of a common language was also explicit in discussion of symbols which at that time were often abbreviations of words in the different vernacular languages of Europe. A committee recommended the substitution of symbols for letters but for wind directions it chose the initial letters in the English language.

The first international congress of geology in 1878 had some points in common with both meteorology and botany. Indeed the botanists' concern in Paris in 1867 with nomenclature had led at least one geologist (the Spaniard Vilanova) to suggest a discussion of geological nomenclature. But the case of geology differs from that of the other sciences in that most of the concern for an international meeting came not from a European country but from the United States. American geologists had encountered strata which had no name in the old world; in different American states they had given different names to the same thing and had finally settled for a pragmatic nomenclature relating to place, leaving the problem of rationalisation to the future. James Hall took the initiative at a meeting of the American Association for the Advancement of Science in Buffalo in 1876 to suggest the calling of an international congress of geologists in Paris at the time of the 1878 Paris exhibition. Although many other subjects beside nomenclature were discussed, this was one of the principal motives for the meeting. Also discussed was the question of agreement on the choice of colour for geological maps. It was not long before the question of a unified nomenclature for palaeontology was raised at these meetings (Berlin, 1885). Thus, after much early confusion, geologists made a major and continuing use of international conferences to achieve greater uniformity in language and cartographic conventions.

A concluding comment on the whole question of standardisation of nomenclature is that the importance of agreement on names and units tended to override considerations of national prestige and priority. It provided a strong incentive for scientists to meet and work out some agreement. When scientists had become accustomed to such international meetings they found that there were other advantages, both intellectual and social, in such meetings and they therefore tended to become permanently established.

THE METRIC SYSTEM

International collaboration in the metric system is sometimes considered to date from the conference of 1875 known as the *Conférence diplomatique du mètre*. Such a view gives a misleading perspective, since the events of the 1870s were really the culmination of activities that had been going on for nearly a century. The possibility of standardising weights and measures had been raised in the British House of Commons in 1790 and, as similar ideas had been current in France, Talleyrand, French minister of foreign affairs, suggested collaboration between the two countries through their principal scientific societies, the Paris Académie des Sciences and the Royal Society. Unfortunately political events in France made collaboration more and more difficult and it was the Académie des Sciences alone which examined the problem in detail. The Academy recommended a new unit of length, the metre, related to the size of the earth. The metre was thus international in theory. In practice it related to the meridian through Paris. The practical geodesic measurements and the construction of standards occupied several years. When the time came to make decisions on a definitive metre and a definitive kilogram some scientists in the Institute persuaded the French government to invite scientists from other countries to collaborate in the final steps. The arrival of representatives from nine allied and neutral countries in Paris in September 1798 has some claim to mark the beginning of the first international scientific conference. The presence of the foreign scientists may have been little more than a gesture but it did help to provide a little more authority for the metric system than a uniquely French determination would have done.

Nevertheless, the metric system was usually considered in other countries to be French and the initiative for further action in the propagation of the system lay, therefore, with that country. In 1841 the French government sent copies of metre standards to a large number of European countries, and metric standards were also sent as far as the United States, South America and Japan. It was, however, international exhibitions which did most to propagate the advantages of the metric system. The 1851 exhibition revived an interest in Britain in an international system, and at the 1855 Paris exhibition members of the panels of judges representing many different countries undertook to make representations to their respective governments on the advantages of a universal system of weights and measures. This raised the obvious question of whether the metric system could be accepted as that universal system. The metric system had already been accepted in several countries of Europe and Central and South America but the spread of the system emphasised the need for uniformity of standards. There was a very real danger in the mid-nineteenth century that different countries would use metric standards which were slightly different. An international conference was, therefore, called in Paris, and 24 foreign countries agreed to send representatives. The date chosen for the conference, August 1870, was unfortunate because, by the time the conference met, France was at war with Prussia. The meeting, nevertheless, took place and did some useful preliminary work, which prepared the way for the *Conférence diplomatique du mètre* and the establishment of an International Bureau of weights and measures.

ASTRONOMY

One could single out astronomy as an example of a science that has had a long and distinguished history of international collaboration. As an observational science on a large scale, astronomy came to involve the whole earth as a base. Most European astronomers of the seventeenth century were content with observation of the stars in northern latitudes but the Englishman Edmond Halley was able with royal support to secure a passage to St. Helena in 1677 to enable him to draw up the first catalogue of stars of the southern sky. Three quarters of a century later Lacaille, with French government support, was able to extend his star catalogue of the southern hemisphere from 350 stars to nearly 10,000.

A second motive for astronomical expeditions arose out of the two transits of Venus predicted for 1761 and 1769. In order to use this rare event to calculate the distance of the sun, different groups of astronomers needed to plan to observe from different parts of the world the exact moment of ingress and egress of the planet across the sun's disc. Le Gentil managed to obtain French government approval to observe the transit of Venus in India and the Paris Academy of Sciences responded to an invitation from the St. Petersburg Academy by sending Chappe to head a Russian party in Siberia. The most ambitious of the British plans was the famous voyage of Captain Cook, which combined an observation from Tahiti with exploration of the South Pacific. International science here overlapped with colonial interests. The expeditions were planned by colonial powers and, in so far as there was any collaboration, it was between European astronomers.

Although the arrangements made to observe the eighteenth-century transits of Venus established something of a precedent, the expeditions were very exceptional. In the 1820s John Herschel and Friedrich Bessel each put forward proposals for a general survey of the stars, but it was the Berlin Academy which took responsibility for most of the work. The first international cooperation in the nineteenth century took place in solar rather than in sidereal astronomy. Improvement in communications made it feasible to make plans on a world-wide scale for the observation of solar eclipses, a phenomenon which was of course strictly local so that the vantage point was crucial. One of the first solar eclipses to attract attention on an international scale was that which crossed central and southern Europe on 8 July 1842. The early eclipse expeditions were often a matter of individual decision, and observers later compared results. They did not collaborate so much as share a common purpose. However, considering that the important phenomenon of a solar corona hardly lasted more than one minute, that was an obvious incentive to make the best use of those precious seconds. In the solar eclipse of 17 May 1882, when astronomers from many countries converged on Sohay in Upper Egypt, each observer was assigned a special task and the advantages of a strict division of labour became evident.

The new technique of photography provided both a permanent record of astronomical observations and an incentive for further international collaboration. The British astronomer, David Gill, who held the position of Her Majesty's astronomer at the Cape of Good Hope, was the first to chart accurately and measure star position by means of photography. His *Cape Photographic Durchmeisterung*, which included nearly half a million southern stars, was to be the forerunner of a general photographic survey of the heavens. Gill corresponded with astronomers in Paris and the result was an international congress in 1887, with representatives from 17 countries, who were to plan the great *Carte du ciel*. By the early twentieth century the need had been felt for a Permanent Commission of the Carte du ciel. The institutional response to the need to centralise a speedy service of astronomical information had been met in 1881 by the creation of a European central office for astronomical telegrams. Another permanent institution was the Commission Internationale de l'Heure, proposed in 1906 with a bureau in Paris.

NEW SCIENCES

A branch of science which claimed increasing attention in the late nineteenth century was psychology. The propagation of physiological psychology through the work of Wundt in Germany brought the subject under the banner of experimental science. The president of the first international congress was the psychologist Ribot, who, as editor of the *Revue philosophique,* had championed the cause of experimental psychology in France. Moreover, as the author of works on the state of psychology in Germany and Britain, Ribot was fully in touch with the latest developments outside his own country. When Ribot opened the first international congress of physiological psychology in Paris in 1889, he explained that, up till then, all psychologists had had in common had been vague aspirations and limited correspondence. By holding such a congress, they were giving their subject some institutional unity. They could not hope at their first meeting to resolve problems but they could decide on the questions they were to ask. Although they had first favoured the term 'physiological' psychology, many people, including the British, preferred to use the wider term 'experimental' and the 1889 Congress agreed that the following Congress, to be held in London in 1892, would be known as the Second International Congress of Experimental Psychology.

Another branch of knowledge with a claim to be scientific, which developed in the nineteenth century, was sociology. I mention this briefly as an example of the institutionalisation of a new area of study selfconsciously related to the natural sciences. One of the key figures, René Worms, was a young entrepreneur, who in 1893 at the age of 24 founded both a journal, which he called the *Revue internationale de sociologie,* and the Institut International de sociologie. The use of the term international was justified by the participation of a few sociologists from Britain, Germany, Austria and the U.S.A. Both the journal and the institute were under the control of an editorial board of which Worms was secretary. Within a year of their foundation they announced the first of a series of international congresses to be held in Paris: 1894, 1895, 1897, 1900. Although there were no more than twenty members at the early congresses, papers were sent in from outside and members of the public were admitted to fill the meetings. The whole enterprise was financed by members but Worms aimed to obtain both government recognition and financial support. Official status was achieved by 1909, although they had to wait until 1931 for a government subsidy. The claim of these early sociological congresses to international status was in tune with the aspirations of the age. By the beginning of the twentieth century the international congress had become a new fashion.

CONCLUSION

One of the benefits of studying science from an international viewpoint is that it provides a truly wide perspective. It is always useful to look at a problem from more than one point of view and, although science is often the

14

product of research in different national contexts, the growth in communication led to the rapid diffusion of science on a global scale.

The transmission of information and skills from one country to another is still a largely unexamined problem. Although I have not chosen to develop the connection, there is an obvious relevance to scientific publications. Many of the leading journals in the nineteenth century became more explicitly international, setting up editorial boards with representatives from different countries. International negotiation in scientific matters may have less dramatic consequences than negotiation on political issues but it is important that scientists should all use the same technical language and the same units. The history of the metre is a part of the history of international science. I have suggested that the importance of a standard nomenclature and the precedents of diplomatic and technological meetings were major factors in the growth of international science.

There are still a few historians of science who feel that the internal exploration of a branch of science is sufficient without any regard to context. But others recognise that institutions can encourage, guide, direct (though not determine) the course of science. If this is so, the historian does not have to stop with the study of a local scientific society or a national institution. He should be prepared to consider the institutional structure of science on an international scale.

By 1900 the international scientific conference had arrived. The study of twentieth-century international cooperation is, therefore, a rather different sort of activity, with the basic data more easily accessible and the main precedents already established. For the nineteenth century further research is required. We still need to study the conditions which brought scientists together before collaboration became a well-established routine. We need to investigate the part played by the leading scientific societies and we need to look at the organisation of individual sciences.

GENERAL BIBLIOGRAPHY

It would not be appropriate to give a long list of the various proceedings of international conferences which have been used as a basis for this preliminary sketch. Apart from these and the *Annual Reports of the British Association for the Advancement of Science* a few works of general relevance are listed below.

Blaisdell, D.C. *International organisation*, New York, 1966.

Crosland, M.P. "The Congress on definitive metric standards, 1798-1799: the first international scientific conference?" *Isis, 60* (1969), 226-231.

Eijkman, P.H. *L'Internationalisme scientifique*, The Hague, 1911.

Gregory, W. (ed.) *International Congresses and Conferences, 1840-1937*, New York, 1938.

Lyons, F.S.L. *Internationalism in Europe, 1815-1914*, London, 1963.

Mangone, G.J. *A Short History of International Organisation*, New York, 1954.

Schroeder, Brigitte. "Caractéristiques des relations scientifiques internationales, 1870-1914", *Journal of World History, 10* (1966-67), 161-177.

Speeckaert, G.P. *The 1,978 International Organizations founded since the Congress of Vienna*, Brussels, 1957.

Stratton, F.J.M. "International Co-operation in Astronomy". *Monthly Notices of the Royal Astronomical Society, 94*, No.4., Feb. 1934, pp.361-372.

Union des Associations Internationales. *Les Congrès Internationaux de 1681 à 1899. Liste complète*, Brussels, 1960.

University of Cambridge, Union Catalogue of Scientific Libraries. *Scientific Conference Proceedings, 1644-1972*, Cambridge, 1975.

INDEX